CONCEPTS OF ELEMENTARY PARTICLE PHYSICS

OXFORD MASTER SERIES IN PHYSICS

The Oxford Master Series is designed for final year undergraduate and beginning graduate students in physics and related disciplines. It has been driven by a perceived gap in the literature today. While basic undergraduate physics texts often show little or no connection with the huge explosion of research over the last two decades, more advanced and specialized texts tend to be rather daunting for students. In this series, all topics and their consequences are treated at a simple level, while pointers to recent developments are provided at various stages. The emphasis is on clear physical principles like symmetry, quantum mechanics, and electromagnetism which underlie the whole of physics. At the same time, the subjects are related to real measurements and to the experimental techniques and devices currently used by physicists in academe and industry. Books in this series are written as course books, and include ample tutorial material, examples, illustrations, revision points, and problem sets. They can likewise be used as preparation for students starting a doctorate in physics and related fields, or for recent graduates starting research in one of these fields in industry.

CONDENSED MATTER PHYSICS

1. M.T. Dove: *Structure and dynamics: an atomic view of materials*
2. J. Singleton: *Band theory and electronic properties of solids*
3. A.M. Fox: *Optical properties of solids, second edition*
4. S.J. Blundell: *Magnetism in condensed matter*
5. J.F. Annett: *Superconductivity, superfluids, and condensates*
6. R.A.L. Jones: *Soft condensed matter*
17. S. Tautz: *Surfaces of condensed matter*
18. H. Bruus: *Theoretical microfluidics*
19. C.L. Dennis, J.F. Gregg: *The art of spintronics: an introduction*
21. T.T. Heikkilä: *The physics of nanoelectronics: transport and fluctuation phenomena at low temperatures*
22. M. Geoghegan, G. Hadziioannou: *Polymer electronics*

ATOMIC, OPTICAL, AND LASER PHYSICS

7. C.J. Foot: *Atomic physics*
8. G.A. Brooker: *Modern classical optics*
9. S.M. Hooker, C.E. Webb: *Laser physics*
15. A.M. Fox: *Quantum optics: an introduction*
16. S.M. Barnett: *Quantum information*
23. P. Blood: *Quantum confined laser devices*

PARTICLE PHYSICS, ASTROPHYSICS, AND COSMOLOGY

10. D.H. Perkins: *Particle astrophysics, second edition*
11. Ta-Pei Cheng: *Relativity, gravitation and cosmology, second edition*
24. G. Barr, R. Devenish, R. Walczak, T. Weidberg: *Particle physics in the LHC era*
26. M. E. Peskin: *Concepts of elementary particle physics*

STATISTICAL, COMPUTATIONAL, AND THEORETICAL PHYSICS

12. M. Maggiore: *A modern introduction to quantum field theory*
13. W. Krauth: *Statistical mechanics: algorithms and computations*
14. J.P. Sethna: *Statistical mechanics: entropy, order parameters, and complexity, second edition*
20. S.N. Dorogovtsev: *Lectures on complex networks*
25. R. Soto: *Kinetic theory and transport phenomena*
27. M. Maggiore: *A modern introduction to classical electrodynamics*

Concepts of Elementary Particle Physics

Michael E. Peskin

SLAC National Accelerator Laboratory, Stanford University

OXFORD
UNIVERSITY PRESS

OXFORD
UNIVERSITY PRESS

Great Clarendon Street, Oxford, OX2 6DP,
United Kingdom

Oxford University Press is a department of the University of Oxford.
It furthers the University's objective of excellence in research, scholarship,
and education by publishing worldwide. Oxford is a registered trade mark of
Oxford University Press in the UK and in certain other countries

© Michael E. Peskin 2019

The moral rights of the author have been asserted

First Edition published in 2019

Reprinted with corrections in 2023

Published in the United States of America by Oxford University Press
198 Madison Avenue, New York, NY 10016, United States of America

British Library Cataloguing in Publication Data
Data available

Library of Congress Control Number: 2019930484
Data available

ISBN 978–0–19–881218–0 (hbk.)
ISBN 978–0–19–881219–7 (pbk.)

DOI: 10.1093/oso/9780198812180.001.0001

Printed and bound by
CPI Group (UK) Ltd, Croydon, CR0 4YY

Links to third party websites are provided by Oxford in good faith and
for information only. Oxford disclaims any responsibility for the materials
contained in any third party website referenced in this work.

Preface

This is a textbook of elementary particle physics, intended for students who have a secure knowledge of special relativity and have completed an undergraduate course in quantum mechanics.

Particle physics has now reached the end of a major stage in its development. The primary forces that act within the atomic nucleus, the strong and weak interactions, now have a fundamental description, with equations that are similar in form to Maxwell's equations. These forces are summarized in a compact mathematical description, called the Standard Model of particle physics. The purpose of this book is to explain what the Standard Model is and how its various ingredients are required by the results of elementary particle experiments.

Increasingly, there is a gap between the study of elementary particles and other areas of physical science. While other areas of physics seem to apply directly to materials science, modern electronics, and even biology, particle physics describes an increasingly remote regime of very small distances. Physicists in other areas are put off by the sheer size and expense of elementary particle experiments, and by the esoteric terms by which particle physicists explain themselves. Particle physics is bound up with relativistic quantum field theory, a highly technical subject, and this adds to the difficulty of understanding it.

Still, there is much to appreciate in particle physics if it can be made accessible. Particle physics contains ideas of great beauty. It reveals some of the most deep and surprising ideas in physics through direct connections between theory and experimental results. In this textbook, I attempt to present particle physics and the Standard Model in a way that brings the key ideas forward. I hope that it will give students an entryway into this subject, and will help others gain a better understanding of the intellectual value of our recent discoveries.

The presentation of elementary particle physics in this book has been shaped by many years of discussion with experimental and theoretical physicists. Particle physicists form a global community that brings together many different points of view and different national styles. This diversity has been a key source of new ideas that have driven the field forward. It has also been a source of intuitive pictures that make it possible to visualize physical processes in the distant and abstract domain of the subnuclear forces. I have tried to bring as many of these pictures as possible into my discussion here. My own way of thinking about particle physics has been shaped by my connection with the great laboratories at Cornell University and SLAC. I am indebted to many colleagues at these laboratories for central parts of the development given here.

I have been reminded often during the writing of this book that many of the great figures responsible for the formulation of the Standard Model have passed on to that symposium in the beyond. In only the past few years, we have lost Sidney Drell, Martin Perl, Richard Taylor, Kenneth Wilson, and, most recently, Burton Richter. All of these people influenced me personally and profoundly affected my thinking about particle physics. It is a challenge for us who follow them not only to finish their work but also to open new chapters in the development of fundamental physics. I hope that this book will provide useful background for those who wish to do so.

The core of this presentation was developed as a set of lectures for CERN summer students in 1997; I thank Luis Alvarez-Gaumé for the invitation to present these lectures. I have presented parts of this material at a number of summer schools and courses, in particular, the course on elementary physics at the Perimeter Scholars International program at the Perimeter Institute. Most recently, I have polished this material by my teaching of the course Physics 152/252 at Stanford University. I am grateful to Patricia Burchat for giving me this opportunity, and for much advice on teaching a course at this level. I thank the students in all of these courses for their patience with preliminary versions of this book and their attention to errors they contained. I thank Sonke Adlung, Harriet Konishi, Sal Moore, and their team at Oxford University Press for their interest in this project. I thank Roy Brener, Tim Cohen, Serge Dendas, Caterina Doglioni, Christopher Hill, Andre Hoang, Sunghoon Jung, Andrew Larkoski, Aaron Pierce, Daniel Schroeder, Bruce Schumm, and André David Tinoco for valuable comments on the presentation, and Jongmin Yoon for an especially careful reading of the manuscript. Most of all, I thank my colleagues in the SLAC Theory Group for their advice and criticism that has benefited my understanding of elementary particle physics.

Michael E. Peskin
Sunnyvale, CA
August, 2018

Figure permissions

We are grateful to the following colleagues, organizations, and publishers, who retain the original copyright, for permission to include figures in this book. The figures are labeled below by our figure numbers. The precise citation for each figure is given in the figure caption.

American Physical Society. Figures 5.3, 5.4 (left and right), 9.2, 9.3, 15.2, 15.3, 19.2, 19.4, 19.6, 20.4, 20.5, and 20.6 are reproduced from *Physical Review Letters*. Figures 5.6, 10.6, 10.7, 13.7, 13.8, 20.2, 21.8, and the figure in Exercise 7.2(m) are reproduced from *Physical Review* **D**. Figures 5.8 and 5.9 are reproduced from *Reviews of Modern Physics*.

Annual Reviews. Figure 4.1 is reproduced from the *Annual Review of Nuclear and Particle Science*.

John Campbell. Figure 13.5 is used by permission of the author.

CDF Collaboration. Figures 13.2, 13.3, and 13.4, which are screen shots from the CDF trigger, are used by permission of the collaboration.

CERN. The event displays on the cover and Figures 6.5, 6.6, 13.9, 17.1, 17.2, 21.1, 21.3, 21.5, 21.7, 21.9, and 21.10 are reproduced from materials and reports published on the CERN web site and are used by permission of the CERN Press Office.

CKMfitter Collaboration. Figure 19.5 is reproduced from the web site of the CKMfitter Collaboration and is used by permission of the collaboration.

Elsevier. Figure 15.1 is reproduced from *Nuclear Physics* **A**. Figure 12.1 is reproduced from *Nuclear Physics* **B**. Figures 10.2, 10.5, 19.1, 19.3, 21.4, 21.6, and the figure in Exercise 7.2(l) are reproduced from *Physics Letters* **B**. Figures 17.5 and 17.8 are reproduced from *Physics Reports*.

Macmillan. Figure 5.1 is reproduced from *Nature*, ©1947, and is used by permission of Macmillan Publisher Ltd.

Lawrence Berkeley Laboratory. Fig. 5.2, ©2014, is reproduced by permission of the regents of the University of California through the Lawrence Berkeley National Laboratory

Kevin McFarland. Figures 15.4, from the web site of Kevin McFarland, is used by his permission and that of the NuTeV Collaboration.

Particle Data Group. Figures 6.1, 6.2, 6.3, 6.4, 8.1, 9.5, 10.1, 11.3, 13.1, and 21.2 are taken from editions of the *Review of Particle Physics* and are used by permission of the Particle Data Group.

Physics Institute of the University of Bonn. Figures 5.5 and 5.7 are reproduced from the proceedings of the 1981 International Symposium on Lepton and Photon Internations at High Energy and used by permission of the institute.

Toshinori Mori. Figure 17.4 is used by permission of the author.

Aldo Serenelli. Figure 20.3 is used by permission of the author.

SLAC. Figures 6.7, 8.2, 9.1, 10.3, and 10.4 are reproduced from the web site of the SLAC National Accelerator Laboratory and are used by permission of the laboratory achivist.

Springer. Figures 10.9, 12.2, 12.3, 17.3, 17.6, and 20.1 are reproduced from the *European Journal of Physics* **C**. Figures 8.3 and 15.5 are reproduced from Zeitschrift für Physik **C**.

Taylor & Francis. Figures 9.4 and 11.2 are reproduced from *An Introduction to Quantum Field Theory*, by Michael E. Peskin and Daniel V. Schroeder, ©1995, by permission of Westview Press, an imprint of Perseus Books, LLC. This book was recently acquired by the Taylor & Francis Group, a subsidiary of Informa UK Ltd. and is now published under the imprint of CRC Press.

ZEUS Collaboration. Figure 9.6, from the ZEUS Collaboration web site, is used by permission of the collaboration.

Contents

Part I

Preliminaries and Tools

Introduction

1

The aim of this book is to describe the interactions of nature that act on elementary particles at distances of the size of an atomic nucleus.

At this time, physicists know about four distinct fundamental interactions. Two of these are macroscopic—gravity and electromagnetism. Gravity has been known since the beginning of history and has been understood quantitatively since the time of Newton. Electrical and magnetic phenomena have also been known since ancient times. The unified theory of electromagnetism was given its definitive form by Maxwell in 1865. Through all of these developments, there was no sign that there could be additional fundamental forces. These would appear only when physicists could probe matter at very small distances.

The first evidence for additional interactions of nature was Becquerel's discovery of radioactivity in 1896. In 1911, Rutherford discovered that the atom consists of electrons surrounding a very tiny, positively charged nucleus. As physicists learned more about atomic structure, it became increasingly clear that the known macroscopic forces of nature could not give the full explanation. By the middle of the 20th century, experiments had revealed a series of questions that could not be resolved without new particles and interactions. These included:

These simple questions give the starting point for the exploration of subnuclear physics.

- What is radioactivity? Why do some atomic nuclei emit high-energy particles? What specific reactions are responsible? What are the particles that are emitted in radioactive decay?

- What holds the atomic nucleus together? The nucleus is made of positively charged protons and neutral neutrons. Electromagnetic forces destabilize the nucleus—as we see from the fact that heavy nuclei are unstable with respect to fission. What is the counterbalancing attractive force?

- What are protons and neutrons made of? These particles have properties that indicate that they are not elementary pointlike particles. What gives them structure? What kinds of particles are inside?

Experiments designed to study these issues produced more confusion before they produced more understanding. The proton and the neutron turned out to be the first of hundreds of particles interacting through

Concepts of Elementary Particle Physics. Michael E. Peskin.
© Michael E. Peskin 2019. Published in 2019 by Oxford University Press.
DOI: 10.1093/oso/9780198812180.001.0001

the nuclear force. The electron turned out to be only one of three apparently pointlike particles with electric charge but no strong interactions. All of these particles were observed to interact with one another through a web of new, short-ranged interactions. Finally, as the 1960's turned to the 1970's, the new interactions were sorted into two basic forces—called the strong and the weak interaction—and simple mathematical expressions for these forces were constructed. Today, physicists refer to these expressions collectively as "the Standard Model of particle physics".

Sometimes, authors or lecturers present the table of elementary particles of the Standard Model and imply that this is all there is to the story. It is not. The way that the forces of nature act on the elementary particles is beautiful and intricate. Often, the telling details of these interactions show up through remarkable aspects of the data when we examine elementary particle behavior experimentally.

These ideas elicit a related question: Of all the ways that nature could be built, how do we know that the Standard Model is the correct one? It seems hardly possible that we could pin down the exact nature of new fundamental interactions beyond gravity and electromagnetism. All of the phenomena associated with the new forces occur at distances smaller than an atomic nucleus, and in a regime where both special relativity and quantum mechanics play an essential role.

It is important to remember that the theory of particle physics must be studied together with the understanding of how experiments are done and how their results are interpreted.

In this book, I will explain the answers to these questions. It turns out that the new forces have common properties and can be built up from simple ingredients. The presence of these ingredients is revealed by well-chosen experiments. The dynamics of the new interactions becomes more clear at higher energies. With the benefit of hindsight, we can begin our study today by studying these dynamical ingredients in their simplest form, working out the consequences of these laws, and comparing the resulting formulae to data from high energy accelerator experiments that illustrate the correctness of these formulae in a very direct way.

Our quest for a fundamental theory of nature is far from complete. In the final chapter of the book, I will discuss a number of issues about fundamental forces for which we still have no understanding. It is also possible, as we probe more deeply into the structure of nature, that we will uncover new interactions that work at even smaller distances than those currently explored. But, at least, one chapter of the story, open since 1896, is now finished. I hope that, working through this book, you will not only understand how to work with the underlying theories describing the strong and weak interactions, but also that you will be amazed at the wealth of evidence that supports the connection of these theories to the real world.

Outline of the book.

The book is organized into three Parts. Part I introduces the basic materials that we will use to probe the nature of new forces at short distances. Parts II and III use this as a foundation to build up the Standard Model theories of the strong and weak interactions.

Part I

Part I begins with basic theory that underlies the subject of particle physics. Even before we attempt to write theories of the subnuclear forces, we expect that those theories will obey the laws of quantum mechanics and special relativity. I will provide some methods for using these important principles to make predictions about the outcome of elementary particle collisions.

In addition, I will describe the types of matter in the theories of strong and weak interactions, the basic elementary particles that interact through these forces. It turns out that there are two types of matter particles that are elementary at the level of our current understanding. Of these, one type, the *leptons*, are seen in our experiments as individual particles. There are six known leptons. Three have electric charge: the electron (e), the muon (μ), and the tau lepton (τ). The other three are the *neutrinos*, particles that are electrically neutral and extremely weakly interacting. Despite this, the evidence for neutrinos as ordinary relativistic particles is very persuasive; I will discuss this in Part III.

particles of the Standard Model: the leptons

Matter particles of the other type, the *quarks*, are hidden from view. Quarks appear as constituents of particles such as protons and neutrons that interact through the strong interaction. There are many known strongly interacting particles, collectively called *hadrons*. I will explain the properties of the most prominent ones, and show that they are naturally considered in families. On the other hand, no experiment has ever seen an isolated quark. It is actually a prediction of the Standard Model that quarks can never appear singly. This makes it especially challenging to learn their properties. One piece of evidence that the description of quarks in the Standard Model is correct is found from the fact it gives a simple explanation for the quantum numbers of observed hadrons and their assortment into families. I will discuss this also in Part I. In the process, I will give names to the hadrons that appear most often in experiments, so that we can discuss experimental methods more concretely.

particles of the Standard Model: the quarks

In a relativistic quantum theory, forces are also associated with particles that can be thought to transmit them. The Standard Model contains four types of such particles. These are the *photon*, the carrier of the electromagnetic interaction, the *gluon*, the carriers of the strong interaction, the W and Z bosons, the carriers of the weak interaction, and the *Higgs boson*, which plays a more subtle role. You will have already encountered the photon in your study of quantum mechanics. I will introduce the gluon in Part II and the W, Z, and Higgs bosons in Part III.

particles of the Standard Model: the bosons

To understand experimental findings about elementary particles, we will need to know at least the basics of how experiments on elementary particles are done, and what sorts of quantities describing their properties are measureable. I will discuss this material also in Part I.

Part II begins with a discussion of the most important experiments that give insight into the underlying character of the strong interaction. One might guess intuitively that the most convincing data on the strong interaction comes from the study of collisions of hadrons with other hadrons. That is incorrect. The experiments that were most crucial in understanding the nature of strong interaction involved electron scatter-

Part II

ing from protons and the annihilation of electrons and positrons at high energy. This latter process has an initial state with no hadrons at all. I will begin Part II with a discussion of the features of these processes at high energy. Our analysis will introduce the concept of the *current-current interaction*, which is an essential part of the physics of both the strong and weak interactions. Starting from the surprisingly simple features of these electromagnetic probes of strongly interacting particles, we will develop a series of arguments that pass back and forth between theory and experiment and eventually lead to a unique proposal for the underlying theory of the strong interaction. We will then describe tests of this theory of increasing sophistication, finally encompassing recent results from experiments at the Large Hadron Collider.

The final chapter of Part II presents our current understanding of the masses of quarks. At first sight, it might seems that it is straightforward to measure the mass of a quark, although the fact that quarks cannot be observed individually makes this more difficult. In fact, it turns out that the question of the quark masses brings in a number of new, subtle concepts. In particular, we will need to understand the idea of *spontaneous symmetry breaking*, a concept that will also prove to be an essential part of the theory of the weak interaction.

Part III Part III presents the description of the weak interaction. Here I will begin from a proposal for the nature of the weak interaction that again is based on a current-current interaction. I will present some quite counterintuitive, and even startling, predictions of that proposal and show that they are actually reproduced by experiment. From this starting point, again in dialogue between theory and experiment, we will build up the full theory. My discussion will include the precision study of the carriers of the weak interaction, the W and Z bosons, and the newest ingredients in this theory, the masses of neutrinos and the properties of the Higgs boson.

In comparing theory and experiment, I will generally present derivations of the formulae that give the most important theoretical predictions. To fully understand this book, it is important that you work through these derivations rather than just skimming the mathematics. For each one, take a large sheet of blank paper and carry out the calculation yourself. The details of the calculation will reveal insights that cannot by obtained from a purely qualitative discussion. You will find that these insights accumulate as you go through the book, as successive derivations cover similar ground from new perspectives.

Still, these derivations will be simplified with respect to a complete treatment of the Standard Model of particle physics. Most of the processes that I will consider will be studied in the limit of very high energies, where the mathematical analysis can be reduced as much as possible and made more transparent. This will be sufficient to discover the equations of the Standard Model and understand their basic tests. But a full discussion of the implications of the Standard Model would cover a more complete list of reactions, including some whose theoretical analysis is quite complex. A treatment of particle physics at that level is beyond the scope of this book.

In particular, many aspects of the theory of elementary particles cannot be understood without a deep understanding of quantum field theory. This book will explain those aspects of quantum field theory that are absolutely necessary for the presentation, but will omit any sophisticated discussion of this subject. A full description of the properties of elementary particles needs more.

For students who would like to study further in particle physics, there are many excellent references written from different and complementary points of view. I have put a list of the most useful texts at the beginning of the References.

A particularly useful reference work is the *Review of Particle Physics* assembled by the Particle Data Group (Patrignani *et al.* 2016). This volume compiles the basic properties of all known elementary particles and provides up-to-date reviews of the major topics in this subject. All elementary particle masses and other physical quantities quoted in this book but not explicitly referenced are taken from the summary tables given in that source.

Symmetries of Space-Time

2

We do not have complete freedom in postulating new laws of nature. Any laws that we postulate should be consistent with well-established symmetries and invariance principles. On distance scales smaller than an atom, space-time is invariant with respect to translations of space and time. Space-time is also invariant with respect to rotations and boosts, the symmetry transformations of special relativity. Many aspects of experiments on elementary particles test the principles of energy-momentum conservation, rotational invariance, the constancy of the speed of light, and the special-relativity relation of mass, momentum, and energy. So far, no discrepancy has been seen. So it makes sense to apply these powerful constraints to any proposal for elementary particle interactions.

Perhaps you consider this statement too strong. As we explore new realms in physics, we might well discover that the basic principles applied in more familiar settings are no longer valid. In the early 20th century, real crises brought on by the understanding of atoms and light forced physicists to abandon Newtonian space-time in favor of that of Einstein and Minkowski, and to abandon the principles of classical mechanics in favor of the very different tools of quantum mechanics. By setting relativity and quantum mechanics as absolute principles to be respected in the subnuclear world, we are making a conservative choice of orientation. There have been many suggestions of more radical approaches to formulating laws of elementary particles. Some of these have even led to new insights: The *bootstrap* of Geoffrey Chew, in which there is no fundamental Hamiltonian, is still finding new applications in quantum field theory (Simmons-Duffin 2017); *string theory*, which radically modifies space-time structure, is a candidate for the overall unification of particle interactions with quantum gravity (Zwiebach 2004, Polchinski 2005). However, the most successful routes to the theory of subnuclear interactions have taken translation invariance, special relativity, and standard quantum mechanics as absolutes. In this book, I will make the assumption that special relativity and quantum mechanics are correct in the realm of elementary particle interactions, and I will use their principles in a strong way to organize my exploration of elementary particle forces.

This being so, it will be useful to formulate the constraints from space-time symmetries in such a way that we can apply them easily. We would like to use the actual transformation laws associated with these

Concepts of Elementary Particle Physics. Michael E. Peskin.
© Michael E. Peskin 2019. Published in 2019 by Oxford University Press.
DOI: 10.1093/oso/9780198812180.001.0001

symmetries as little as possible. Instead, we should formulate questions in such a way that the answers are expressions *invariant* under space-time symmetries. Generally, there will be a small and well-constrained set of possible invariants. If we are lucky, only one of these will be consistent with experiment.

2.1   Relativistic particle kinematics

As a first step in simplifying the use of constraints from special relativity, I will discuss the kinematics of particle interactions. Any isolated particle is characterized by an energy and a vector momentum. In special relativity, these are unified into a 4-vector. I will write energy-momentum 4-vectors in energy units and notate them with an index $\mu = 0, 1, 2, 3$,

Representation of the energy and momentum of a particle in 4-vector notation.

$$p^\mu = (E, \vec{p}c)^\mu \ . \tag{2.1}$$

I will now review aspects of the formalism of special relativity. Probably you have seen these formulae before in terms of rulers, clocks, and moving trains. Now we will need to use them in earnest, because elementary particle collisions generally occur at energies at which it is essential to use relativistic formulae.

Under a boost by v along the $\hat{3}$ direction, the energy-momentum 4-vector transforms as $p \to p'$, with

$$E' = \frac{1}{\sqrt{1 - v^2/c^2}}(E + \frac{v}{c}p^3 c) \ , \quad p^{3\prime}c = \frac{1}{\sqrt{1 - v^2/c^2}}(p^3 c + \frac{v}{c}E) \ ,$$
$$p^{1,2\prime}c = p^{1,2}c \ . \tag{2.2}$$

It is convenient to write this as a matrix transformation

$$p' = \Lambda p \quad \text{with} \quad \Lambda = \begin{pmatrix} \gamma & 0 & 0 & \gamma\beta \\ 0 & 1 & 0 & 0 \\ 0 & 0 & 1 & 0 \\ \gamma\beta & 0 & 0 & \gamma \end{pmatrix} \ , \tag{2.3}$$

where

$$\beta = \frac{v}{c} \quad \gamma = \frac{1}{\sqrt{1 - \beta^2}} \ . \tag{2.4}$$

In multiplying matrices and vectors in this book, I will use the convention that repeated indices are summed over. Then, for example, I will write (2.3) as

$$p'^\mu = \Lambda^\mu{}_\nu \, p^\nu \ , \tag{2.5}$$

In this book, unless it is explicitly indicated otherwise, repeated indices are summed over. This convention is one of Einstein's lesser, but still much appreciated, innovations.

omitting the explicit summation sign for the index ν. Lorentz transformations leave invariant the Minkowski space vector product

$$p \cdot q = E_p E_q - \vec{p} \cdot \vec{q} \ . \tag{2.6}$$

To keep track of the minus sign in this product, I will make use of raised and lowered Lorentz indices. Lorentz transformations preserve the metric tensor

$$
\eta_{\mu\nu} = \begin{pmatrix} 1 & 0 & 0 & 0 \\ 0 & -1 & 0 & 0 \\ 0 & 0 & -1 & 0 \\ 0 & 0 & 0 & -1 \end{pmatrix} , \qquad \eta^{\mu\nu} = \begin{pmatrix} 1 & 0 & 0 & 0 \\ 0 & -1 & 0 & 0 \\ 0 & 0 & -1 & 0 \\ 0 & 0 & 0 & -1 \end{pmatrix} . \quad (2.7)
$$

Using this matrix, and the summation convention, we can write (2.6) as

$$
p \cdot q = p^\mu \eta_{\mu\nu} q^\nu . \tag{2.8}
$$

Alternatively, let q with a lowered index be defined by

$$
q_\mu = \eta_{\mu\nu} q^\nu = (E_q, -\vec{q})_\mu . \tag{2.9}
$$

The invariant product of p and q is written

$$
p \cdot q = p^\mu q_\mu . \tag{2.10}
$$

I will use raised and lowered Lorentz indices to keep track of the minus sign in the Minkowski vector product. Please pay attention to the position of indices—raised or lowered—throughout this book.

To form an invariant, we always combine a raised index with a lowered index. As the equations in this book become more complex, we will find this trick very useful in keeping track of the Minkowski space minus signs.

A particularly important Lorentz invariant is the square of a momentum 4-vector,

$$
p \cdot p \equiv p^2 = E^2 - |\vec{p}|^2 c^2 . \tag{2.11}
$$

Being an invariant, this quantity is independent of the state of motion of the particle. In the rest frame

$$
p^\mu = (E_0, \vec{0})^\mu . \tag{2.12}
$$

I will define the mass of a particle as its rest-frame energy

$$
(mc^2) \equiv E_0 . \tag{2.13}
$$

The mass of a particle is a Lorentz-invariant quantity that characterizes that particle in any reference frame.

Since p^2 is an invariant, the expression

$$
(mc^2)^2 = p^2 = E^2 - |\vec{p}|^2 c^2 \tag{2.14}
$$

is true in any frame of reference.

In this book, I will write particle momenta in two standard ways

$$
p^\mu = (E_p, \vec{p}c)^\mu \quad \text{or} \quad p^\mu = mc^2 \gamma (1, \vec{\beta})^\mu , \tag{2.15}
$$

where

Definitions of the quantities E_p, β, γ associated with relativistic particle motion.

$$
E_p = c(|\vec{p}|^2 + (mc)^2)^{1/2} , \qquad \beta = \frac{|\vec{p}|c}{E_p} , \qquad \gamma = (1 - \beta^2)^{-1/2} . \tag{2.16}
$$

Especially, the symbol E_p will always be used in this book to represent this standard function of momentum and mass. I will refer to a 4-vector with $E = E_p$ as being "on the mass shell".

To illustrate these conventions, I will now work out some simple but important exercises in relativistic kinematics. Imagine that a particle of mass M, at rest, decays to two lighter particles, of masses m_1 and m_2. In the simplest case, both particles have zero mass: $m_1 = m_2 = 0$. Then, energy-momentum conservation dictates that the two particle energies are equal, with the value $Mc^2/2$. Then, if the final particles move in the $\hat{3}$ direction, we can write their 4-vectors as

$$p_1^\mu = (Mc^2/2, 0, 0, Mc^2/2)^\mu \qquad p_2^\mu = (Mc^2/2, 0, 0, -Mc^2/2)^\mu \ . \quad (2.17)$$

The next case, which will appear often in the experiments we will consider, is that with m_1 nonzero but $m_2 = 0$. In the rest frame of the original particle, the momenta of the two final particles will be equal and opposite. With a little algebra, one can determine

$$p_1^\mu = (E_p, 0, 0, pc)^\mu \ , \qquad p_2^\mu = (pc, 0, 0, -pc)^\mu \quad (2.18)$$

These kinematic formulae will be used very often in this book.

(for motion in the $\hat{3}$ direction), where

$$E_p = \frac{M^2 + m_1^2}{2M} c^2 \ , \qquad pc = \frac{M^2 - m_1^2}{2M} c^2 \ . \quad (2.19)$$

It is easy to check that these formulae satisfy the constraints of total energy-momentum conservation and that p_1^μ satisfies the mass-shell constraint (2.14).

Finally, we might consider the general case of nonzero m_1 and m_2. Here, it takes a little more algebra to arrive at the final formulae

$$p_1^\mu = (E_1, 0, 0, pc)^\mu \ , \qquad p_2^\mu = (E_2, 0, 0, -pc)^\mu \quad (2.20)$$

with

$$E_1 = \frac{M^2 + m_1^2 - m_2^2}{2M} c^2 \ , \qquad E_2 = \frac{M^2 - m_1^2 + m_2^2}{2M} c^2 \ , \quad (2.21)$$

and

$$p = \frac{c}{2M} (\lambda(M, m_1, m_2))^{1/2} \ , \quad (2.22)$$

where the kinematic λ function is defined by

$$\lambda(M, m_1, m_2) = M^4 - 2M^2(m_1^2 + m_2^2) + (m_1^2 - m_2^2)^2 \ . \quad (2.23)$$

These three sets of formulae apply equally well to reactions with two particles in the initial state and two particles in the final state. It is only necessary to replace Mc^2 with the center of mass energy E_{CM} of the reaction.

2.2   Natural units

In the discussion of the previous chapter, I needed to introduce many factors of c in order to make the treatment of energy, momentum, and mass more uniform. This is a fact of life in the description of high energy particles. Ideally, we should take advantage of the worldview of relativity to pass seamlessly among these concepts. Equally well, our discussions of particle dynamics will take place in a regime in which quantum mechanics plays an essential role. To make the best use of quantum concepts, we should be able to pass easily between the concepts of momentum and wavenumber, or energy and frequency.

To make these transitions most easily, I will, in this book, adopt *natural units*,

$$\hbar = c = 1 \ . \tag{2.24}$$

That is, I will measure momentum and mass in energy units, and I will measure distances and times in inverse units of energy. For convenience in discussing elementary particle physics, I will typically use the energy units MeV or GeV. This will eliminate a great deal of unnecessary baggage that we would otherwise need to carry around in our formulae.

The conventions that define natural units.

For example, to write the mass of the electron, I will write

$$\text{not} \quad m_e = 0.91 \times 10^{-27} \text{g} \quad \text{but rather} \quad m_e = 0.51 \text{ MeV} \ . \tag{2.25}$$

An electron with a momentum of the order of its rest energy has, according to the Heisenberg uncertainty principle, a position uncertainty

$$\frac{\hbar}{m_e c} = 3.9 \times 10^{-11} \text{ cm} \ , \tag{2.26}$$

which I will equally well write as

$$\frac{1}{m_e} = (0.51 \text{ MeV})^{-1} \ . \tag{2.27}$$

Natural units make it very intuitive to estimate energies, lengths, and times in the regime of elementary particle physics. For example, the lightest strongly interacting particle, the π meson, has a mass

Natural units are useful for estimation.

$$m_\pi c^2 = 140 \text{ MeV} \ . \tag{2.28}$$

This corresponds to a distance

$$\frac{\hbar}{m_\pi c} = 1.4 \times 10^{-13} \text{ cm} \tag{2.29}$$

and a time

$$\frac{\hbar}{m_\pi c^2} = 0.47 \times 10^{-22} \text{ sec} \ . \tag{2.30}$$

These give—within a factor 2 or so—the size of the proton and the lifetimes of typical unstable hadrons. So, the use of m_π gives a good first estimate of all dimensionful strong interaction quantities. To obtain an estimate in the desired units—MeV, cm, sec—we would decorate the simple expression m_π with appropriate factors of $\hbar$ and c and then evaluate as above.

The material in this book will be easier to grasp if you make yourself comfortable with the use of natural units. This will both simplify formulae and simplify many estimates of energies, distances, and times.

It may make you uncomfortable at first to discard factors of $\hbar$ and c. Get used to it. That will make it much easier for you to perform calculations of the sort that we will do in this book. Some useful conversion factors for moving between distance, time, and energy units are given in Appendix B.

One interesting quantity to put into natural units is the strength of the electric charge of the electron or proton. In this book, the constant e will be a positive quantity equal to the electric charge of the proton. The electric charge of the electron will be $(-e)$. More generally, I will write the charge of a particle as Qe, with $Q = +1$ for a proton and $Q = -1$ for an electron.

The Coulomb potential between charges 1 and 2 is given in standard notation by

$$V(r) = \frac{Q_1 Q_2 e^2}{4\pi\epsilon_0 r} \ . \tag{2.31}$$

I will use units for electromagnetism in which also

$$\epsilon_0 = \mu_0 = 1 \ . \tag{2.32}$$

Then the Coulomb potential reads

$$V(r) = \frac{Q_1 Q_2 e^2}{4\pi} \frac{1}{r} \ . \tag{2.33}$$

Since r, in natural units, has the dimensions of $(\text{energy})^{-1}$, the value of the electric charge must have a form in which it is dimensionless. Indeed,

$$\alpha \equiv \frac{e^2}{4\pi\epsilon_0 \hbar c} \tag{2.34}$$

is a dimensionless number, called the *fine structure constant*, with the value

$$\alpha = 1\,/\,137.036 \ . \tag{2.35}$$

The intrinsic strengths of the basic elementary particle interactions are not apparent from the size of their effects—or from their names. Here is a preview.

There are two remarkable things about this equation. First, it is surprising that there is a dimensionless number α that characterizes the strength of the electromagnetic interaction. Second, that number is small, signalling that the electromagnetic interaction is a weak interaction. One of the goals of this book will be to determine whether the strong and weak subnuclear interactions can be characterized in the same way, and whether these interactions—looking beyond their names—are intrinsically strong or weak. I will discuss estimates of the strong and weak interaction coupling strengths at appropriate points in the course. It will turn out that the strong interaction is weak, at least when measured under the correct conditions. It will also turn out that the weak interaction is also weak in dimensionless terms. It is weaker than the strong interactions, but not as weak as electromagnetism.

Group theory plays an important role in quantum mechanics, and this importance extends to the study of elementary particle physics. You have encountered group theory concepts in your quantum mechanics course, but it is likely that those arguments did not make explicit reference to group theory. In particle physics, we lean much more heavily on group theory, and so it is best to discuss these concepts formally and give them their proper names. Please, then, study Sections 2.3 and 2.4 carefully, especially if you are uncomfortable with mathematical abstraction. With careful reading, you will see that the concepts I describe generalize physical arguments that are already familiar to you.

2.3   A little theory of discrete groups

Group theory is a very important tool for elementary particle physics. In this section and the next, I will review how group theory is used in quantum mechanics, and I will discuss some properties of groups that we will meet in this book. To do this, I will mainly discuss systems that you have already studied in your quantum mechanics course, giving a new description of these systems using a more formal and abstract mathematical language,. Please do not be put offby this. The mathematical terms and concepts that I will introduce will generalize to and, I hope, illuminate, many new systems that we will study in this book.

In quantum mechanics, we deal with groups on two levels. First, there are abstract groups. In mathematics, a *group* is a set of elements $G = \{a, b, \ldots\}$ with a multiplication law defined, so that ab is defined and is an element of G. The multiplication law satisfies the three properties

Here are the axioms that define a group.

(1) Multiplication is associative: $a(bc) = (ab)c$.

(2) G contains an *identity element* 1 such that, for any element of G, $1a = a1 = a$.

(3) For each a in G, there is an *inverse* element a^{-1} in G such that $aa^{-1} = a^{-1}a = 1$.

Every symmetry of nature normally encountered in physics satisfies these axioms and is described by an abstract group.

Next, we need to relate the abstract group to operators that act on the space of states in a quantum mechanics problem. A group transformation is a symmetry of a quantum-mechanical system if it leaves the dynamics of that system invariant. This will be true if the transformations commute with the Hamiltonian H. It is useful to state this relationship more precisely.

In quantum mechanics, the quantum states are vectors in a Hilbert space. Symmetries act to convert one of these states into another. A symmetry transformation carries each state into another one in such as way that the whole Hilbert space is mapped into itself, preserving norms, that is, preserving quantum mechanical probabilities. Thus it must act as a unitary transformation of the Hilbert space. A group G is then described in a quantum mechanics problem as a set of unitary transformations $\{\mathcal{U}(a)\}$, one for each element of G, that obey the multiplication law of G. That is, if a, b, c are elements of G and $ab = c$, the transformations in the set should satisfy $\mathcal{U}(a)\mathcal{U}(b) = \mathcal{U}(c)$. Such a set of unitary transformations is called a *unitary representation* of G.

The group G will be a *symmetry* of the quantum mechanics problem if, for all elements a of G, the corresponding unitary transformations commute with the Hamiltonian,

The action of a group on the Hilbert space of states in quantum mechanics is described through unitary representations of the group, that is, through unitary matrices with the same multiplication law as the corresponding group elements. Thus, unitary group representations will be used in many aspects of the physics discussed in this book.

$$[\mathcal{U}(a), H] = 0 . \tag{2.36}$$

A direct consequence of (2.36) is that, if $|\psi\rangle$ is an eigenstate of H with energy E, $\mathcal{U}(a)|\psi\rangle$ will also be an eigenstate of H with the same energy. The inverse operation to a will be represented by the inverse transformation: $\mathcal{U}(a^{-1}) = \mathcal{U}(a)^{-1} = \mathcal{U}(a)^\dagger$.

Without reference to quantum mechanics, but just thinking about the pure mathematics of groups, we can ask the following question: Given a group G with elements $\{a\}$, can we find a set of finite-dimensional unitary matrices U_a that obey the multiplication law of the group, that is, such that $U_a U_b = U_c$? This is a classical mathematics problem, and mathematicians have classified the possible answers for all of the groups commonly encountered in physics, and for more complex examples, including one called the "Monster" group. These *finite-dimensional unitary representions* of groups will appear in quantum mechanics problems as tranformations among a finite number of eigenstates of the Hamiltonian that implement a symmetry of the problem.

The theory of group representations is a deep subject that can become quite technically complex. It is certainly not necessary to master this subject in order to study elementary particle physics. In this book, we will make use of only the simplest examples. However, it is often good to recognize that a physics problem of interest involves the representations of a relevant symmetry group. This gives a starting point to analyze the problem and connects the solution to those of other problems with which we might be more familiar.

Here is a simple example: Consider the abstract group called Z_2 that contains two elements $\{1, -1\}$ satisfying the multiplication law

$$1 \cdot 1 = (-1)(-1) = 1 , \quad 1 \cdot (-1) = (-1) \cdot 1 = (-1) . \tag{2.37}$$

We can represent Z_2 in a quantum mechanics problem by 2×2 unitary matrices. To make this concrete, consider a system with two particles π^+ and π^-. Define the operator C to transform

$$C \left| \pi^+ \right\rangle = \left| \pi^- \right\rangle , \qquad C \left| \pi^- \right\rangle = \left| \pi^+ \right\rangle . \tag{2.38}$$

The action of C on this 2-dimensional space is given by the matrix

$$\begin{pmatrix} 0 & 1 \\ 1 & 0 \end{pmatrix} \qquad \text{acting on} \qquad \begin{pmatrix} \left| \pi^+ \right\rangle \\ \left| \pi^- \right\rangle \end{pmatrix} . \tag{2.39}$$

If H is the Hamiltonian for this quantum-mechanical system and $[C, H] = 0$, that would imply that the masses and decay rates of π^+ and π^- must be equal. On the same Hilbert space, we can define the trivial operation

$$\mathbf{1} \left| \pi^+ \right\rangle = \left| \pi^+ \right\rangle , \qquad \mathbf{1} \left| \pi^- \right\rangle = \left| \pi^- \right\rangle . \tag{2.40}$$

This is given by

$$\begin{pmatrix} 1 & 0 \\ 0 & 1 \end{pmatrix} \qquad \text{acting on} \qquad \begin{pmatrix} \left| \pi^+ \right\rangle \\ \left| \pi^- \right\rangle \end{pmatrix} . \tag{2.41}$$

The unitary matrices $\{\mathbf{1}, C\}$ form a 2-dimensional unitary representation of the group Z_2. If these matrices commute with H, we say that H has Z_2 symmetry.

We could have discussed the relation of C to H and its eigenstates without making explicit reference to the fact that the unitary matrix C is part of a group representation. However, using the language of group theory connects this example to others that we might have studied. Not all groups are as simple to understand as Z_2. The more complicated the group, the more useful this connection is.

A group G is called *Abelian* if, for all a, b in G, $ab = ba$. A unitarity representation of an Abelian group G consists of unitary matrices that commute with one another. This means that they can be simultaneously diagonalized. The operation of the group is then reduced to simple numbers. In the example above, the matrices (2.41) and (2.39) are diagonalized in a common basis. It is conventional to use C also as a symbol for the eigenvalue of C on one of its eigenstates. In this case, the eigenstates are

An Abelian group is described by its eigenstates and their eigenvalues. The eigenvalues are precisely what physicists call the *quantum numbers* of a state.

$$C = +1 \; : \; [\,|\pi^+\rangle + |\pi^-\rangle]/\sqrt{2}$$
$$C = -1 \; : \; [\,|\pi^+\rangle - |\pi^-\rangle]/\sqrt{2} \; . \tag{2.42}$$

Because $C^2 = 1$, operating twice with the matrix C must give back the original state: $C \cdot C \, |\psi\rangle = |\psi\rangle$. This must, in particular, be true for an eigenstate. Then the eigenvalues of C can only be ± 1. We say that the first state in (2.42) has $C = +1$ and the second has $C = -1$.

Symmetries of the Hamiltonian may involve transformations of space-time coordinates, such as the special relativity transformations discussed in Section 2.1. These are called *space-time symmetries*. In examples like the one above, the symmetries relate different particles or quantum states without reference to space-time. These are called *internal symmetries*. A given abstract group such as Z_2 may describe a space-time or an internal symmetry.

Space-time symmetries vs. internal symmetries.

If G contains two elements a, b that do not commute, $ab \neq ba$, it is called a *non-Abelian* group. If G is non-Abelian, and $\{U_a\}$ is a finite-dimensional unitary representation of G, it is generally not possible to simultaneously diagonalize all of the unitary matrices in $\{U_a\}$. However, by a change of basis, we can reduce these matrices to a common block-diagonal form

$$U_R \rightarrow \begin{pmatrix} U_1 & 0 & 0 \\ 0 & U_2 & 0 \\ 0 & 0 & U_3 \end{pmatrix} , \tag{2.43}$$

where the blocks $U_1, U_2, U_3, \cdots$ are as small as possible. These minimal-size unitary transformations representing G are called *irreducible unitary representations of G*. For an irreducible representation $\{U_i\}$, the size of the matrices is called the *dimension d_i* of the representation. The notion of irreducible representations is probably more familiar to you in the context of continuous groups. I will put your knowledge of the rotation group into this context in the next section.

The concept of an *irreducible* group representation. Many physics problems in quantum mechanics are solved by breaking up a larger Hilbert space into irreducible representations of an appropriate symmetry group.

Mathematicians have shown that, for each discrete group G with a finite number of elements, there is only a limited number of inequivalent finite-dimensional irreducible representations. Any other matrix representation of the group is reducible, in the sense of (2.43), into a sum of these basic representations. It can be proved that, for a discrete group G with n elements, the inequivalent unitary transformations satisfy

$$\sum_i d_i^2 = n \; . \tag{2.44}$$

Once we have found irreducible representations that add up to (2.44), we have completely determined the structure of the possible representations of G.

An example is given by the group of Π_3 of permutations on three elements. We can represent such a permutation as the result of transforming the set of labels [123] to a set of labels in another order. With this representation, the group has 6 elements that can be written

$$\{ \ [123] \ , \ [231] \ , \ [312] \ , \ [132] \ , \ [321] \ , \ [213] \ \} \ . \tag{2.45}$$

Permutations multiply $a \cdot b = c$ by composition, for example,

$$[231] \cdot [231] = [312]$$
$$[132] \cdot [312] = [321] \ . \tag{2.46}$$

That is, applying the two permutations in order (right to left) gives the resulting permutation as shown.

The 6 permutations in (2.45) can be associated with 6 states in a Hilbert space. In this representation, the representation matrices are 6×6 matrices with entries 0 and 1. It can be shown that this is a reducible representation. It contains two 1-dimensional irreducible representations. One of these is the trivial representation that multiplies each element by 1. Another is the representation that multiplies a state by +1 for an even or cyclic permutation—the first three elements of (2.45)—and multiplies a state by −1 for an odd permutation—the last three elements of (2.45). There is also one 2-dimension representation, presented in Exercise 2.3. These three irreducible representations together satisfy (2.44).

2.4   A little theory of continuous groups

The concepts reviewed in the previous section extend to the situation of groups with a continous set of elements. Important examples are the basic space-time symmetries: the group of spatial translations, the group of spatial rotations, and the group of Lorentz transformations, which includes rotations and boosts.

The group of space translations has the simplest structure. All translations commute with one another. You learned in quantum mechanics that translations are implemented by unitary transformations. For translations by a in one dimension

$$U(a) = \exp[-iaP] \tag{2.47}$$

where P is the operator measuring the total momentum of the system. This is made most clear by considering the wavefunction of a plane wave of momentum p,

The action of a space translation in quantum mechanics gives a simple example of a unitary representation of an Abelian group.

$$\langle x | p \rangle = e^{ipx} \ . \tag{2.48}$$

Acting on the state $|p\rangle$ with (2.47), we find

$$\langle x | \, U(a) \, | p \rangle = e^{ip(x-a)} \ , \tag{2.49}$$

which is the same wavefunction displaced by a. Using the language introduced in the previous section, we say that the set of unitary operators $\{U(a)\}$ is a unitary representation of the group of space translations.

The expression of each $U(a)$ as an exponential implies a relation between the group of translations and the Hermitian operator P. We describe this relationship by saying that P is the *generator* of $\{U(a)\}$ or the generator of the group of translations.

The statement that P is Hermitian is equivalent to the statement that the $U(a)$ are unitary,

$$U(a)^\dagger = \exp[+iaP^\dagger] = \exp[+iaP] = U(a)^{-1} \ . \tag{2.50}$$

Then, continuous unitary transformations are generated by Hermitian operators. In quantum mechanics, Hermitian operators correspond to observables.

Observables have time-independent values if the corresponding operators commute with the Hamiltonian of the quantum mechanics problem. In this example, momentum is conserved if $[P, H] = 0$. Through the correspondence (2.47), this statement is exactly equivalent to the statement that $[U(a), H] = 0$, that is, that the equations of motion of the system are invariant under translations. This relation is completely general. If Q is a Hermitian operator on the Hilbert space, the statement that Q is a conserved quantity,

In quantum mechanics, every symmetry that leaves the Hamiltonian invariant is associated with a conserved quantity. This follows from the connection between Hermitan operators and unitary symmetry transformations.

$$[Q, H] = 0 \ , \tag{2.51}$$

is equivalent to the statement that Q generates a symmetry of the equations of motion,

$$[U_Q(a), H] = 0 \quad \text{for} \quad U_Q(a) = \exp[-iaQ] \ . \tag{2.52}$$

This is the quantum-mechanical version of *Noether's theorem* in classical mechanics: Every symmetry of the equations of motion is associated with a conservation law, and vice versa.

The group of translations is an Abelian group, since all translations commute with one another. This implies that all of the matrices $U(a)$ can be simultaneously diagonalized. Actually, for every $U(a)$, the eigenstates of $U(a)$ are the eigenstates of P, that is, states of definite momentum. Each eigenstate of P gives a one-dimensional unitary representation of the translation group.

A non-Abelian continuous group that should be familiar to you is the rotation group in 3 dimensions. In quantum mechanics, rotations are implemented on the Hilbert space by the unitary operators

The action of rotations in quantum mechanics gives an example of the unitary representation of a non-Abelian group.

$$U(\vec{\alpha}) = \exp[-i\vec{\alpha} \cdot \vec{J}] \tag{2.53}$$

where $\vec{\alpha}$ gives the axis and angle of the rotation and $\vec{J}$ are the operators of angular momentum. These operators satisfy the commutation relation

As in the previous example, the conservation law of angular momentum is associated with the symmetry of invariance under rotations.

$$[J^i, J^j] = i\epsilon^{ijk} J^k \ . \tag{2.54}$$

It can be shown that, if Hermitian operators J^i satisfy (2.54), the unitary operators constructed from them satisfy the composition rules of 3d rotations. That is, if

$$U(\vec{\beta})U(\vec{\alpha}) = U(\vec{\gamma}) , \qquad (2.55)$$

then the rotation $\vec{\gamma}$ is the one that results from rotating first through $\vec{\alpha}$ and then through $\vec{\beta}$. The operators J^i are thus the generators of rotations. In fact the complete structure of the group of rotations is specified by the commutation relation (2.54).

In quantum mechanics, finite-dimensional matrix representations of the rotation group play an important role. The quantum states of atoms are organized into multiplets of definite angular momentum, for example, the 2P or 3D states of the hydrogen atom. States of definite angular momentum give the finite-dimensional irreducible matrix representations of the rotation group.

Through the correspondence (2.53), a finite-dimensional representation of the rotation group is generated by a set of finite-dimensional matrices that satisfy (2.54). The simplest such representations are the trivial, 1-dimensional representation

$$J^i = 0, \qquad (2.56)$$

the 2-dimensional representation

$$J^1 = \frac{\sigma^1}{2} , \quad J^2 = \frac{\sigma^2}{2} , \quad J^1 = \frac{\sigma^3}{2} , \qquad (2.57)$$

where σ^i are the Pauli sigma matrices

$$\sigma^1 = \begin{pmatrix} 0 & 1 \\ 1 & 0 \end{pmatrix} , \quad \sigma^2 = \begin{pmatrix} 0 & -i \\ i & 0 \end{pmatrix} , \quad \sigma^3 = \begin{pmatrix} 1 & 0 \\ 0 & -1 \end{pmatrix} , \qquad (2.58)$$

and the 3-dimensional representation

$$J^1 = \begin{pmatrix} 0 & 0 & 0 \\ 0 & 0 & -i \\ 0 & i & 0 \end{pmatrix} , \quad J^2 = \begin{pmatrix} 0 & 0 & i \\ 0 & 0 & 0 \\ -i & 0 & 0 \end{pmatrix} , \quad J^3 = \begin{pmatrix} 0 & -i & 0 \\ i & 0 & 0 \\ 0 & 0 & 0 \end{pmatrix} . \qquad (2.59)$$

It is instructive to check explicitly that (2.57) and (2.59) satisfy (2.54). The three representations given here are those of spin 0, spin $\frac{1}{2}$, and spin 1. We will meet these representations again and again in the applications I will discuss in this book. Similarly, for every integer or half-integer value j, there is a set of three $(2j+1) \times (2j+1)$ matrices satisfying these commutation relations. This is the spin j representation of the rotation group.

One of the standard problems in atomic physics is to decompose a set of quantum states into irreducible representations of the rotation group. For example, states of an atom may be labelled by orbital angular momentum ℓ and spin angular momentum s. This gives a set of states with $(2\ell+1)(2s+1)$ elements. The total angular momentum j takes values

$$|\ell - s| \leq j \leq (\ell + s) . \qquad (2.60)$$

The reduction of a set of states of an atom with orbital and spin angular momenta (ℓ, s) into states of total angular momentum j is an example of the reduction of a reducible representation of a continuous group—in this case, the rotation group—into a sum of irreducible representations.

Since $[\vec{J}, H] = 0$, each value of j gives a set of $(2j + 1)$ states with the same energy. In Section 4.1, we will translate this group theory exercise into a statement about the energy levels of the hydrogen atom.

We can consider the group of rotations in 3 dimensions as an abstract group whose multiplication law is defined by the composition of rotations. This group is called $SO(3)$. Similarly, there is an abstract group of rotations in d dimensions, called $SO(d)$. The case $d = 2$ is simple; it is the group of rotations of a circle, an Abelian group of translations of an angle ϕ, with ϕ identified with $(\phi + 2\pi)$. This abstract group is the same one that we meet when we consider the group of phase transformations

$$e^{-i\phi} \to e^{-i\alpha}e^{-i\phi} \ . \tag{2.61}$$

This is a transformation by a 1×1 unitary matrix, so we also call this group $U(1)$.

General $n \times n$ unitary matrices form a representation of an abstract group called $U(n)$. Any $n \times n$ unitary matrix can be written in the form of (2.47) as generated by a set of $n \times n$ Hermitian matrices

$$U = \exp[-i\alpha^a t^a] \ . \tag{2.62}$$

The sum over a runs over a basis of $n \times n$ Hermitian matrices, which contains n^2 elements. One of these elements is the unit matrix,

$$t^0 = 1 \tag{2.63}$$

This matrix commutes with all of the other t^a. If we omit this element from the set of Hermitian matrices, we obtain a non-Abelian group of matrices with $n^2 - 1$ generators, the $n \times n$ Hermitian matrices with zero trace. The group generated by these $n^2 - 1$ matrices is called $SU(n)$. It is the group of $n \times n$ unitary matrices with determinant 1.

Definition of the group $SU(n)$.

For $n = 2$, the Pauli sigma matrices (2.58) form a basis for the 2×2 traceless Hermitian matrices. Thus, $SO(3)$ and $SU(2)$ are names for the same abstract group. (Mathematicians make a distinction between these groups, but the difference will not be relevant to the calculations done in this book.) This abstract group describes rotations in three dimensions, but it will also describe some internal symmetries of elementary particles that we will meet in the course of our discussion.

A continuous group of transformations generated by Hermitian matrices, in the form (2.62), is called a *Lie group*. The commutation algebra of the generators t^a,

$$[t^a, t^b] = i f^{abc} t^c \tag{2.64}$$

This equation, which expresses the non-commuting nature of the generators of a Lie group, contains the full information about the representations and the geometry of the group.

is called the *Lie algebra* of the group. The constants f^{abc} are called the *structure constants* of the Lie algebra. It can be shown that we can always choose a basis for the t^a such that the structure constants f^{abc} are completely antisymmetric in $[abc]$. These definitions straightforwardly generalize the presentation that I have given of the rotation group in 3 dimension. In the case of the rotation group,

$$f^{abc} = \epsilon^{abc} \ . \tag{2.65}$$

In the same way as for the rotation group, the Lie algebra of the generators determines the multiplication law of any two elements of the group.

In this book, we will meet only special cases of Lie groups. The particular groups $U(1) = SO(2)$, $SU(2) = SO(3)$, and $SU(3)$ will have important roles in our story. Still, the abstract properties of Lie groups will be useful to us in understanding how to apply these groups to particle physics. I will introduce some further formalism of Lie groups when we will need it in Chapter 11.

2.5 Discrete space-time symmetries

The symmetries of special relativity include the continuous symmetries of rotations and Lorentz transformations. But they also include two distinct space-time transformations that leave the metric tensor (2.7) invariant but cannot be constructed as a product of continuous rotations and boosts. This will turn out to be an important issue for elementary particle physics. According to Noether's theorem, conservation of energy-momentum is equivalent to the invariance of the equations of motion with respect to space-time translations, and the conservation of angular momentum is equivalent to the invariance of the equations of motion with respect to rotations and boosts. However, there is no fundamental principle that requires that extra, discrete space-time transformations must be symmetries of the Hamiltonian or that quantities associated with these extra discrete symmetries must be conserved. This is a separate question that can only be answered by experiment. We will see in Part III that the answer given to this question is quite surprising.

Minkowski space has two extra space-time symmetries: *parity P* and *time-reversal T*.

The two space-time transformations that are not part of the continuous Lorentz group are *parity* (P) and *time reversal* (T). These space-time operations satisfy

$$P^2 = 1 \qquad T^2 = 1 \tag{2.66}$$

In quantum mechanics, these transformations are implemented by operators with eigenvalues ± 1. I will also refer to the eigenvalue of a quantum state as the value P or T for that state. Continuous Lorentz invariance does not imply that these values P and T are conserved. However, P and T are observed to be conserved in electromagnetism and atomic physics. The study of energy levels of nuclei confirms that P and T are also conserved by the strong nuclear interaction.

Parity is defined as the operation on 4-vectors

$$x^\mu = (x^0, \vec{x})^\mu \to (x^0, -\vec{x})^\mu \ . \tag{2.67}$$

A rotation matrix, for example,

$$\Lambda = \begin{pmatrix} 1 & 0 & 0 & 0 \\ 0 & \cos\theta & -\sin\theta & 0 \\ 0 & \sin\theta & \cos\theta & 0 \\ 0 & 0 & 0 & 1 \end{pmatrix} , \tag{2.68}$$

or, indeed, any matrix that implements a continuous Lorentz transformation, has

$$\det \Lambda = +1 \ . \tag{2.69}$$

But (2.67) is implemented by a matrix with $\det \Lambda = -1$. Thus, this matrix cannot be generated as a product of continuous rotations. Time reversal is defined similarly as the operation

$$x^\mu = (x^0, \vec{x})^\mu \to (-x^0, \vec{x})^\mu \ . \tag{2.70}$$

By the same logic, time reversal cannot be continuously generated.

In quantum mechanics, an isolated particle can also have an *intrinsic parity*. That is, under parity, its quantum state of momentum $\vec{k}$ can transform as

$$P \left| A(\vec{k}) \right\rangle = + \left| A(-\vec{k}) \right\rangle \quad \text{or} \quad - \left| A(-\vec{k}) \right\rangle \ . \tag{2.71}$$

We refer to these two cases as intrinsic parity $(+1)$ or (-1). A particle can also have an intrinsic quantum number under time reversal.

A quantum particle can have *intrinsic parity* +1 or −1.

In quantum mechanics, time reversal is implemented by an *anti-unitary* operator. In this book, I will avoid detailed analysis of time-reversal properties as much as possible.

There is one more discrete transformation that is closely related to these space-time operations. As we will see in the next chapter, quantum field theory implies that, for each particle in nature, there must exist an *antiparticle* with the same mass and opposite values of all conserved charges. We can then define an operation called *charge conjugation* (C) that converts each particle to its antiparticle and vice versa. C then also naturally satisfies

It is useful to consider *charge conjugation C* as a discrete space-time transformation on the same level as P and T.

$$C^2 = 1 \ . \tag{2.72}$$

Quantum states can have intrinsic values of C equal to $+1$ or -1. C is observed to be conserved in electromagnetic and strong nuclear reactions.

I have already explained that it is a question for experiment whether P, C, and T are conserved by all interactions in nature. However, it is a theorem in quantum field theory that the combination CPT must be a symmetry of all particle interactions. This statement can be tested experimentally and, so far, it holds up. We will take up the issue of the separate conservation of P, C, and T in our discussion of the weak interaction in Part III.

Exercises

(2.1) Consider the decay of a particle of mass M, at rest, into two particles with masses m_1 and m_2, both nonzero. With an appropriate choice of axes, the momentum vectors of the final particles can be written

$$p_1 = (E_1, 0, 0, p) \qquad p_2 = (E_2, 0, 0, -p) \qquad (2.73)$$

with $E_1^2 = p^2 + m_1^2$, $E_2^2 = p^2 + m_2^2$.

(a) Show that

$$p = \left[M^4 - 2M^2(m_1^2 + m_2^2) + (m_1^2 - m_2^2)^2 \right]^{1/2} / 2M \qquad (2.74)$$

(b) Take the limit $m_2 \to 0$ and show that this formula reproduces the result (2.19) for the decay into one massive and one massless particle.

(c) Find formulae for E_1 and E_2 in terms of M, m_1, m_2.

(2.2) Using natural units, estimate the following quantities:

(a) If the photon has a mass, the electric fields generated by charges will fall off exponentially at distances larger than the photon Compton wavelength. It is possible to obtain limits on the photon mass by looking for this effect in the solar system. For example, the magnetic field of Jupiter is found to be a conventional dipole field out to many times the radius of the planet. Estimate the corresponding upper limit on the photon mass in MeV.

(b) The range of the weak interaction is given by Compton wavelength of the W boson, which has a mass of 80.4 GeV. Estimate this length in cm.

(c) If the electron is a composite particle with a nonzero size, that will affect the observed rate for electron-electron and electron-positron scattering. Given that these rates are in good agreement with the predictions for pointlike electrons up to a center of mass energy of 200 GeV, estimate the upper limit on the size of the electron, in cm.

(2.3) Show that the following are unitary representations of the permutation group Π_3 by verifying that they satisfy the multiplication law of Π_3:

(a) The 1-dimension representation in which all six permutations in (2.45) are represented by 1.

(b) The 1-dimension representation in which [123], [231], and [312] are represented by 1 and [213], [321], and [132] are represented by -1.

(c) The 2-dimensional representation that assigns

$$[123] \to \begin{pmatrix} 1 & 0 \\ 0 & 1 \end{pmatrix} \qquad [231] \to \begin{pmatrix} 0 & -1 \\ 1 & -1 \end{pmatrix}$$

$$[312] \to \begin{pmatrix} -1 & 1 \\ -1 & 0 \end{pmatrix} \qquad [213] \to \begin{pmatrix} 0 & 1 \\ 1 & 0 \end{pmatrix}$$

$$[321] \to \begin{pmatrix} 1 & -1 \\ 0 & -1 \end{pmatrix} \qquad [132] \to \begin{pmatrix} -1 & 0 \\ -1 & 1 \end{pmatrix}$$

$$(2.75)$$

(2.4) This problem explores the non-Abelian nature of the Lorentz group.

(a) The 4×4 matrix $\Lambda_3(\beta)$ that represents a boost by β in the $\hat{3}$ direction is given by (2.3). Write the corresponding 4×4 matrix $\Lambda_1(\beta)$ that represents a boost by β in the $\hat{1}$ direction.

(b) Compute the composite Lorentz transformation $\Lambda_C = \Lambda_1(\beta) \Lambda_3(\beta)$. The component $\Lambda_C{}^0{}_0$ of this matrix should be the composite boost γ_C. From this, compute the new velocity β_C.

(c) By acting Λ_C on the 4-vector $(1, 0, 0, 0)$, show that the elements $\Lambda_C{}^i{}_0$ give the direction of the boost. Show from this that the new velocity is $\vec{\beta}_C = (\beta, 0, \beta/\gamma)$. Show that the magnitude of this vector agrees with the result of part (b).

(d) The matrix Λ_C is not symmetric, so it cannot be a pure boost. It is, in fact, a combination of a boost and a rotation. To understand this better, expand the elements of Λ_C in powers of β for small β, keeping terms up to order β^2.

(e) Write the Λ matrix for a pure boost to the velocity $\vec{\beta}_C$. This matrix should be symmetric. The space-space part should be

$$\delta^{ij} + (\gamma_C - 1) \frac{\beta_C^i \beta_C^j}{\beta_C^2} . \qquad (2.76)$$

(f) Expand the matrix found in part (e) to order β^2. Show that it explains the symmetric part of the result found in part (d). Identify the remaining antisymmetric part as an infinitesimal rotation in the $\hat{3}$-$\hat{1}$ plane. The rotation that results from the non-commuting nature of boosts in different directions is called a *Wigner rotation*.

Relativistic Wave Equations

3

In the previous chapter, I developed some simple rules for the treatment of special relativity that will aid us in our search for the laws of elementary particle interaction. In this chapter, I will discuss some of the concepts that we will need to use quantum mechanics effectively.

The standard treatment of 1-particle quantum mechanics will not be adequate for our purposes. First of all, the Schrödinger equation is not Lorentz-invariant. In that equation, time and space appear asymmetrically. In a relativistic theory, the wavefunctions of quantum particles must obey wave equations in which time and space appear symmetrically in accord with special relativity. In this chapter, I will discuss three of the most important of these equations.

Standard quantum mechanics is inadequate in another way. In elementary particle reactions, the number of particles can change as individual particles are created and destroyed. We have already noted in the previous chapter that every particle must have an antiparticle with the same mass. Typically, elementary particle interactions allow the creation of a particle together with its antiparticle, or the annihilation of a particle with its antiparticle. Then, quantum mechanics must be generalized to a multiparticle theory.

> The Schrödinger equation is not adequate to describe elementary particles. We need a theoretical framework that is relativistic, and that allows particles to be created and destroyed.

Both generalizations are accomplished in relativistic quantum field theory. However, there is no space in this small book for a complete description of quantum field theory, or even for a derivation of its major implications. Instead, I will use this chapter to explain some essential points of quantum field theory that will be needed for our analysis. In Chapter 7, I will explain how we use quantum field theory to make predictions for elementary particle reactions, and I will give some shortcuts and heuristics that will allow us to apply these ideas easily.

3.1 The Klein-Gordon equation

A wave equation is said to be invariant under a group of symmetries if, for any solution, the symmetry transform of that solution is another solution of the wave equation. For a scalar field, the Lorentz transform

Concepts of Elementary Particle Physics. Michael E. Peskin.
© Michael E. Peskin 2019. Published in 2019 by Oxford University Press.
DOI: 10.1093/oso/9780198812180.001.0001

of a waveform is the same waveform evaluated at Lorentz-transformed points. In an equation,

$$\phi(x) \to \phi'(x) = \phi(\Lambda^{-1}x) \ . \tag{3.1}$$

Canonically, Λ^{-1} appears in this formula so that, if $\phi(x)$ has a maximum at $x = a$, $\phi'(x)$ will have a maximum at the Lorentz-transformed point $x = \Lambda a$. A Lorentz-invariant theory of waves should have the property that, if $\phi(x)$ solves the wave equation, then $\phi'(x)$ in (3.1) does also.

The simplest equation satisfying this property is the Klein-Gordon equation

$$\left(\frac{\partial^2}{\partial t^2} - \nabla^2 + m^2 \right)\phi(t, \vec{x}) = 0 \ . \tag{3.2}$$

In this equation, t and $\vec{x}$ appear in a symmetric way. The 4-gradient

$$\partial_\mu = (\frac{\partial}{\partial t}, \frac{\partial}{\partial \vec{x}})_\mu \tag{3.3}$$

transforms under Lorentz transformations as a 4-vector with a lowered index. That is, the quantities

$$p \cdot \partial = (E\frac{\partial}{\partial t} + \vec{p} \cdot \vec{\nabla}) \quad \text{and} \quad \partial^2 = (\frac{\partial^2}{\partial t^2} - \nabla^2) \tag{3.4}$$

are Lorentz-invariant operators. Using (3.4), we can write the Klein-Gordon equation (3.2) in a more manifestly Lorentz-invariant form,

$$(\partial^2 + m^2)\phi(x) = 0 \ . \tag{3.5}$$

We can also see the invariance of (3.5) by examining the solutions of this equation explicitly. These are

$$\phi(x) = e^{-iEt+i\vec{p}\cdot\vec{x}} = e^{-ip\cdot x} \ , \tag{3.6}$$

where $p^\mu = (E, \vec{p})^\mu$ is a 4-vector satisfying

$$E^2 - |\vec{p}|^2 = p^2 = m^2 \ . \tag{3.7}$$

This criterion is Lorentz-invariant. The Lorentz-invariance of the 4-vector product is the statement that

$$p \cdot x = (\Lambda p) \cdot (\Lambda x) \quad \text{or} \quad (\Lambda p) \cdot x = p \cdot (\Lambda^{-1}x) \ . \tag{3.8}$$

Then the boost of the solution (3.6) is

$$\phi'(x) = e^{-ip \cdot \Lambda^{-1}x} = e^{-i(\Lambda p) \cdot x} \ , \tag{3.9}$$

which is also a solution of the equation.

The Klein-Gordon equation has the odd feature, from the point of view of a quantum-mechanical interpretation, that it has solutions corresponding both to positive and negative energy. Solving (3.7) for E, we find that both solutions

$$E = \pm E_p \tag{3.10}$$

are acceptable. This is a common property of all relativistic wave equations. Quantum field theory gives an attractive way to understand the negative energy solutions, which I will explain below.

Another way to derive the relativistic invariance of the Klein-Gordon equation is to write a variational principle that gives rise to this equation. You might be used to the variational principles of Lagrangian mechanics. In that formalism, we write an action functional S

$$S[x(t), \dot{x}(t)] = \int dt L(x, \dot{x}) . \tag{3.11}$$

The principle that S is stationary with respect to all variations of the solution $x(t)$ yields the equation of motion of the system. Mathematically, if $x(t) \to x(t) + \delta x(t)$, then we can write δS in the form

$$\delta S[x(t), \dot{x}(t)] = \int dt \, \delta x(t) \left[\mathcal{E}[x(t), \dot{x}(t), \ddot{x}(t)] \right] . \tag{3.12}$$

Then the equation of motion is $\mathcal{E} = 0$.

To obtain a relativistic equation of motion, we start with a relativistically invariant expression for the action S. The action S should be a function of the waveform $\phi(x)$. Instead of an integral over t only, I will integrate symmetrically over all of Minkowski space. Then the action principle takes the form

$$S[\phi(x)] = \int d^4x \, \mathcal{L}(\phi, \partial_\mu \phi) \tag{3.13}$$

By choosing an action S in this form, we guarantee that the action is relativistically invariant. Then the equation of motion following from the variational principle must be a relativistic field equation.

The function $\mathcal{L}$ is called the *Lagrange density*. I will choose the Lagrange density to be relativistically invariant. Then S is the invariant integral of a invariant function and thus is guaranteed to be Lorentz-invariant.

To illustrate how we apply this formalism, I will propose a simple form for $\mathcal{L}$. Consider, then,

Lagrangian formulation of the Klein-Gordon equation.

$$\mathcal{L} = \frac{1}{2} \left(\partial^\mu \phi \partial_\mu \phi - m^2 \phi^2 \right) . \tag{3.14}$$

There are no uncontracted 4-vector indices. Thus, this expression, and, by extension its integral S over all space-time, is Lorentz-invariant. The variation of $\mathcal{L}$ with respect to $\phi(x)$ is

$$\delta \mathcal{L} = \partial_\mu \delta\phi \, \partial^\mu \phi - m^2 \delta\phi \, \phi \tag{3.15}$$

Putting this under the integral d^4x and integrating by parts in the first term, we find

$$\delta S = \int d^4x \, \delta\phi(x) \left[(-\partial^2 - m^2)\phi(x) \right] . \tag{3.16}$$

The variational principle states that the Lagrangian equation of motion is the condition that δS vanishes for an arbitrary variation of $\phi(x)$. In (3.16), this condition implies that the quantity in brackets must vanish. This gives exactly the Klein-Gordon equation (3.5).

The Lagrangian formalism guarantees that the transform of any solution of the equation of motion is equally well a solution of the equation of motion. The logic is quite transparent: A solution ϕ of the wave equation is a stationary point of $S[\phi]$. But, if S is invariant under Lorentz transformations, then the Lorentz transform ϕ' of this solution will have the same value: $S[\phi] = S[\phi']$. This will also be true of other, nearby, field configurations. Thus, ϕ' will also be a stationary point of S. Then ϕ' also will be a solution to the wave equation.

The principle that a relativistic field theory is described by a relativistically invariant Lagrange density is a very powerful one. This principle will allow us to turn general ideas about the nature of new particle interactions into concrete proposals for the equations of motion. I will elaborate this variational approach further, by stages, in Sections 3.3 and 3.4, and later, in Chapters 11, 14 and 16. At the end of this development, we will have a mathematical formalism that will allow us to write the equations for the strong and weak interactions in a compact and, I hope, persuasive, form.

The Lagrangian formalism guarantees that, if $\phi(x)$ solves the wave equation, any boost or rotation of $\phi(x)$ also solves the wave equation.

3.2   Fields and particles

In principle, we could use the Klein-Gordon equation as a single-particle quantum theory in which the Klein-Gordon wave replaces the Schrödinger wavefunction. However, as I have explained above, a theory of relativistic particles should actually have the capability to discuss many particles, as many as we wish. To accomplish this, we need a different strategy.

It can be shown that this is accomplished by writing the Hamiltonian that leads to the Klein-Gordon equation and then quantizing that Hamiltonian. The resulting quantum theory has a Lorentz-invariant ground state, called the *vacuum state*, and excited states with the energy-momentum of particles with mass m. In this section, I will describe some important properties of this quantum theory. These properties are common to quantum field theories based on relativistic wave equations.

In the solution of the Klein-Gordon quantum theory, the field $\phi(x)$ becomes an operator that can create and destroy particles. Let $|0\rangle$ be the ground state of the Hamiltonian for the Klein-Gordon theory. This is the *vacuum state*; it contains zero particles. Let $|\varphi(p)\rangle$ be a state with one particle of momentum p. This is a state of higher energy, with energy E_p above the energy of the vacuum. The operator $\phi(x)$ has a nonzero matrix element corresponding to destruction of the particle,

In quantum field theory, $|0\rangle$ denotes the *vacuum state*, the state of empty space with no particles. $|p\rangle$ denotes a state with one particle of momentum p. This is an excited state with higher energy than the vacuum state.

$$\langle 0| \, \phi(x) \, |\varphi(p)\rangle = e^{-ip\cdot x} \; . \tag{3.17}$$

The field operator $\phi(x)$ satisfies the Klein-Gordon equation, and so the right-hand side of (3.17) must be a solution to the Klein-Gordon equation. So, indeed, the right-hand side must be of the form of (3.6). The 0 component of p^μ is the positive energy solution from (3.10), with $p^0 = +E_p$. The field operator $\phi(x)$ destroys the particle φ at the space-time point x. We should then interpret the right-hand side of (3.17)

The matrix element representing the destruction of a spin 0 particle by its quantum field.

as the wavefunction that the particle occupied at the moment it was destroyed. This is a Schrödinger wavefunction of standard form, with momentum $\vec{p}$ and energy $+E_p$.

The complex conjugate of the equation (3.17) is an equation

$$\langle \varphi(p)| \, \phi^\dagger(x) \, |0\rangle = e^{+ip\cdot x} \; , \qquad (3.18)$$

The matrix element representing the creation of a spin 0 particle by its quantum field.

Now the negative energy solution of the Klein-Gordon equation appears on the right-hand side. This is natural also, because the particle φ now appears in a bra vector, so that the right-hand side would be the complex conjugate of the Schrödinger wavefunction into which the field creates the particle at the point x.

The field $\phi(x)$ can be either real- or complex-valued. That is, we have the two choices

$$\phi^\dagger(x) = \phi(x) \quad \text{or} \quad \phi^\dagger(x) \neq \phi(x) \; . \qquad (3.19)$$

In the second case, the positive energy solutions for $\phi^\dagger(x)$ give us a new matrix element

$$\langle 0| \, \phi^\dagger(x) \, |\varphi'(p)\rangle = e^{-ip\cdot x} \; , \qquad (3.20)$$

where φ' is a new particle distinct from φ. We will see below, in Section 3.4, that if $\phi(x)$ carries electric charge Q, φ has charge Q while φ' has charge $-Q$. The particles φ and φ' have the same mass, because their associated fields satisfy the same Klein-Gordon equation. We say that φ' is the *antiparticle* of φ.

If the field $\phi(x)$ is real-valued, the particles φ and φ' can be identical. In this case, the particle destroyed by $\phi(x)$ can be its own antiparticle.

The formulae for creation and destruction of particles by field operators will play an important role in all of the calculations done in this book. I summarize the equations (3.17), (3.18), which apply to particles of spin 0, and the corresponding formulae for particles of spin $\frac{1}{2}$ and spin 1, in Appendix C.

3.3 Maxwell's equations

The particles φ that appeared in the previous section carried no quantum numbers except for energy and momentum. From nonrelativistic quantum mechanics, we know that some particles can carry intrinsic angular momentum. For example, electrons carry an intrisic spin of $\frac{1}{2}\hbar$ in addition to their orbital angular momentum. Similarly, photons carry an intrisic spin of $1 \cdot \hbar$. I will now discuss how particles with these properties can be described by quantum field theory.

Begin with the case of spin 1. Spin 1 is the vector representation of angular momentum. To encode this, consider a 3-vector field $V^i(x)$, that is, a field that transforms under rotations R according to

$$V^i(x) \rightarrow V^{i\prime}(x) = R_{ij} V^j(R^{-1}x) \; . \qquad (3.21)$$

Illustration of the transformation of a vector field as in (3.21):

In this equation, a rotation moves the coordinate of the field in the same way as in (3.1) but also changes the orientation of the field by the same rotation.

If the field $V^i(x)$ is an operator in the quantum theory and destroys a particle v, the matrix element corresponding to that operation would have the form

$$\langle 0| V^i(x) |v(p,\epsilon)\rangle = \epsilon^i e^{-ip\cdot x} , \qquad (3.22)$$

The matrix element representing the destruction of a spin 1 particle by its quantum field. Note that the vector index of the field is carried on the right-hand side by the polarization vector of the particle.

The part of the Schrödinger wavefunction representing momentum must be the same as in the Klein-Gordon case, but now there must be another element to carry the index i and represent the orientation under rotations. So we need a 3-vector associated with the particle. For this, I have introduced the 3-component vector ϵ^i. Under a rotation, the left-hand side of this equation transforms with an overall matrix R_{ij}. For consistency, ϵ^i must transform as

$$\epsilon^i \to R_{ij}\epsilon^j . \qquad (3.23)$$

Definition of the *polarization vector* of a spin-1 particle.

Then the particle v carries a real 3-vector that rotates as the state $|v(p,\epsilon)\rangle$ is rotated. This vector is called the *polarization vector* of the particle.

In a similar way, we can construct fields corresponding to any spin j representation of the rotation group. Let $\mathcal{R}^{(j)}_{ab}$ be a rotation matrix in the spin j representation. This would be a $(2j+1)\times(2j+1)$ matrix. A spin-j field would transform under rotations according to

$$W^a(x) \to W^{a\prime}(x) = \mathcal{R}^j_{ab}W^b(R^{-1}x) . \qquad (3.24)$$

The field $W^a(x)$ would destroy a particle w according to the matrix element

$$\langle 0| W^a(x) |w(p,\eta)\rangle = \eta^a e^{-ip\cdot x} , \qquad (3.25)$$

The particle w would then carry a $(2j+1)$-dimensional polarization vector η^a which would transform as a spin-j vector under rotations. For $j=\frac{1}{2}$, η would be a 2-component spinor.

The equations (3.22), (3.25) are not yet relativistically invariant. In fact, it is subtle to construct relativistic wave equations for particles of nonzero spin. In the remainder of this section, I will discuss the case of spin 1.

There is an obvious generalization of (3.22) to a relativistic equation. This is

There is a problem in constructing a relativistic quantum theory for fields of spin 1.

$$\langle 0| V^\mu(x) |v(p,\epsilon)\rangle = \epsilon^\mu e^{-ip\cdot x} , \qquad (3.26)$$

where ϵ^μ is now a 4-vector. But there is a problem. To preserve Lorentz invariance of this state, the norm of the 1-particle state should be proportional to

$$-\epsilon^\mu \epsilon_\mu . \qquad (3.27)$$

I have put a minus sign here so that the expression is positive when ϵ is a spacelike unit vector, as we would expect for the vector ϵ in (3.22). However, if ϵ^μ could also be a timelike vector—for example,

$\epsilon = (1, 0, 0, 0)$—the state would have negative norm and, formally in quantum mechanics, negative probability. The spin 1 wave equation must somehow forbid timelike ϵ from appearing.

Photons are spin 1 particles, and so we can ask if their wave equation solves this problem. Indeed, it is so. Maxwell's equations can be written as equations for the 4-vector vector potential $A_\mu(x)$. But, as you learned in your electrodynamics course, Maxwell's equations have propagating solutions with only two possible polarization vectors,

$$\vec{A}(x) = \vec{\epsilon}\, e^{-iEt+i\vec{p}\cdot\vec{x}} \ , \tag{3.28}$$

where $E = |p|$ (zero mass) and

$$\vec{\epsilon} \cdot \vec{p} = 0 \ . \tag{3.29}$$

We can view (3.28) as the wavefunction of a photon. Under a Lorentz rotation, the form of (3.28) is preserved. Under a Lorentz boost, the exponential is unchanged, but the polarization vector, while remaining spacelike, acquires a component in the time direction. It can be shown that this Lorentz transformation of $\vec{\epsilon}$ has no effect on the photon's interactions, as long as the electromagnetic current is conserved. Through this logic, we obtain a description of the states that is Lorentz-invariant, though not quite manifestly so. This makes it possible to quantize the Maxwell field consistently using only states of positive norm.

Maxwell's equations solve that problem (in a way that is not quite obvious).

This solution does not generalize in a simple way to spin 1 fields with a vector boson mass. For those, we will need a special construction called the *Higgs mechanism*, which I will explain in Chapter 16.

It will be instructive to write the variational principle for the Maxwell field. I will start with the 4-vector potential

$$A^\mu(x) = (-\phi(x), -\vec{A}(x))^\mu \ ; \tag{3.30}$$

here, $\phi(x)$ and $\vec{A}(x)$ are the conventional scalar and vector potentials of classical electrodynamics. The electric and magnetic field strengths are contained in the tensor

$$F^{\mu\nu} = \partial^\mu A^\nu - \partial^\nu A^\mu \ . \tag{3.31}$$

Working out the components of this tensor carefully, remembering that the notation $\partial^\mu = (\partial/\partial t, -\vec{\nabla})^\mu$ leads to some extra minus signs, we find

$$F^{0i} = -\nabla^i \phi - \partial_t A^i = E^i$$
$$F^{ij} = \nabla^i A^j - \nabla^j A^i = \epsilon^{ijk} B^k \ . \tag{3.32}$$

A simple, manifestly Lorentz-invariant Lagrange density for this system is

$$\mathcal{L} = -\frac{1}{4} F^{\mu\nu} F_{\mu\nu} + J^\mu A_\mu \ . \tag{3.33}$$

Lagrangian formulation of Maxwell's equations.

I have added a new field $J^\mu(x)$ to represent an external electromagnetic current. This object should satisfy the equation of current conservation. With

$$J^\mu = (\rho, \vec{J})^\mu \ , \tag{3.34}$$

that relation is

$$0 = \partial_\mu J^\mu = \partial_t \rho + \vec{\nabla} \cdot \vec{J} \,. \tag{3.35}$$

Now vary (3.33) with respect to $A^\mu(x)$. The variation is

$$\delta\mathcal{L} = -\frac{1}{2}(\partial_\mu\,\delta A_\nu - \partial_\nu\,\delta A_\mu)F^{\mu\nu} + \delta A_\nu\,J^\nu \,. \tag{3.36}$$

Gathering terms under the integral $\int d^4x$ and integrating by parts, we find

$$\delta S = \int d^4x\,\delta A_\nu \left[\partial_\mu F^{\mu\nu} - J^\nu\right] \,. \tag{3.37}$$

Thus, the field equations following from (3.33) are

$$\partial_\mu F^{\mu\nu} = -J^\mu \,. \tag{3.38}$$

Carefully inserting (3.32), (3.34), we see that these equations are precisely

$$\nabla^i E^i = \rho \qquad \epsilon^{ijk}\nabla^j B^k - \partial_t E^i = J^i \,, \tag{3.39}$$

which are exactly the inhomogeneous Maxwell equations written in natural units $\epsilon_0 = \mu_0 = c = 1$. It should be familiar to you, and it is easily checked, that the formulae (3.32) automatically satisfy the homogeneous Maxwell equations. Since Maxwell's equations follow from the relativistically invariant action principle (3.33), they must automatically be Lorentz invariant wave equations in the sense that I have discussed in Section 3.1.

3.4   The Dirac equation

To describe spin $\frac{1}{2}$ particles such as the electron, we need to construct a relativistic wave equation for a spin $\frac{1}{2}$ field. Dirac solved this problem by constructing a special set of matrices that, viewed appropriately, transform as a 4-vector. To construct these matrices, Dirac suggested using *anticommutation* rather than commutation relations: $\{A, B\} = AB + BA$. In particular, he introduced matrices γ^μ that satisfy the *Dirac anticommutation algebra*

$$\{\gamma^\mu, \gamma^\nu\} = 2\eta^{\mu\nu} \,, \tag{3.40}$$

where the right-hand side is the metric (2.7) of special relativity. It can be shown that there are no sets of 2×2 or 3×3 matrices satisfying this algebra. The smallest such matrices are 4×4. Here is a representation of the algebra,

This representation of the Dirac algebra is useful for studying problems in which spin $\frac{1}{2}$ particles move nonrelativistically.

$$\gamma^0 = \begin{pmatrix} 1 & 0 \\ 0 & -1 \end{pmatrix} \qquad \gamma^i = \begin{pmatrix} 0 & \sigma^i \\ -\sigma^i & 0 \end{pmatrix} \,, \tag{3.41}$$

where the elements of these matrices are 2×2 blocks and the σ^i are the 2×2 Pauli sigma matrices. Here is another representation

This representation of the Dirac algebra is useful for studying problems in which spin $\frac{1}{2}$ particles move at speeds close to the speed of light. We will use this representation most often in our study of the fundamental structure of particle interactions.

$$\gamma^0 = \begin{pmatrix} 0 & 1 \\ 1 & 0 \end{pmatrix} \qquad \gamma^i = \begin{pmatrix} 0 & \sigma^i \\ -\sigma^i & 0 \end{pmatrix} \,. \tag{3.42}$$

It can be shown that all 4×4 representations of the Dirac algebra are equivalent by unitary transformations.

Using γ^μ as if it were a 4-vector, we can write the Dirac field equation

$$\left(i\gamma^\mu \partial_\mu - m\right)\Psi = 0 \ . \tag{3.43}$$

I will now prove three properties of this equation: First, any solution of the Dirac equation is a solution of the Klein-Gordon equation. Then the exponential part of the solution, at least, has the form

$$\Psi \sim e^{-ip \cdot x} \tag{3.44}$$

and is Lorentz invariant. Second, the solutions of the Dirac equation for $\vec{p} = 0$ are precisely two positive-energy and two negative-energy solutions. Third, a field satisfying the Dirac equation natually has a vector current which is conserved. For electrons, this operator would be interpreted as the electric charge current. To fully prove that the Dirac equation is relativistic, we would have to define the Lorentz transformation properties of γ^μ carefully and prove that these matrices transform as a 4-vector. That discussion is beyond the level of this book. You can find it in any textbook of quantum field theory.

To prove the first property, multiply the Dirac equation by the operator $(i\gamma^\mu \partial_\mu + m)$. We find

$$\begin{aligned} 0 &= (i\gamma^\mu \partial_\mu + m)(i\gamma^\mu \partial_\mu - m)\Psi \\ &= \left[-\gamma^\mu \gamma^\nu \partial_\mu \partial_\nu + im\gamma^\nu \partial_\nu - im\gamma^\mu \partial_\mu - m^2\right]\Psi \ . \end{aligned} \tag{3.45}$$

The two derivatives in the first term commute, so we can replace $\gamma^\mu \gamma^\nu$ by $\frac{1}{2}\{\gamma^\mu, \gamma^\nu\}$. Then (3.45) takes the form

$$0 = \left[-\eta^{\mu\nu} \partial_\mu \partial_\nu - m^2\right]\Psi \tag{3.46}$$

or finally

$$[-\partial^\mu \partial_\mu - m^2]\Psi = 0 \ . \tag{3.47}$$

So, indeed, all solutions of the Dirac equation satisfy the Klein-Gordon equation.

To prove the second property, first look for solutions of the Dirac equation with zero 3-momentum. These would be of the form

$$\Psi = \begin{pmatrix} a \\ b \\ c \\ d \end{pmatrix} e^{-iEt} \ . \tag{3.48}$$

Use the first matrix representation given above. Then the Dirac equation reads

$$(\gamma^0 E - m) \begin{pmatrix} a \\ b \\ c \\ d \end{pmatrix} = 0 \ , \tag{3.49}$$

or

$$\left[\begin{pmatrix} E & \\ & -E \end{pmatrix} - \begin{pmatrix} m & \\ & m \end{pmatrix}\right]\begin{pmatrix} \begin{pmatrix} a \\ b \end{pmatrix} \\ \begin{pmatrix} c \\ d \end{pmatrix} \end{pmatrix} = 0 \ . \tag{3.50}$$

There are two solutions with $E = +m$,

$$\Psi = \begin{pmatrix} \xi \\ 0 \end{pmatrix} e^{-imt} \ , \tag{3.51}$$

where ξ is an arbitrary 2-component spinor, identified with the electron spin. Similarly, there are two solutions with $E = -m$

$$\Psi = \begin{pmatrix} 0 \\ \eta \end{pmatrix} e^{+imt} \ , \tag{3.52}$$

where η is another arbitrary 2-component spinor. These negative energy solutions are identified with the *antiparticles* of the Dirac fermions. For the Dirac equation describing electrons, the negative energy solutions describe *positrons*. The spinor η is identified with the *opposite* of the positron spin.

For nonzero $\vec{p}$, the Dirac equation reads

$$\begin{pmatrix} E - m & -\vec{\sigma} \cdot \vec{p} \\ \vec{\sigma} \cdot \vec{p} & -E - m \end{pmatrix}\begin{pmatrix} \begin{pmatrix} a \\ b \end{pmatrix} \\ \begin{pmatrix} c \\ d \end{pmatrix} \end{pmatrix} e^{-iEt + i\vec{p}\cdot\vec{x}} = 0 \ . \tag{3.53}$$

The solutions of this equation are constructed in Exercise 3.4. These solutions are conventionally written

$$\Psi = U^s(p) \ e^{-ip\cdot x} \tag{3.54}$$

for the two positive energy solutions ($s = 1, 2$) and

$$\Psi = V^s(p) \ e^{+ip\cdot x} \tag{3.55}$$

for the two negative energy solutions ($s = 1, 2$). For (3.55), note that the energy and momentum in the exponent are the opposites of the antiparticle energy and momentum. Similarly the spin orientation in $V^s(p)$ is the opposite of the antiparticle spin orientation.

For nonzero values of p/m, the components of the Dirac fields that are zero in (3.51), (3.52) become filled in. For example,

$$\Psi = \begin{pmatrix} \xi \\ \frac{\vec{\sigma}\cdot\vec{p}}{2m}\xi \end{pmatrix} e^{-iE_p t + i\vec{p}\cdot\vec{x}} \ . \tag{3.56}$$

to first order in $|\vec{p}|$, with $E_p = (p^2 + m^2)^{1/2} \approx m + p^2/2m$. I will discuss the form of $U^s(p)$ and $V^s(p)$ at higher momentum, and, especially, for extremely relativistic energies, in Chapter 8.

In a quantum field theory of Dirac particles (for definiteness, electrons e^-), the basic 1-particle states are states $|e(p, s)\rangle$ in which the electron has a definite momentum p and spin s. The matrix elements for destroying and creating one electron are

$$\langle 0| \, \Psi_a(x) \, |e^-(p, s)\rangle = U^s(p)e^{-ip\cdot x} \,,$$
$$\langle e^-(p, s)| \, \Psi_a^\dagger(x) \, |0\rangle = U^{\dagger s}(p)e^{+ip\cdot x} \,. \tag{3.57}$$

In this theory, the electron has a spin-$\frac{1}{2}$ antiparticle e^+, the positron. The matrix elements for destroying and creating a positron are

$$\langle 0| \, \Psi_a^\dagger(x) \, |e^+(p, s)\rangle = V^{\dagger s}(p)e^{-ip\cdot x} \,,$$
$$\langle e^+(p, s)| \, \Psi_a(x) \, |0\rangle = V^s(p)e^{+ip\cdot x} \,. \tag{3.58}$$

To construct the electric current, note first that the Dirac matrices for $\mu = i = 1, 2, 3$ are anti-Hermitian

$$(\gamma^0)^\dagger = \gamma^0 \qquad (\gamma^i)^\dagger = -\gamma^i \,. \tag{3.59}$$

To form Hermitian operators from the Dirac field, it is convenient to note that

$$(\gamma^0 \gamma^\mu)^\dagger = (\gamma^\mu)^\dagger (\gamma^0)^\dagger = +\gamma^0 \gamma^\mu \tag{3.60}$$

for all four cases $\mu = 0, 1, 2, 3$. Then the square of the Dirac field

$$\Psi^\dagger \Psi \tag{3.61}$$

is more conveniently written as the $\mu = 0$ component of

$$\Psi^\dagger \gamma^0 \gamma^\mu \Psi \,. \tag{3.62}$$

This quantity is Hermitian,

$$(\Psi^\dagger \gamma^0 \gamma^\mu \Psi)^\dagger = \Psi^\dagger (\gamma^0 \gamma^\mu)^\dagger \Psi = \Psi^\dagger \gamma^0 \gamma^\mu \Psi \,, \tag{3.63}$$

and, it can be shown, it transforms as a 4-vector under Lorentz transformations. From now on, I will write

$$\overline{\Psi} = \Psi^\dagger \gamma^0 \tag{3.64}$$

so that the operator in (3.62) appears as

$$j^\mu = \overline{\Psi} \gamma^\mu \Psi \,. \tag{3.65}$$

It would be wonderful if this operator, which is now written manifestly as a 4-vector, would turn out to be the operator that represents the electromagnetic current of the electron field.

We can work out the equation of motion of this 4-vector. For this, we need the complex conjugate of the Dirac equation. This is

$$-i\partial_\mu \Psi^\dagger (\gamma^\mu)^\dagger - m\Psi^\dagger = 0 \,. \tag{3.66}$$

Multiplying on the right by γ^0, we find the simpler form

$$-i\partial_\mu \overline{\Psi} (\gamma^\mu) - m\overline{\Psi} = 0 \,. \tag{3.67}$$

The matrix element representing the destruction of a spin-$\frac{1}{2}$ particle by its quantum field. Note that the spinor index of the field is carried on the right-hand side by the U^s spinor of the particle.

The matrix element representing the destruction of a spin-$\frac{1}{2}$ antiparticle by its quantum field. Note that the spinor index of the field is carried on the right-hand side by the V^s spinor of the antiparticle.

The vector current operator of the Dirac theory.

Then

$$\partial_\mu(\overline{\Psi}\gamma^\mu\Psi) = (\partial_\mu\overline{\Psi})\gamma^\mu\Psi + \overline{\Psi}\gamma^\mu\partial_\mu\Psi$$
$$= im\overline{\Psi}\Psi - im\overline{\Psi}\Psi$$
$$= 0 \ . \tag{3.68}$$

So, indeed, (3.65) satisfies the standard equation of conservation of a current

$$\partial_\mu j^\mu = 0 \ . \tag{3.69}$$

The Dirac equation is coupled to an external electromagnetic field using the 4-vector potential, generalizing the prescription used to couple the electromagnetic field to the Schrödinger equation in non-relativistic quantum mechanics.

The Dirac equation can be coupled to an electromagnetic field in the way that is standard in quantum mechanics, by introducing the electromagnetic vector potential into the derivatives. Relativistically and in natural units, this replacement is written

$$\partial_\mu \to D_\mu = (\partial_\mu + ieA_\mu) \tag{3.70}$$

so that the Dirac equation becomes

$$[i\gamma^\mu(\partial_\mu + ieA_\mu) - m]\Psi = 0 \ . \tag{3.71}$$

Remembering that $A^0 = -\phi$, for the hydrogen atom

$$A^0 = -\frac{e}{4\pi r} \ , \qquad \vec{A} = 0 \ . \tag{3.72}$$

Then the Dirac equation for an electron in the hydrogen atom reads

$$\left[\gamma^0(i\frac{\partial}{\partial t} + \frac{\alpha}{r}) + i\vec{\gamma}\cdot\vec{\nabla} - m\right]\Psi = 0 \ . \tag{3.73}$$

I will have much more to say about the logic of this principle for coupling the Dirac equation to the electromagnetic field in Chapter 11.

Finally, I will write a variational principle for the Dirac equation. The Lagrange density is

Lagrangian formulation of the Dirac equation.

$$\mathcal{L} = \overline{\Psi}(i\gamma^\mu\partial_\mu - m)\Psi(x) \tag{3.74}$$

Given that $\overline{\Psi}\gamma^\mu\Psi$ transforms under Lorentz transformations as a 4-vector, this formula is manifestly Lorentz invariant. The variation of (3.74) with respect to $\overline{\Psi}(x)$ gives the Dirac equation (3.43); the variation with respect to $\Psi(x)$, after integration by parts, gives (3.67). These equations then must be relativistically invariant wave equations.

It is one short step beyond this to write a Lagrangian for the interaction of electrons and photons. To do this, we only need to combine Maxwell's equations and the Dirac equation in a consistent way. This is automatically accomplished by writing the Lagrangian containing both ingredients,

Lagrangian of Quantum Electrodynamics (QED).

$$\mathcal{L} = -\frac{1}{4}F^{\mu\nu}F_{\mu\nu} + \overline{\Psi}(i\gamma^\mu D_\mu - m)\Psi \ , \tag{3.75}$$

where $D_\mu = \partial_\mu + ieA_\mu$ as in (3.70). You can easily check that the principle that action integral should be stationary leads to Maxwell's equations plus the Dirac equation. Notice that there is a term in the Dirac Lagrangian that involves A_μ,

$$\mathcal{L} = \cdots + i\overline{\Psi}\gamma^\mu(+ieA_\mu)\Psi = \cdots - A_\mu[e\overline{\Psi}\gamma^\mu\Psi] . \qquad (3.76)$$

Comparing to (3.36), we see that the electromagnetic current that is the source for the Maxwell fields is exactly

$$J^\mu = -ej^\mu = -e\overline{\Psi}\gamma^\mu\Psi . \qquad (3.77)$$

This completes the identification of (3.65) with the electromagnetic current.

The quantum field theory based on (3.75) is called *Quantum Electrodynamics* or QED. This theory is in extraordinary agreement with the actual properties of electrons and photons. The predictions of this theory agree with the observed magnetism of the electron and the measured properties of hydrogen atomic states to the accuracy of parts per billion (Kinoshita 1990).

3.5 Relativistic normalization of states

The equations such as (3.17), (3.22), and (3.57) for the creation and destruction of particles depend on a detail that I have not yet discussed. The right-hand sides of these equations can have the simple Lorentz transformation laws shown only if the particle states are normalized in a Lorentz-invariant way. I will now explain how this must be done.

In nonrelativistic quantum mechanics, we typically normalize momentum states according to the convention

$$\langle p_1|p_2\rangle = (2\pi)^3\delta^{(3)}(\vec{p_1} - \vec{p_2}) . \qquad (3.78)$$

But this normalization is not relativistically invariant. To remedy this, we must use the *relativistic normalization*

$$\langle p_1|p_2\rangle = 2E_{p_1}(2\pi)^3\delta^{(3)}(\vec{p_1} - \vec{p_2}) . \qquad (3.79)$$

In this book, all particle states are normalized using this relativistically invariant prescription.

I will now check that (3.79), rather than (3.78), is Lorentz invariant. We boost a momentum vector p in the $\hat{3}$ direction, using the formulae (2.2)

$$E'_p = \gamma(E_p + \beta p^3) \qquad p^{1\prime} = p^1$$
$$p^{3\prime} = \gamma(p^3 + \beta E_p) \qquad p^{2\prime} = p^2 . \qquad (3.80)$$

Then the normalization equation (3.79) transforms to

$$\langle p'_1|p'_2\rangle = 2E'_{p_1}(2\pi)^3\delta^{(3)}(\vec{p'_1} - \vec{p'_2})$$
$$= 2\gamma(E_{p_1} + \beta p_1^3)(2\pi)^3\delta(\gamma(p_1^3 + \beta E_{p_1}) - p_2^{3\prime})$$
$$\delta(p_1^{1\prime} - p_2^{1\prime})\delta(p_1^{2\prime} - p_2^{2\prime}) . \qquad (3.81)$$

In the last two delta functions, we can use (3.80) to replace p' by p. To simplify the first delta function, we must recall the transformation of a delta function for an argument that vanishes at $x = 0$,

$$\delta(g(x)) = \frac{1}{|dg/dx|}\delta(x) \tag{3.82}$$

Using this formula, we can rewrite the first delta function and find

$$\langle p_1'|p_2'\rangle = 2\gamma(E_{p_1} + \beta p_1^3)(2\pi)^3\delta(p_1^1 - p_2^1)\delta(p_1^2 - p_2^2)$$
$$\cdot \frac{1}{\gamma(1 + \beta\, dE_{p_1}/dp_1^3)}\delta(p_1^3 - p_2^3) \ . \tag{3.83}$$

Evaluating the term in the second line,

$$1 + \beta\frac{dE_{p_1}}{dp_1^3} = 1 + \beta\frac{p_1^3}{\sqrt{p_1^2 + m_1^2}} = \frac{E_{p_1} + \beta p_1^3}{E_{p_1}} \ . \tag{3.84}$$

Assembling the pieces, we find

$$\langle p_1'|p_2'\rangle = 2E_{p_1}\,(2\pi)^3\delta^{(3)}(\vec{p}_1 - \vec{p}_2) \ . \tag{3.85}$$

that is, the normalization condition (3.79) is invariant under Lorentz transformations.

Another way to understand the relativistic invariance of the normalization condition just given is to consider the related integral over phase space. For the same reasons as described above, the integral

$$\int \frac{d^3p}{(2\pi)^3} \tag{3.86}$$

is not Lorentz-invariant. However, the integral

$$\int \frac{d^4p}{(2\pi)^4}(2\pi)\delta(p^2 - m^2) \tag{3.87}$$

is manifestly relativistically invariant. The integral over p^0 is

$$\int dp^0\delta(p^2 - m^2) = \int dp^0\delta((p^0)^2 - |\vec{p}|^2 - m^2) = \frac{1}{2p^0}\bigg|_{p^0=E_p} \ . \tag{3.88}$$

Then

$$\int \frac{d^4p}{(2\pi)^4}(2\pi)\delta(p^2 - m^2) = \int \frac{d^3p}{(2\pi)^3}\frac{1}{2E_p} \ . \tag{3.89}$$

This gives the relativistically invariant integral over momentum space. The factor $2E_p$ in the denominator is set up to cancel the factor $2E_p$ in (3.79).

With this integral and the relativistically normalized states, the sum over momentum states is written

The sum over a complete set of 1-particle states with relativistic normalization contains the extra factor $1/2E_p$.

$$1 = \int \frac{d^3p}{(2\pi)^3}\frac{1}{2E_p}\,|p\rangle\langle p| \ . \tag{3.90}$$

That is,

$$\mathbf{1}\,|k\rangle = \int \frac{d^3p}{(2\pi)^3} \frac{1}{2E_p}\,|p\rangle\,\langle p|k\rangle = \int \frac{d^3p}{(2\pi)^3} \frac{1}{2E_p}\,|p\rangle\,2E_p(2\pi)^3\delta(\vec{p}-\vec{k}) = |k\rangle \ .$$

(3.91)

In natural units, where all energies, momenta, and inverse distances are written with the dimensions of GeV, the right-hand side of the normalization condition has the dimensions

$$2E_p(2\pi)^3\delta^{(3)}(\vec{p}_1 - \vec{p}_2) \sim (\mathrm{GeV}) \cdot (\mathrm{GeV})^{-3} = (\mathrm{GeV})^{-2} \ .$$

(3.92)

Then relativistically normalized states have the dimensions

$$|p\rangle \sim (\mathrm{GeV})^{-1} \ .$$

(3.93)

Relativistically normalized 1-particle states have the mass dimension $(\mathrm{GeV})^{-1}$.

It is only with these conventions that the matrix elements such as (3.17) can be correct in all reference frames. In the remainder of this book, I will normalize all states relativistically so that we can use simple formulae that take maximum advantage of Lorentz invariance.

3.6 Spin and statistics

In quantum field theory, all particles appear as quantum states found in the quantization of relativistic wave equations. This blurs the distinction between the particles that make up matter, the quarks and leptons, and the particles such as the photon that mediate forces. Each type of particle has an associated wave equation. The only differences between these equations come as a consequence of the differing spins of the particles or fields.

However, it turns out that there is profound difference between particles with integer spin and particles with half-integer spin. In 1940, Pauli proved *the connection between spin and statistics*. It has since been shown that this result, like the theorem of *CPT* conservation, can be proved from a basic system of axioms for quantum field theory (Streater and Wightman 2000). The theorem states:

- A field with integer spin creates and destroys particles with integer spin that obey Bose-Einstein statistics
- A field with half-integer spin creates and destroys particles with half-integer spin that obey Fermi-Dirac statistics.

This theorem implies that integer spin particles can come together to form macroscopic fields of force. Half-integer spin particles obey the Pauli exclusion principle and thus form rigid structures that we call matter. Otherwise, both types of particles and fields are treated identically within quantum field theory.

The connection between spin and statistics is a consequence of quantum field theory. This theorem explains why particles of matter and forces, for example, electrons and photons, can have qualitatively different properties while arising in essentially the same way from the formalism of quantum fields.

In this section, we have discussed wave equations for fields of spins 0, $\frac{1}{2}$, and 1. In the spin 1 case, we saw that there were nontrivial barriers to the formulation of appropriate equations. These barriers become more formidable as the spin increases. For spin $\frac{3}{2}$, there is an equation for the massless case, called the Rarita-Schwinger equation. The extension of this equation to massive particles requires the same special circumstances as for spin 1. For spin 2, the only possible field equation for an interacting particle is Einstein's field equation for gravity. Beyond spin 2, there are no known field equations for interacting fields of fixed spin, although some consistent systems of equations are known that include families of particles with an infinite number of different spins (Vasiliev 1990).

For the Standard Model of particle physics, we will need only fields of spin 0, $\frac{1}{2}$, and 1. So we now have a foundation to use in searching for that theory.

Exercises

(3.1) Consider a quantum field theory defined by the Lagrangian

$$\mathcal{L} = \frac{1}{2}(\partial_\mu \varphi)^2 - \frac{1}{2}m^2\phi^2 - \frac{\lambda}{4}\varphi^4 . \tag{3.94}$$

This model field theory is called φ^4-theory.

(a) Use the Lagrangian to derive the field equation for $\varphi(x)$.

(b) I claim that the Hamiltonian of this theory is

$$H = \int d^3x \left[\frac{1}{2}(\frac{\partial}{\partial t}\varphi)^2 + \frac{1}{2}(\vec{\nabla}\varphi)^2 + \frac{1}{2}m^2\varphi^2 + \frac{\lambda}{4}\varphi^4 \right] . \tag{3.95}$$

Use the equation of motion derived in (a) to show that H is conserved: $(d/dt)H = 0$, if boundary terms at infinity can be ignored.

(c) I claim that the total momentum of the field in this theory is

$$\vec{P} = \int d^3x \left[\frac{\partial}{\partial t}\varphi \, \vec{\nabla}\varphi \right] . \tag{3.96}$$

Use the equation of motion derived in (a) to show that $\vec{P}$ is conserved.

(3.2) The Dirac matrices γ^μ, $\mu = 0, 1, 2, 3$, are 4×4 matrices that satisfy the algebra

$$\{\gamma^\mu, \gamma^\nu\} = 2\eta^{\mu\nu} , \tag{3.97}$$

where $\eta^{\mu\nu}$ is the metric tensor of special relativity.

(a) Show that the matrices (3.41) satisfy the Dirac algebra.

(b) Show that the matrices (3.42) satisfy the Dirac algebra.

(c) Show that these representations are equivalent. That is, write a 4×4 unitary matrix U such that

$$\gamma_B^\mu = U\gamma_A^\mu U^\dagger$$

(3.3) Define the matrix γ^5 by

$$\gamma^5 = i\gamma^0\gamma^1\gamma^2\gamma^3 . \tag{3.98}$$

(a) Show that γ^5 anticommutes with all of the matrices γ^μ.

(b) Work out the form of γ^5 using the representation (3.41).

(c) For a wave with momentum $\vec{p} = p\hat{3}$ and energy E_p, find the solution of the Dirac equation (3.53) in which $U(p)$ has $(a, b) = (1, 0)$. This solution represents an electron moving in the $\hat{3}$ direction with spin up.

(d) Take the limit of the solution in (c) as $p \to \infty$. Show that it is an eigenstate of γ^5.

(e) For a wave with momentum $\vec{p} = p\hat{3}$ and energy E_p, find the solution of the Dirac equation (3.53) in which $U(p)$ has $(a, b) = (0, 1)$. This solution represents an electron moving in the $\hat{3}$ direction with spin down.

(f) Take the limit of the solution in (e) as $p \to \infty$. Show that it is an eigenstate of γ^5.

(3.4) Consider an event in which an unstable particle H decays into two photons. Work in the rest frame of the unstable particle. The photons are emitted back-to-back. Take the $\hat{3}$ axis to be aligned with the direction of the photons. Let photon 1 be the one travelling in the $+\hat{3}$ direction and photon 2 be the one travelling the $-\hat{3}$ direction.

(a) Argue that the spin of H must be integer, not half-integer.

(b) Possible polarization vectors for the photon 1 are

$$\vec{\epsilon}_{1R} = \frac{1}{\sqrt{2}}(\hat{1} + i\hat{2}) \qquad \vec{\epsilon}_{1L} = \frac{1}{\sqrt{2}}(\hat{1} - i\hat{2})$$

(3.99)

Rotate these vectors by ϕ about the $\hat{3}$ axis. A state of angular momentum $J^3 = +1$ gets a phase $e^{-i\phi}$. Show that the two choices correspond to photon states of angular momentum $J^3 = +1$ and -1, respectively, about the $\hat{3}$ axis.

(c) Write the corresponding polarization vectors for photon 2, by rotating the vectors in (1) by $180°$ about $\hat{2}$. These have $J^3 = +1, -1$ about the direction of motion of the photon (which is now $-\hat{3}$).

(d) The wavefunction of the 2-photon state is then a sum of terms of the form

$$\vec{\epsilon}_{1X}\vec{\epsilon}_{2Y}$$

(3.100)

where $X, Y = R, L$. There are four possible values for (X, Y). For each, compute the total J^3 for the state (2). Show that, in the states with $X = R, Y = L$ or $X = L, Y = R$, the spin of the original particle H must be ≥ 2.

(e) Consider the state with $X = Y = R$. Show that this state is transformed into itself by a rotation by $180°$ about $\hat{2}$. The same is true for the state $X = Y = L$.

(f) If the original particle H has spin J and decays to the state $X = Y = R$, it must have been in the state $|J0\rangle$, with $J^3 = 0$. How does this state transform when rotated by $180°$ about $\hat{2}$? (The transformation must be the same as that of the spherical harmonic $Y_{J0}(\theta,\phi)$.)

(g) Conclude that an unstable particle of spin 1 may not decay to two photons. This result is called the *Landau-Yang theorem*. (Note that invariance under parity has not been used in this argument.)

The Hydrogen Atom and Positronium

4

Before we begin our study of elementary particle physics proper, we will need one more set of introductory concepts. Physicists were able to grasp the structure of strongly interacting particles because they saw similarities between these particles and the quantum states of the the hydrogen atom and other comparably simple quantum-mechanical systems. Because this analogy will be important to us, I will spend this chapter reviewing important properties of the hydrogen atom. I will also discuss a system very similar to the hydrogen atom, the bound state of an electron and its antiparticle, the positron. Indeed, the study of the electron-positron bound states is already elementary particle physics, but with the well-understood interaction of electromagnetism playing the role of the binding force.

There will be two major themes in this discussion. First, though you certainly will have encountered Schrödinger's solution of the hydrogen atom in your quantum mechanics course, this gives only a part of the physics of the hydrogen atom. Corrections due to relativity and the spin of the electron create small energy splittings within the multiplets of states found by Schrödinger. The analogous effects in the bound states of the strong interaction are much larger. Thus, we will need to understand the full spectrum of states of the hydrogen atom to understand the qualitative features of the spectrum of strongly interacting particles. Second, the space-time symmetry parity (P) plays an important role in hydrogen. The spectrum of electron-positron bound states, *positronium*, is also strongly regulated by charge conjugation symmetry C. The study of these systems will give us some examples of the use of P and C before we apply them to the physics of the strong interaction.

4.1 The ideal hydrogen atom

To begin, recall the basic formulae for the idealized hydrogen atom of quantum mechanics textbooks. Consider the nonrelativistic limit for the electron, and take the proton mass to be very large. The potential felt

Concepts of Elementary Particle Physics. Michael E. Peskin.
© Michael E. Peskin 2019. Published in 2019 by Oxford University Press.
DOI: 10.1093/oso/9780198812180.001.0001

by the electron is

$$V(r) = -\frac{e^2}{4\pi r} = -\frac{\alpha}{r} \ . \tag{4.1}$$

It is a standard topic in quantum mechanics to solve the Schrödinger equation for this problem. The bound state energies are

$$E = -\frac{Ry}{n^2} \ , \tag{4.2}$$

where n is an integer and Ry is the the Rydberg energy

$$Ry = \frac{1}{2}\frac{me^4}{(4\pi)^2} = 13.6 \text{ eV} \ . \tag{4.3}$$

The hydrogen atom's Rydberg energy and Bohr radius take a simple form in natural units.

In natural units,

$$Ry = \frac{1}{2}\alpha^2 m \ . \tag{4.4}$$

From this formula we see that the binding energy of hydrogen is much smaller than the rest energy of the electron precisely because the electromagnetic interactions are weak. The Bohr radius of hydrogen also takes a simple form in natural units

$$a_0 = 1/\alpha m \ . \tag{4.5}$$

The velocity of the electron in the atom is of the order of $v/c \sim \alpha$; this justifies the use of the nonrelativistic approximation.

The bound states of hydrogen are arranged in levels associated with integers $n = 1, 2, 3, \ldots$. Each level contains the orbital angular momentum states

$$\ell = 0, 1, \cdots, n-1 \qquad m = -\ell, \cdots, \ell \ . \tag{4.6}$$

The orbital wavefunctions are the spherical harmonics $Y_{\ell m}(\theta, \phi)$, which are even under spatial reflection for even ℓ and odd for odd ℓ. Then, under P, these states transform as

$$P \ |n\ell m\rangle = (-1)^\ell \ |n\ell m\rangle \ . \tag{4.7}$$

4.2   Fine structure and hyperfine structure

The real hydrogen atom has more structure. First, add to the problem the fact that the electron is a particle with intrinsic spin 1/2. Thus, it has two spin states. Each state of fixed (n, ℓ) then contains $2(2\ell + 1)$ quantum states. In states with $\ell \neq 0$, the spin and orbital motion interact to split the degeneracy of these states. This is called the *spin-orbit interaction*. This interaction has two sources, one dynamical, one kinematic.

The action of P commutes with angular momentum and so preserves the spin direction. It is easy to see that orbital angular momentum $\vec{L} = \vec{r} \times \vec{p}$ commutes with P, since P reverses the direction of both $\vec{r}$ and $\vec{p}$. More pictorially, if you draw a spinning top and then perform a spatial inversion, you will see that the spin direction remains the same. So, states with the same values of L^2, S^2, and total angular momentum $J^3 = L^3 + S^3$ but different values of L^3 and S^3 can mix quantum-mechanically.

The dynamical source of spin-orbit interaction can be seen by considering the force acting on a spinning electron in its rest frame. A spinning charged particle generally has a magnetic moment. A spinning classical distribution of charges has a magnetic moment equal to

$$\vec{\mu} = \frac{Q}{2M}\vec{L} \ . \tag{4.8}$$

However, this formula is not quite correct for the magnetic moment of the electron (or other fundamental spin-$\frac{1}{2}$ particles) due to their intrinsic spin. To correct the formula, it is conventional to include a fudge factor called the *Landé g-factor* and parametrize

$$\vec{\mu} = g\frac{Q}{2M}\vec{S} \ . \tag{4.9}$$

Then, for an electron,

$$\vec{\mu}_e = -\frac{g\,e}{2m}\vec{S} \ . \tag{4.10}$$

Definition of the Landé g-factor that gives the size of the magnetic moment of an elementary particle.

This vector has magnitude $\vec{\mu}_e = (g/2)e\hbar/2m$. For the electron, the g-factor is close to 2. The relativistic theory of the electron field, given by the Dirac equation, predicts $g = 2$ precisely. Corrections due to QED interactions slightly modify this relation to

$$g_e = 2(1 + \frac{\alpha}{2\pi} + \cdots) \ . \tag{4.11}$$

An electron moving with velocity $\vec{v}$ through a Coulomb field feels, in its rest frame, a magnetic field

$$\vec{B}' = -\vec{v} \times \vec{E} = -\vec{v} \times \frac{e\hat{r}}{4\pi r^2}$$
$$= +\frac{e}{4\pi m r^3}\vec{L} \ . \tag{4.12}$$

This field tends to orient the electron spin. The effect is described by a Hamiltonian

$$\Delta H = -\vec{B}' \cdot \vec{\mu} = \frac{g_e}{2m^2 r^3}\frac{e^2}{4\pi}\vec{L} \cdot \vec{S} \ . \tag{4.13}$$

A second influence on the electron spin comes from the dynamics of spin in special relativity. An electron moving in a circle is subject to acceleration toward the center and, thus, to a sequence of boosts. The successive boosts do not commute with one another. Instead, they multiply out to a net rotation. This effect is called *Thomas precession*. It yields a precession of the spin at the angular velocity

$$\vec{\omega} = \frac{\gamma^2}{\gamma + 1}\vec{a} \times \vec{v} \ , \tag{4.14}$$

where, as in (2.16),

$$\gamma = (1 - v^2)^{-1/2} \ . \tag{4.15}$$

A derivation of the formula (4.14) is given in Exercise 4.3. For orbital motion in a Coulomb field, the acceleration is

$$m\vec{a} = -\frac{e^2}{4\pi r^2}\hat{r} \tag{4.16}$$

so the precession angular velocity is

$$\vec{\omega} = -\frac{1}{2}\frac{e^2}{4\pi mr^2}\frac{\vec{L}}{mr} . \tag{4.17}$$

This effect is described by a Hamiltonian

$$\Delta H = -\frac{1}{2m^2r^3}\frac{e^2}{4\pi}\vec{L}\cdot\vec{S} . \tag{4.18}$$

The two effects have opposite sign, leading to a total spin-orbit interaction

$$\Delta H = \frac{g-1}{2}\frac{\alpha}{m^2r^3}\vec{L}\cdot\vec{S} . \tag{4.19}$$

Since $g \approx 2$, the second effect is about half of the first. The sign is such that the state with $\vec{L}$ and $\vec{S}$ opposite in sign has lower energy.

For extremely relativistic motion of an electron in a magnetic field, $\gamma \gg 1$, the Thomas precision actually cancels the direct field-induced precession up to the factor $(g_e - 2)$. The small but nonzero size of the precession can be used to control the spin of polarized electrons in magnetic transport systems.

It is straightforward to diagonalize the operator $\vec{L}\cdot\vec{S}$. Let $\vec{J}$ be the total angular momentum: $\vec{J} = \vec{L} + \vec{S}$. The square of this operator, J^2, commutes with all scalar operators, including L^2, S^2, and $\vec{L}\cdot\vec{S}$. Notice that

$$\vec{L}\cdot\vec{S} = \frac{1}{2}\left((\vec{L}+\vec{S})^2 - L^2 - S^2\right) ; \tag{4.20}$$

that is,

$$\vec{L}\cdot\vec{S} = \frac{1}{2}(J^2 - L^2 - S^2) . \tag{4.21}$$

So, the $2(2\ell+1)$ states with given ℓ split into two levels, each of which has a definite value of $J^2 = j(j+1)$. For example, the 2×3 2P states of hydrogen split into

$$2\mathrm{P}_{j=1/2} \quad (2 \text{ states}) \quad + \quad 2\mathrm{P}_{j=3/2} \quad (4 \text{ states}) . \tag{4.22}$$

In the $j = \frac{1}{2}$ state,

$$\vec{L}\cdot\vec{S} = \frac{1}{2}[\frac{1}{2}\cdot\frac{3}{2} - 1\cdot 2 - \frac{1}{2}\cdot\frac{3}{2}] = -1 . \tag{4.23}$$

In the $j = \frac{3}{2}$ state,

$$\vec{L}\cdot\vec{S} = \frac{1}{2}[\frac{3}{2}\cdot\frac{5}{2} - 1\cdot 2 - \frac{1}{2}\cdot\frac{3}{2}] = +\frac{1}{2} . \tag{4.24}$$

It is important to note that the center of gravity of the full set of states does not change. In this example,

$$2 \cdot (-1) + 4 \cdot (+\frac{1}{2}) = 0 . \tag{4.25}$$

This happens very often in the energy splitting of a multiplet of degenerate states. The phenomenon is most easily understood by writing the sum of the perturbations as proportional to

$$\text{tr}[\vec{L} \cdot \vec{S}] . \tag{4.26}$$

We can evaluate this trace by summing over states with fixed L and S. For example,

$$\sum_{S^3} \langle S^3 | \vec{S} | S^3 \rangle = 0 . \tag{4.27}$$

and similarly for states of fixed L. Then it is easy to see that (4.26) is zero, so that the energies of the states, on average, are not shifted.

The order of magnitude of the spin-orbit interaction is

$$\left\langle \frac{\alpha}{m^2 r^3} \right\rangle \sim \frac{\alpha}{m^2 a_0^3} \sim \alpha^4 m \sim \alpha^2 \, Ry . \tag{4.28}$$

The spin-orbit interaction in hydrogen is an effect of order α^2 relative to the Rydberg energy. It *lowers* the energy of states with smaller J and *raises* the energy of states with larger J.

Thus, this effect is a factor of 10^{-4} smaller than the splitting of the principal levels of hydrogen. These splittings are called the *fine structure* of the hydrogen atom.

More structure appears when we add in the spin of the proton. The proton also has a magnetic moment

$$\vec{\mu}_p = g_p \frac{e}{2m_p} \vec{S} , \tag{4.29}$$

where $g_p = 5.6$. The fact that this number is nowhere near $g = 2$ tells us that the proton is not an elementary Dirac fermion. I will present the explanation for the large g-factor of the proton in Section 5.5.

The magnetic moments of the proton and the electron interact, with the ground state favoring the configuration in which the two spins are opposite. The Hamiltonian has the form

$$\Delta H = C \vec{S}_p \cdot \vec{S}_e , \tag{4.30}$$

where the constant C depends on the electron wavefunction. Then, for example, the 1S state of hydrogen is split into two levels, corresponding to total spin $\vec{J} = \vec{S}_e + \vec{S}_p$ equal to 0 and 1,

$$
\begin{aligned}
J = 1 : & \quad |\uparrow\uparrow\rangle \quad \tfrac{1}{\sqrt{2}}(|\uparrow\downarrow\rangle + |\downarrow\uparrow\rangle) \quad |\downarrow\downarrow\rangle \\
J = 0 : & \quad \tfrac{1}{\sqrt{2}}(|\uparrow\downarrow\rangle - |\downarrow\uparrow\rangle)
\end{aligned}
\tag{4.31}
$$

The value of the splitting is

$$\frac{8}{3} g_p \alpha^2 \cdot Ry \cdot \frac{m_e}{m_p} . \tag{4.32}$$

The spin-spin interaction in hydrogen is an effect of order $\alpha^2 m_e/m_p$ relative to the Rydberg energy. It *lowers* the energy of states with smaller total S and *raises* the energy of states with larger total S.

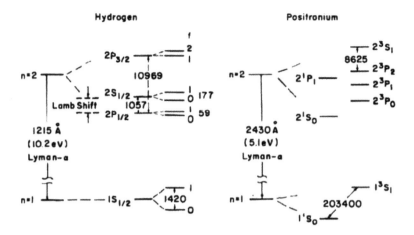

Fig. 4.1 Comparison of the 1S, 2S, and 2P energy levels of hydrogen and positronium, from (Berko and Pendleton 1980).

It is smaller than the fine structure by another factor of 10^{-3}, due to the large mass of the proton. The spin-spin splittings in the spectrum of hydrogen are called the *hyperfine structure*. The transition from the 1S($S = 1$) to the 1S($S = 0$) state of hydrogen by emission of a radio-wavelength photon gives the 21 cm radiation line that plays a central role in radio astrophysics.

4.3   Positronium

These same ideas can be applied to a particle-antiparticle system. The simplest case is *positronium*, the bound state of an electron and its antiparticle, the positron. It is relatively easy to make positronium. While most unstable nuclei emit electrons through beta decay, some emit positrons (β^+ emitters). A positron moving slowly through a gas may pick up an electron and form a positronium atom.

Figure 4.1 compares the spectra of hydrogen and positronium. There are two important differences. First, the two-body problem in positronium involves the reduced mass

$$\mu = \frac{m_1 m_2}{m_1 + m_2} = \frac{1}{2} m_e , \tag{4.33}$$

so the Rydberg in positronium is $\frac{1}{2}$ that in hydrogen. Second, the hyperfine splittings in positronium are roughly the same size as the fine structure splittings, and both are of order $\alpha^4 m_e$.

The eigenstates of positronium can be classified under parity, and also under charge conjugation. Consider first P. The orbital eigenstate $Y_{\ell m}$ gets a factor $(-1)^\ell$ under space inversion. An additional factor comes from an odd property of the Dirac equation (unfortunately, not simply explained). We are free to choose, by convention, that the electron has even intrinsic parity,

$$P \left| e^-(\vec{p}) \right\rangle = + \left| e^-(-\vec{p}) \right\rangle . \tag{4.34}$$

However, its antiparticle must then have odd intrinsic parity

$$P \left| e^+(\vec{p}) \right\rangle = - \left| e^+(-\vec{p}) \right\rangle . \tag{4.35}$$

Combining this sign with the spatial reflection property (4.7), the positronium states have parity

$$P = (-1)^{\ell+1} . \tag{4.36}$$

Quantum number of a positronium state under a parity transformation.

For the transformation properties under C, we must account three effects. C converts the electron to the positron and the positron to the electron. The electron and positron are fermions, and so, when we put the electron and positron back into their original order in the wavefunction, we get a factor (-1). Reversal of the coordinate in the orbital wavefunction gives a factor $(-1)^\ell$. Finally, the electron and positron spins are interchanged. The $S = 1$ state is symmetric in spin

$$S = 1 : \qquad |\uparrow\uparrow\rangle \quad \frac{1}{\sqrt{2}}(|\uparrow\downarrow\rangle + |\downarrow\uparrow\rangle) \quad |\downarrow\downarrow\rangle , \tag{4.37}$$

but the $S = 0$ state is antisymmetric

$$S = 0 : \qquad \frac{1}{\sqrt{2}}(|\uparrow\downarrow\rangle - |\downarrow\uparrow\rangle) \tag{4.38}$$

and so gives another factor (-1). In all, the positronium states have C

$$C = (-1)^{\ell+1} \cdot \begin{cases} 1 & S = 1 \\ -1 & S = 0 \end{cases} . \tag{4.39}$$

Quantum number of a positronium state under a charge conjugation.

The low-lying states of the positronium spectrum then have the J^{PC} values:

$$\tag{4.40}$$

The 2P states 0^{++}, 1^{++}, and 2^{++} arise from coupling the $L = 1$ orbital angular momentum to the $S = 1$ total spin angular momentum.

The photon couples to the electron with charge $(-e)$ and to the positron with charge $(+e)$. In the quantum theory, the amplitude of the photon-electron-electron interaction is proportional to $(-e)$ and the amplitude of the photon-positron-positron interaction is proportional to $(+e)$. Electrodynamics is consistent with C symmetry, then, only if we assign the photon to be odd under C

$$C \left| \gamma(\epsilon, p) \right\rangle = - \left| \gamma(\epsilon, p) \right\rangle . \tag{4.41}$$

Then, in the level diagram above, states of opposite C are linked by photon transitions, *e.g.*

$$2\mathrm{P}(J^{++}) \to 1\mathrm{S}(1^{--}), \qquad J = 0, 1, 2$$
$$2\mathrm{P}(1^{+-}) \to 1\mathrm{S}(0^{-+})$$
$$1\mathrm{S}(1^{--}) \to 1\mathrm{S}(0^{-+}) \tag{4.42}$$

In atomic physics, the strongest photon transitions are the E1 transitions, in which the electromagnetic field couples to the electric dipole moment of the atom. E1 transitions reverse the P of the state. Transitions in which the electromagnetic field couples to the magnetic dipole moment of the atom are called M1 transitions. These have smaller rates, by a factor α, and they do not change P. In (4.42), the first two transitions are E1, the last is M1.

C has strong and surprising implications for the decay of positronium.

Charge conjugation invariance has a striking implication for the decay rates of the ground states of positronium. Positronium, a massive particle, cannot decay to a single photon conserving energy and momentum. In principle, it could decay to 2 photons. However, the 2-photon state has $C = +1$, so only the $S = 0$ state (*para-positronium*) can decay in this way. The $S = 1$ states (*ortho-positronium*) can decay only to 3 photons. The decay rate is then suppressed by a factor of α, and also by some numerical factors. The formulae for the decay rates are: for the $S = 0$ state,

$$1/\tau = \frac{1}{2}\alpha^5 m \qquad \tau = 1.2 \times 10^{-10} \text{ sec}, \tag{4.43}$$

and for the $S = 1$ states,

$$1/\tau = \frac{2}{9\pi}(\pi^2 - 9)\alpha^6 m \qquad \tau = 1.4 \times 10^{-7} \text{ sec}, \tag{4.44}$$

So, when we emit positrons into a gas, 1/4 of the positronium atoms decay in a tenth of a nanosecond, but then we must wait 1000 times longer for the other 3/4 to decay. It is a strange result, but experiment verifies it (Berko and Pendleton 1980).

Exercises

(4.1) Imagine that the electron had spin 1 instead of spin $\frac{1}{2}$. Show that the spin-orbit interaction would split the 2P levels into states with three different energies. Compute the relative energy shifts. Show that the center of gravity of the levels is not changed by this energy splitting.

(4.2) The relativistic equations studied in Chapter 3 generally predict the the corresponding particles have Landé g factor equal to 2. We can explore this for particles of spin $\frac{1}{2}$ using the Dirac equation.

  (a) A field obeying the Dirac equation in the presence of a background electromagnetic field also obeys the second-order equation

  $$(i\gamma^\mu D_\mu + m)(i\gamma^\nu D_\nu - m)\Psi = 0 , \quad (4.45)$$

  where $D_\mu = (\partial_\mu + ieA_\mu)$. Simplify this equation by using the identity

  $$\gamma^\mu \gamma^\nu = \frac{1}{2}\{\gamma^\mu, \gamma^\nu\} + \frac{1}{2}[\gamma^\mu, \gamma^\nu] . \quad (4.46)$$

  and show that it reduces to the Klein-Gordon equation plus one extra term.

  (b) Simplify the new term by proving the identity

  $$[D_\mu, D_\nu] = +ieF_{\mu\nu} . \quad (4.47)$$

  Using the explicit form of the γ^μ matrices, evaluate this term in a background magnetic field for which $F_{ij} = \epsilon_{ijk}B^k$ and $F_{0i} = 0$.

  (c) Act the resulting equation on the Dirac equation solution (3.51). Show that, to first order in B, the energy of the state is shifted by a term of the form of $\Delta E = -\vec{\mu} \cdot \vec{B}$. In the expression for $\vec{\mu}$, identify $g = 2$.

(4.3) This problem explores Thomas precession .

  (a) Using the commutation relation (2.54), show that the Hamiltonian $H = \vec{\omega} \cdot \vec{S}$ and the Heisenberg equation of motion

  $$i\frac{d}{dt}\vec{S} = [\vec{S}, H] \quad (4.48)$$

  lead to the equation of motion

  $$\frac{d}{dt}\vec{S} = \vec{\omega} \times \vec{S} . \quad (4.49)$$

This is a precession of $\vec{S}$ with angular velocity $\vec{\omega}$. Using this relation, we can go back and forth between computed values of the precession frequency and its description by an effective Hamiltonian.

  (b) Now consider a particle moving in the $+\hat{3}$ direction and also being accelerated in the $+\hat{1}$ direction. Write the 4×4 boost matrix Λ that boosts the particle from its rest frame to the velocity $v\hat{3}$.

  (c) In the lab, we observe the particle to be accelerated with acceleration a, so that it is now moving with the velocity $v\hat{3} + a\delta t\hat{1}$. Write the boost matrix $\overline{\Lambda}$ that boosts the particle from its rest frame to this new frame, working to first order in δt. The space-space part of $\overline{\Lambda}$ should be

  $$\delta^{ij} + (\gamma-1)\frac{(v\hat{3} + a\delta t\hat{1})^i (v\hat{3} + a\delta t\hat{1})^j}{v^2} . \quad (4.50)$$

  (d) To understand this transformation, compute $\Lambda^{-1}\overline{\Lambda}$. This matrix gives the effect of the acceleration as seen by the particle in its rest frame. Show this this takes the form of a boost by $\gamma a\delta t$ and a rotation by $(\gamma - 1)a\delta t/v$ with the axis of rotation in the direction $\hat{a} \times \hat{v}$.

  (e) Using (4.15), show that the rotation found here reproduces the formula for the Thomas precession frequency quoted in (4.14).

(4.4) It is possible to solve exactly for the energy eigenvalues of the hydrogen atom problems for the relativistic equations for scalars (Klein-Gordon equation) and electrons (Dirac equation). The solutions are long but actually not so difficult, if you use the tricks that are suggested below at the various stages of the solution. Try it!

The electrostatic potential for an electron in the hydrogen atom is conveniently written

$$V(r) = -\frac{\alpha}{r} . \quad (4.51)$$

Let m be the electron mass. Take the proton to be fixed and infinitely heavy.

  (a) The nonrelativistic Schrödinger equation for a particle in a potential is

  $$\left[-\nabla^2 + 2mV(r) - 2mE\right]\psi = 0 . \quad (4.52)$$

Prepare for the diagonalization of this operator by making r dimensionless with the substitution $\rho = r/a_0 = rm\alpha$ and letting ψ be an eigenstate of L^2 with eigenvalue $\ell(\ell+1)$. Write the resulting equation, an ordinary differential equation for $\psi(\rho)$.

(b) The Klein-Gordon equation for a particle in a potential is

$$\left[(E - V(r))^2 + \nabla^2 - m^2\right]\varphi = 0 \quad (4.53)$$

Expand out this equation, substitute $\rho = rE\alpha$, and let φ be an eigenstate of L^2 with eigenvalue $\ell(\ell+1)$. Note that the form of this equation is the same as that in (a), with small changes in the constant factors.

(c) Look for solutions of these equations of the form

$$\rho^{\nu-1}e^{-k\rho} . \quad (4.54)$$

For the Schrödinger equation, show that $\nu = \ell + 1$, $k = 1/\nu$, and

$$E = -\frac{1}{2}\frac{\alpha^2 m}{\nu^2} . \quad (4.55)$$

For the Klein-Gordon equation, show that $\nu = \lambda + 1$, where λ satisfies

$$\ell(\ell+1) - \alpha^2 = \lambda(\lambda+1) \quad (4.56)$$

Define δ_ℓ by

$$\lambda = \ell - \delta_\ell \quad (4.57)$$

Compute δ_ℓ and show that it is small, of the order of α^2.

(d) The other bound state solutions of the Schrödinger equation are of the form

$$\rho^{\nu-p-1}P_p(\rho)e^{-k\rho} . \quad (4.58)$$

where $P_p(\rho)$ is a polynomial of degree p (an integer). You can look up in your favorite quantum mechanics book that it remains true that $k = 1/\nu$ and that the formula (4.55) for the energy levels still holds. The same formulae apply to the Klein-Gordon equation, except that ν is no longer an integer. It is shifted from an integer value by δ_ℓ, as in part (c). Using these facts, you can immediately write down the eigenvalue of the operator. This gives an equation for E in the Klein-Gordon case, which is an easy quadratic equation for E. Show that the solution is

$$E = \frac{m}{[1 + \alpha^2/(n - \delta_\ell)^2]^{1/2}} , \quad (4.59)$$

where n is an integer, the usual principal quantum number. Note that the energy levels now depend on both n and ℓ.

(e) Expand (4.59) to order α^4. Show that the first two terms give back the nonrelativistic answer, while the order α^4 term gives a relativistic correction that depends on ℓ.

(f) Now turn to the Dirac equation. The Dirac equation in a potential is

$$\left[\gamma^0(E - V(r)) + i\vec{\gamma} \cdot \vec{\nabla} - m\right]\Psi = 0 \quad (4.60)$$

Choose the representation of the Dirac matrices

$$\gamma^0 = \begin{pmatrix} 0 & 1 \\ 1 & 0 \end{pmatrix} \qquad \vec{\gamma} = \begin{pmatrix} 0 & \vec{\sigma} \\ -\vec{\sigma} & 0 \end{pmatrix} \quad (4.61)$$

where all blocks are 2×2 and the $\vec{\sigma}$ are the Pauli sigma matrices. Make this equation second-order by multiplying by the operator

$$\left[\gamma^0(E - V(r)) + i\vec{\gamma} \cdot \vec{\nabla} + m\right] . \quad (4.62)$$

Show that all matrices disappear except for one term in which $\vec{\nabla}$ acts on $V(r)$. This term is proportional to

$$i\vec{\gamma}\gamma^0 = \begin{pmatrix} i\vec{\sigma} & 0 \\ 0 & -i\vec{\sigma} \end{pmatrix} . \quad (4.63)$$

So, the top two components of Ψ obey an independent equation from the bottom two components. Each equation is an equation for a 2-component spinor field. You will find that these equations have the same bound state eigenvalues, so we only have to solve one of these.

(g) Expand the squares and write out the equation as in the Klein-Gordon case. Again, let $\rho = rE\alpha$. You should find the same structure as in the Klein-Gordon case, except that now the coefficient of $1/r^2$ is

$$\ell(\ell+1) - \alpha^2 + i\alpha\vec{\sigma} \cdot \hat{r} , \quad (4.64)$$

where $\hat{r}$ is a unit vector in the radial direction. This is, unfortunately, a 2×2 matrix.

(h) To make (4.64) a number, we need a rather subtle trick: The operator $\vec{\sigma} \cdot \hat{r}$ commutes with the total angular momentum J^2. On the other hand, the factor $\hat{r}$ changes ℓ by ± 1 unit. So this operator must mix pairs of states with the same j and different ℓ. These are states with $\ell = j + \frac{1}{2}$ and states with $\ell = j - \frac{1}{2}$. Show, however, that

$$(\vec{\sigma} \cdot \hat{r})^2 = 1 . \qquad (4.65)$$

This means that, in the basis just described

$$\vec{\sigma} \cdot \hat{r} = \begin{pmatrix} 0 & 1 \\ 1 & 0 \end{pmatrix} . \qquad (4.66)$$

Show that (4.64) can then be written

$$\begin{pmatrix} (j - \frac{1}{2})(j + \frac{1}{2}) - \alpha^2 & i\alpha \\ i\alpha & (j + \frac{1}{2})(j + \frac{3}{2}) - \alpha^2 \end{pmatrix} . \qquad (4.67)$$

(i) Find the eigenvalues of this matrix. Show that they are of the form $\lambda(\lambda + 1)$, where λ is shifted from an integer by

$$\delta_j = (j + \frac{1}{2}) - [(j + \frac{1}{2})^2 - \alpha^2]^{1/2} . \qquad (4.68)$$

(j) Put all of the pieces together and derive

$$E = \frac{m}{[1 + \alpha^2/(n - \delta_j)^2]^{1/2}} . \qquad (4.69)$$

Note that this formula depends only on n and j, not on ℓ.

(k) Expand the formula to order α^4 and find the relativistic corrections to the energy levels of the hydrogen atom according to the Dirac equation. These formulae are in good agreement with experiment (to this order in α).

The Quark Model

5

We now have ample preparation to begin a discussion of particle physics and the strong and weak interactions. The first topic that I will discuss is the nature of the strongly interacting particles created in nuclear reactions. These include the proton and neutron, the π mesons, and many related particles. These particles are collectively called *hadrons*.

We now know that hadrons are not elementary particles. They are bound states of more elementary constituents called *quarks*. However, it is very important to have a sharp qualitative understanding of the hadrons. As we will discuss at the end of this chapter, quarks are never seen as isolated particles but only as constituents of hadrons. This behavior is consistent with the laws of quantum mechanics, but it is very counterintuitive. This has two implications for our study. First, if quarks are not seen directly, the evidence for their existence inside bound states must be especially strong. In this chapter, I will describe how the observable properties of hadrons give a first level of this evidence for quarks. We will see stronger evidence for the quark structure of hadrons in Chapters 8 and 9. Second, if quarks are not seen directly, our experimental measurements on the strong interaction must be done at the level of hadrons. To understand what is actually measured in the experiments I will discuss, you will need to keep in mind the names, identities, and basic properties of the hadrons.

5.1 The discovery of the hadrons

The lightest strongly interacting particles are the π mesons, with masses of

$$\pi^{\pm} \;:\quad 139.57 \text{ MeV} \qquad \pi^0 \;:\quad 134.98 \text{ MeV} \tag{5.1}$$

The history of the discovery of these particles is fascinating. In 1935, Hideki Yukawa showed that, in the quantum theory of the Klein-Gordon equation

$$(\partial^2 + m^2)\varphi = 0 \tag{5.2}$$

the interaction of the field with static sources leads to a potential (the *Yukawa potential*)

$$V(r) = \frac{g^2}{4\pi r}\, e^{-mr} \;. \tag{5.3}$$

Concepts of Elementary Particle Physics. Michael E. Peskin.
© Michael E. Peskin 2019. Published in 2019 by Oxford University Press.
DOI: 10.1093/oso/9780198812180.001.0001

Fig. 5.1 Charged particle tracks in photographic emulsion, produced by cosmic rays, showing the decay of a π meson to a muon, from (Lattes, Occhialini, and Powell 1947).

The Yukawa potential is universally attractive, with characteristic range

$$\frac{1}{m} \quad \text{or} \quad \frac{\hbar}{mc} \, . \tag{5.4}$$

Comparing to the range of the nuclear forces, he concluded (Yukawa 1935) that these forces would be explained by a Klein-Gordon particle, the *meson*, of mass

$$m \sim 200 \text{ MeV} \, . \tag{5.5}$$

Shortly thereafter, a particle of mass about 100 MeV was discovered in cosmic rays. It was quickly concluded that this was the Yukawa meson. However, it was then found that this particle was extremely penetrating, with a range in matter of tens of meters. Theorists set to work inventing reasons why the basic particle of the nuclear force did not in fact interact with nuclei. Then, in 1947, Lattes, Ochialini, and Powell exposed photographic emulsion at high altitude, including at the observatory at Mt. Chacaltaya in Bolivia. They found that another particle was visible there and decayed to the supposed meson. They observed that the first particle was produced in nuclear interactions. One of the figures from their paper is shown in Fig. 5.1 (Lattes, Occhialini, and Powell 1947).

After this discovery, the former meson was demoted to the μ, which turned out on closer investigation to be a lepton. The new meson was the π. This put us on the correct road to an understanding of the strong interaction.

Over the next fifteen years, more strongly interacting particles, *hadrons*, were discovered, first, in cosmic ray observations and, beginning in the 1950's, in experiments at particle accelerators. Figs. 5.2 and 5.3 show displays of collision events that give evidence for some of these particles. The pictures were made using a device called a *bubble chamber*, a volume of liquid in which the passage of a relativistic charged particle will produce a line of tiny bubbles, which can then be photographed. The momenta of the particles responsible for the various tracks can be measured, as we will discuss in Section 6.4, and specific charged particles can be identified. Neutral particles are gaps where no bubbles are

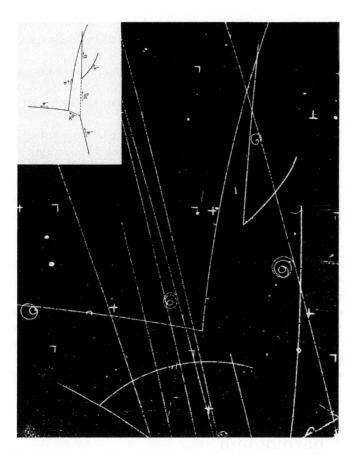

Fig. 5.2 A 1957 photograph from the 10-inch bubble chamber at Lawrence Berkeley National Laboratory, showing the reaction $\pi^- p \to K^0 \Lambda$, with subsequent decays of the K^0 and the Λ into two charged particles (figures courtesy of LBNL).

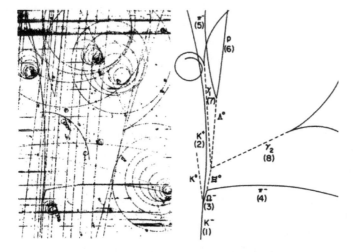

Fig. 5.3 A bubble chamber photograph from Brookhaven National Laboratory that proved the existence of the Ω^- baryon, from (Barnes 1964).

visible, ending in a decay to two or more charged particles. Adding the momenta of these charged particles, one can find evidence for neutral particles of definite mass. The specific particles that appear in these figures will be introduced in Sections 5.3 and 5.5. Eventually, over a hundred strongly interacting particles were discovered. It appeared that the π mesons were not fundamental particles but rather bound states made of some more elementary constituents.

The discovery and characterization of the hadrons is a fascinating chapter in the history of physics, but in this book I will simply present the final understanding of this subject. Here, this subject will serve as a starting point for the investigation of the strong interaction described in Part II. Very informative accounts of the discovery of the hadrons may be found in the historical account by Pais (1986) and in older textbooks such as (Källén 1964) and (Gasiorowicz 1966).

By studying the systematics of the hadrons, physicists tried to guess how they could be built up from more basic states. In 1964, Murray Gell-Mann and George Zweig proposed the *quark model* (Gell-Mann 1964, Zweig 1964), which I will describe in the following sections of the chapter. It was not at all obvious at the time that the quark model was correct. In fact, a large part of the high-energy physics community did not accept the quark model until ten years later, when remarkable new evidence for it was found. In this textbook account, I will start with this evidence and then work backward to the description of the hadrons known in 1964.

5.2   Charmonium

Electron-positron annihilation provides a simple setting to study elementary particles created in the annihilation process. This is an expecially powerful way to study strongly interacting particles.

A beautifully simple way to create any particle, together with its antiparticle, is to annihilate electrons and positrons at high energy. The annihilation results in a short-lived excited state of electromagnetic fields. This state can then re-materialize into any particle-antiparticle pair that couples to electromagnetism and has a total mass less than the total energy of the annihilating e^+e^- system.

In the 1960's, a number of e^+e^- colliders were constructed around the world. At energies below 2 GeV, e^+e^- annihilation shows a series of resonances, which become increasingly broad and blend together to give an annihilation rate that depends smoothly on energy. For example, the ρ', at 1450 MeV, has a width of about 400 MeV. At higher energies, the annihilation rate was expected to decrease as $1/E_{CM}^2$. The actual measurements showed an increasing, and oddly inconsistent, rate.

Discovery of the J/ψ resonance in e^+e^- annihilation and production at a mass of 3.1 GeV.

In 1971, the e^+e^- collider SPEAR began operating at the SLAC National Laboratory in California. In November 1974, the SPEAR experimenters discovered an enormous, very narrow, resonance, at about 3.1 GeV (Augustin *et al.* 1974). This resonance would correspond to a new strongly interacting particle. This particle decayed most often to pions and other hadrons, but also, about 6% of the time, to electron-positron pairs and the same to $\mu^+\mu^-$ pairs. When they announced

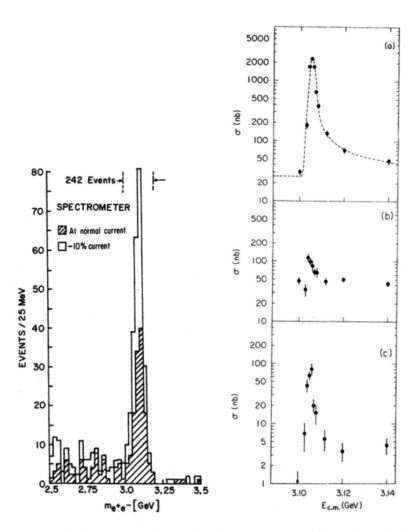

Fig. 5.4 Left: Accumulation of events from the reactions $pp \to e^+ e^- + X$ at a fixed value of the $e^+ e^-$ invariant mass, proving the existence of the narrow resonance $J\psi$, from (Aubert *et al.* 1974). Right: Observation of a resonance in $e^+ e^-$ annihilation near 3.1 GeV, proving the existence of the J/ψ, from (Augustin *et al.* 1974). From top to bottom, the three plots show the production rate of final states with (a) multiple hadrons, (b) $e^+ e^-$, and (c) pairs of μ, π, K.

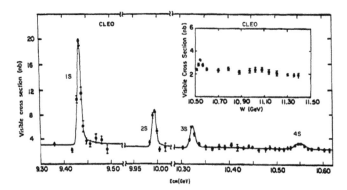

Fig. 5.5 Resonances in e^+e^- annihilation to hadrons corresponding to the S states of the Υ family, from (Silverman 1981). The inset extends the dataset to higher center of mass energy.

this discovery, they learned that the group of Samuel Ting, working at Brookhaven National Laboratory in Upton, New York, had also observed this new particle (Aubert *et al.* 1974). Ting's group had studied the reaction

$$p + p \to e^+e^- + X \tag{5.6}$$

where the particles X are not observed. They had looked for a resonance in the mass spectrum of the e^+e^- pair, which would indicate a state of definite mass that decayed to e^+e^-. Again, an enormous enhancement appeared. The two discovery papers appeared back-to-back in *Physical Review Letters*. I reproduce the key plots in Fig. 5.4. This particle is now called the J/ψ. The discovery shocked everyone because the resonance was so narrow. As we will discuss in Section 7.1, the lifetime τ of a particle is related to its observed width in energy Γ by $\tau = \hbar/\Gamma$. Thus, a small width corresponds to a long lifetime, longer by three orders of magnitude than the lifetimes of typical hadrons of mass 3 GeV. A few weeks later, the SPEAR group discovered a second narrow resonance at 3686 MeV, the ψ' (Abrams *et al.* 1974).

Another group of narrow resonances is found in e^+e^- annihilation at higher energy. The lightest state of this family, called the Υ, has a mass of 9460 MeV. It was discovered by the group of Leon Lederman in the reaction $pp \to \mu^+\mu^- + X$ at the Fermilab proton accelerator (Herb *et al.* 1977). The full family of resonances was later uncovered at the Cornell University e^+e^- collider CESR. In Fig. 5.5, I show the data from the CLEO experiment at CESR, showing evidence for 4 clear resonances (Silverman 1981). Two more resonances, which are barely visible in this plot, have since been established.

The physics of the e^+e^- annihilation process allows us to determine the quantum numbers of these particles. In the process $e^+e^- \to$ hadrons, the highest rate reactions are those in which the e^+e^- pair is annihilated by the electromagnetic current $\vec{j} = \overline{\psi}\vec{\gamma}\psi$ through the matrix element

$$\langle 0| \, \vec{j}(x) \, |e^+e^-\rangle \,. \tag{5.7}$$

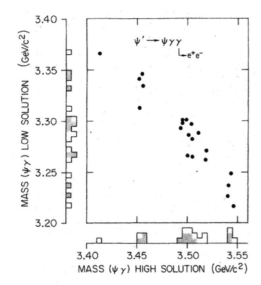

Fig. 5.6 Distribution of $\psi\gamma$ invariant masses observed in the reaction $\psi' \to \gamma\gamma\, J/\psi$ by the Mark I experiment at SPEAR, from (Tannenbaum *et al.* 1978).

The current has spin 1, $P = -1$, and $C = -1$. These must also be properties of the annihilating e^+e^- state, and of the new state that is produced. So, all of the ψ and Υ states must have $J^{PC} = 1^{--}$. The current creates or annihilates a particle and antiparticle at a point in space. So, if these particles are particle-antiparticle bound states, the wavefunctions in these bound states must be nonzero at the origin. Most probably, they would be the 1S, 2S, *etc.* bound states of a potential problem.

If this guess is correct, the states with higher L must also exist. They might be produced in radiative decays of the ψ and Υ states. Indeed, in the summer of 1975, the SPEAR group and the DASP group at the DESY Laboratory in Hamburg, Germany, observed 2-photon decays of the ψ' (Braunschweig *et al.* 1975, Feldman *et al.* 1975)

$$\psi' \to \gamma + \chi$$
$$\to \gamma + J/\psi \, . \tag{5.8}$$

Three intermediate χ states were observed. Because the transitions involve emission of a photon, these three states must have $C = +1$. Some later evidence for these transitions, from the SPEAR experiment, is shown in Fig. 5.6. Notice that the lower energy photons have a somewhat narrow energy spread, while the higher energy photons are broadened in energy. The broadening would naturally happen for the photons from the χ decays to J/ψ, since the χ recoils against the first photon and so its decay products are boosted in a direction uncorrelated with the photon emission.

Because these transitions are the strongest photon decays of the ψ', it is tempting to identify them with E1 transitions, as are seen in atomic

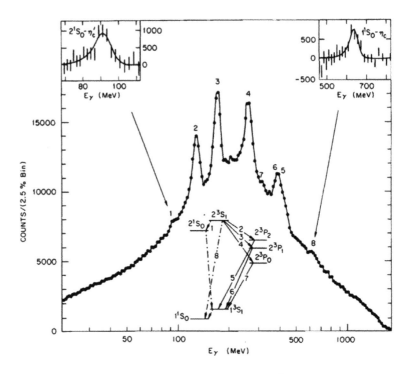

Fig. 5.7 Energy spectrum of photons observed at the ψ' resonance by the Crystal Ball experiment, from (Scharre 1981).

physics and in positronium. In a bound state, E1 transitions reverse the value of P and add orbital angular momentum $L = 1$. In this case, the new states χ in (5.8) would have the quantum numbers $J^{PC} = 0^{++}, 1^{++}, 2^{++}$.

We now have the pattern of states,

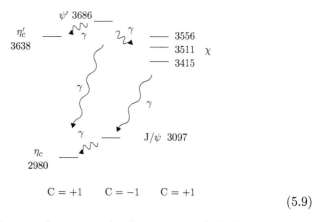

$$ (5.9) $$

The spectrum of states of the J/ψ and its partners reproduces the form of the positronium spectrum.

Remarkably, this reproduces exactly the pattern of the lowest-energy states of positronium. In the positronium spectrum, there are also 1S and 2S 0^{-+} states. These states, called η_c and η_c', were discovered through weaker M1 photon transitions from ψ'. Fig. 5.7 shows the

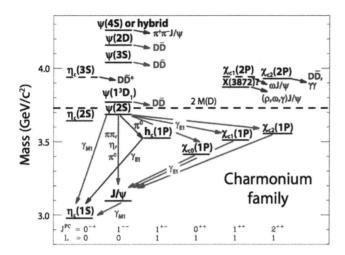

Fig. 5.8 Observed states and transitions of the J/ψ system, from (Eichten, Godfrey, Mahlke, and Rosner 2008). Note that the lowest P and D states are labelled 1P and 1D in this figure, whereas in the hydrogen atom they would be called 2P and 3D. The dashed horizontal line marks the threshold for pair production of D mesons, described in Section 5.4.

amazing photon spectrum observed by the Crystal Ball experiment at SPEAR (Scharre 1981), with peaks at the photon energies associated with almost all of these transitions. The positronium spectrum also contains a 2P state with quantum numbers 1^{+-}. The analogous state in charmonium, called h_c, has $C = -1$, and so it cannot be reached from the ψ particles by a photon. It was discovered later, at Fermilab, in the reaction $p\bar{p} \to h_c \to \gamma\eta_c \to 3\gamma$ (Andreotti *et al.* 2005), and by the CLEO experiment at Cornell, in the reaction $\psi' \to \pi^0 h_c \to \pi^0\gamma\eta$ (Rosner *et al.* 2005).

The complete set of ψ and Υ family states now known is shown in Figs. 5.8 and 5.9, from a recent review of Eichten, Godfrey, Mahlke, and Rosner (2007). This makes even more clear that the analogy to positronium is precise. These states are bound states of a spin $\frac{1}{2}$ fermion and its antiparticle. In the case of the ψ family, the fermion is called the *charm quark* (c); this quark has a mass of about 1.8 GeV. In the case of the Υ family, the fermion is called the *bottom quark* (b); this quark has a mass of about 5 GeV.

It is worth noting that the P states in these spectra lie below the corresponding S states. In the hydrogen atom, the 2P and 2S states are almost degenerate. This is a special property of the $1/r$ potential. Other possible potentials give these states different energies. A potential that increases at large distances can lift the 2S states above the 2P states. The reason for this is that the the 2P states are smaller radially than the 2S states,

$$\langle r \rangle_{2P} < \langle r \rangle_{2S} , \tag{5.10}$$

because the 2S state must be orthogonalized to the 1S state and therefore

The spin $\frac{1}{2}$ particles whose particle-antiparticle bound states form the J/ψ and Υ systems of states are called *quarks*.

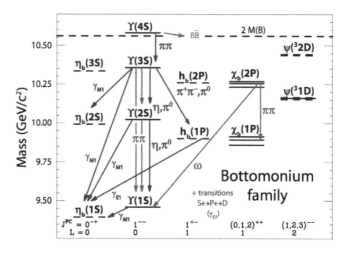

Fig. 5.9 Observed states and transitions of the Υ system, from (Eichten, Godfrey, Mahlke, and Rosner 2008). The dashed horizontal line marks the threshold for pair production of B mesons, described in Section 5.4.

must extend further out from the origin. Potentials of the form

$$V(r) = A\log r \quad \text{or} \quad V(r) = -\frac{A}{r} + Br \tag{5.11}$$

give a good fit to the observed energies of the states of the ψ and Υ systems. With a potential of this form, quarks cannot escape from their bound states but rather are permanently confined. This raises issues of physics that I will discuss later in the course. It also makes it ambiguous what we mean by the quark mass; again, we will return to this question later.

> It is difficult to understand why the J/ψ and Υ are so long-lived. But, the analogy with positronium suggests an explanation.

With the information provided so far, it still seems a mystery why the J/ψ and Υ states are so narrow. If these particles decay by strong interactions, we might expect that the quark and antiquark inside these bound states would annihilate rapidly through the strong interaction, leading to a width of hundreds of MeV. But the measured widths of these resonances are:

$$J/\psi : \quad 93 \text{ keV} \qquad \Upsilon : \quad 54 \text{ keV} . \tag{5.12}$$

The widths of the η_c states are

$$\eta_c : \quad 30 \text{ MeV} \qquad \eta_c' : \quad 10 \text{ MeV} , \tag{5.13}$$

also very small compared to the expected values. We have seen the ratio of about 1000 between the 0^{-+} and 1^{--} lifetimes in the previous chapter; this is just the ratio of the ortho- and para-positronium lifetimes. The long lifetimes of the J/ψ and Υ would be explained if these states could only decay to weakly coupled spin 1, $C = -1$ bosons. I would like to suggest here that the decays are to spin 1 particles, called *gluons*, that are the basic quanta of the strong interaction,

$$\eta_c \to 2g \qquad J/\psi \to 3g . \tag{5.14}$$

It still seems strange that the gluons would be weakly coupled to heavy quarks. I will take up that issue later in Chapters 10 and 11.

5.3   The light mesons

Now we can go back to the π mesons and other relatively light hadrons. There are three π mesons, π^+, π^0, π^-. The lightest states of the ψ and Υ families are the η_c and η_b, with the quantum numbers 0^{-+}. We have seen that this is natural for the $S = 0$, 1S bound states of a system of spin $\frac{1}{2}$ particles. By detailed study of their interactions, it was determined that the π mesons also had $J^P = 0^-$. The π^0 decays to 2 photons, so it is $C = +1$. All of this is consistent with the interpretation of the pions as spin-$\frac{1}{2}$ fermion-antifermion bound states.

In fact, it was found that the lightest-mass hadrons all have the J^P quantum numbers either 0^- or 1^-. There are 9 relatively light 0^- hadrons, the *pseudoscalar mesons*,

$$
\begin{array}{ccccc}
& & \underline{\eta'} & & 958 \\[4pt]
& & \underline{\eta} & & 548 \\[4pt]
\underline{K^-} & \underline{\bar{K}^0}\;\underline{K^0} & & \underline{K^+} & 498 \\[4pt]
& \underline{\pi^-}\quad\underline{\pi^0}\;\underline{\pi^+} & & & 140
\end{array}
\tag{5.15}
$$

The family of light pseudoscalar (0^-) mesons.

and 9 somewhat heavier 1^- hadrons, the *vector mesons*,

$$
\begin{array}{ccccc}
& & \underline{\phi^0} & & 1020 \\[4pt]
\underline{K^{*-}} & \underline{\bar{K}^{*0}}\;\underline{K^{*0}} & & \underline{K^{*+}} & 892 \\[4pt]
\rho^-\!\underline{\quad\quad} & \underline{\overline{\;\omega^0\;}} & \!\underline{\quad\quad}\rho^+ & & \begin{array}{c}781 \\ 770\end{array}
\end{array}
\tag{5.16}
$$

The family of light vector (1^-) mesons.

The numbers in the figures give the masses of the particles in MeV. The K and K^* states are not produced singly in strong interactions. They are only produced together with one another, or with special excited states of the proton. For example, we see the reactions

$$
\pi^- p \to n K^+ K^- \qquad \pi^- p \to \Lambda^0 K^0 \;,
\tag{5.17}
$$

where Λ^0 is a heavy excited state of the proton, but we do not see the reaction

$$
\pi^- p \to n K^0 \;.
\tag{5.18}
$$

For this reason, the K mesons and the Λ^0 baryon became known as the *strange particles*. It was found that the rules for K and K^* production can be expressed simply by saying that the strong interaction preserves a quantum number called *strangeness*, with K^0, K^+, K^{*0}, and K^{*+}

having strangeness $S = -1$, their antiparticles having $S = +1$, and the Λ^0 having $S = +1$.

The π^0, η, and η' decay to 2 photons, so these are $C = +1$ states. The ρ^0, ω, and ϕ decay to $\gamma\pi^0$, so these are $C = -1$ states. These observations favor an interpretation of these states as bound states of spin-$\frac{1}{2}$ fermions and their corresponding antiparticles. The fermions must be of three types, called the *up*, *down*, and *strange quarks*,

$$u, d, s . \tag{5.19}$$

For example, we can model the π^+ and ρ^+ as the $u\bar{d}$ bound states with spin 0 and spin 1, respectively. States with strangeness $+1$ will be assigned one s quark, and states with strangeness -1 will have one $\bar{s}$ antiquark.

The mass pattern of the 0^- states is not so clear, but the mass pattern of the 1^- states is quite obvious. The K^* states, with strangeness ± 1, are about 120 MeV heavier than the ρ and ω mesons, and the ϕ is about 120 MeV heavier than the K^* states. We can then interpret the 9 1^- states as bound states of

$$(\bar{u}, \bar{d}, \bar{s}) \times (u, d, s) , \tag{5.20}$$

The spectrum of light mesons can be built of as bound states of the three quarks u, d, s with the corresponding antiquarks. The s quark is somewhat heavier than the others and carries the conserved quantum number *strangeness*.

with the s quark carrying strangeness $S = -1$ and being about 120 MeV heavier than the u, d. The u quark should have an electric charge

$$Q_u = Q_d + 1 = Q_s + 1 . \tag{5.21}$$

The ϕ is interpreted as an $\bar{s}s$ bound state. Quite properly, its main decay modes are

$$\phi \to K^+K^-, K^0\overline{K}^0 . \tag{5.22}$$

In this model, the near mass degeneracy of the π^+, π^-, π^0 and of the K^+, K^0 would be a consequence of near mass degeneracy of the u and d quarks. It is tempting to guess that these quarks are exactly degenerate, up to small corrections due to electromagnetic effects from their different electric charges. This point of view is not actually correct, but it will serve for the moment. I will return to this set of issues in Section 14.4. There I will explain the form of the mass spectrum of the pseudoscalar mesons, and we will find a way to evaluate the masses of the u and d quarks.

It is tempting to guess that the strong interactions are invariant under the discrete symmetry

$$u \leftrightarrow d \tag{5.23}$$

and, approximately, under the additional discrete symmetries

$$u \leftrightarrow s \qquad d \leftrightarrow s . \tag{5.24}$$

In fact, the symmetry of the strong interaction is larger. In the 1930's, Heisenberg suggested that the nuclear forces are invariant under a continuous symmetry, called *isospin* (Heisenberg 1932). In this viewpoint, the proton and neutron are viewed as a 2-component system that can be rotated by 2×2 unitary transformations. The group of transformations is $SU(2)$, appearing here as an internal symmetry of the strong interaction. Isospin turned out to be a powerful symmetry constraint on the properties of nuclear energy levels. The group $SU(2)$ is isomorphic to the rotation group in 3 dimensions, and so the representations of isospin are exactly the familiar ones of angular momentum. If we assign the u and d quarks to a spin $\frac{1}{2}$ representation of isospin

$$\begin{pmatrix} u \\ d \end{pmatrix} \to U(\alpha) \begin{pmatrix} u \\ d \end{pmatrix} = e^{-i\vec{\alpha}\cdot\vec{\sigma}/2} \begin{pmatrix} u \\ d \end{pmatrix} , \qquad (5.25)$$

then the combinations $\bar{u}u$, $\bar{u}d$, $\bar{d}u$, $\bar{d}d$ form the $\frac{1}{2} \otimes \frac{1}{2}$ representation, that is, a sum of spin 0 and spin 1. The isospin 1 representation has three degenerate states, and we can identify these with the three π and the three ρ mesons. The isospin 0 states can be identified with the η and ω mesons. The continuous isospin symmetry is a larger symmetry group than the discrete replacement symmetry and it is more powerful. For example, isospin symmetry relates different elementary particle processes by $SU(2)$ Clebsch-Gordan coefficients. I give some examples of isospin predictions of this type in Exercise 5.2.

It is sometimes useful to ignore the mass difference between the s quark and the lighter quarks and treat the three quarks (u, d, s) as having identical strong interactions. These three states are transformed by an $SU(3)$ continuous symmetry. In this book, I will make references to this symmetry, but I will avoid situations in which we need to compute $SU(3)$ Clebsch-Gordan coefficients.

In my discussion of the C of the 0^- and 1^- states above, it was awkward to treat states like the π^+ and π^-. These states are interchanged by C

$$C \left| \pi^+ \right\rangle = \left| \pi^- \right\rangle \qquad (5.26)$$

and thus are not eigenstates of C. To restore a simple transformation law, we define G-*parity* by

$$G = C\,e^{i\pi I_2} = CR_2(\pi) . \qquad (5.27)$$

G-parity is a consequence of C and isospin symmetry that is easier than C to apply to reactions of the π meson.

This is a rotation by $180°$ about the $\hat{2}$ axis of isospin, followed by charge conjugation. G will be a good symmetry to the extent that both C and isospin are conserved by the strong interactions.

Since

$$R_2(\pi)\,(\hat{1} + i\hat{2}) = -(\hat{1} - i\hat{2}) \qquad R_2(\pi)\,\hat{3} = -\hat{3} , \qquad (5.28)$$

we find that

$$R_2(\pi) \left| \pi^+ \right\rangle = -\left| \pi^- \right\rangle \qquad R_2(\pi) \left| \pi^0 \right\rangle = -\left| \pi^0 \right\rangle . \qquad (5.29)$$

Then

$$G \left| \pi^+ \right\rangle = -\left| \pi^+ \right\rangle \qquad G \left| \pi^- \right\rangle = -\left| \pi^- \right\rangle \qquad G \left| \pi^0 \right\rangle = -\left| \pi^0 \right\rangle . \quad (5.30)$$

Then all three pions have $G = -1$. This leads to useful selection rules. For example, when the J/ψ decays directly to pions, it always decays to an *odd* number of pions. The J/ψ has $I = 0$ and $C = -1$, thus, $G = -1$, so this rule is explained by G-parity.

According to the quark model, all mesons — bosonic hadrons — are described as quark-antiquark bound states. In general, this description works well. The situation is not completely clear because the pion is so light. Mesons heavier than about 1300 MeV can decay rapidly by emitting pions. They may have large resonance widths and large interference or mixing between states. This leaves room for additional non-$\bar{q}q$ states. If the strong interactions are indeed due to spin 1 gluons, we would expect to see hadrons that are gg bound states. There is some evidence for such states, but the question is not yet settled even today.

> (Almost) all known mesons can be identified with $q\bar{q}$ bound states.

5.4   The heavy mesons

If the heavy quarks b and c are the same type of particle as the light quarks u, d, s, we also expect to find heavy mesons with one light and one heavy quark. Two families of these are known. First, near 2 GeV, we have the long-lived 0^- states

$$\underline{D_s^+} \quad 1968$$
$$\underline{D^0 \quad D^+} \quad 1869 \tag{5.31}$$

> The 0^- and 1^- meson states containing one c quark.

and the 1^- states that decay to these by photon emission

$$\underline{D_s^{*+}} \quad 2112$$
$$\underline{D^{*0} \quad D^{*+}} \quad 2010 \tag{5.32}$$

These states are explained as $c\bar{u}$, $c\bar{d}$, and $c\bar{s}$ bound states. Each state has a corresponding antiparticle. The 3S and higher states of the ψ spectrum decay to pairs of these D mesons. Similarly, near 5 GeV, we have the long-lived 0^- states

$$\underline{B_s^0} \quad 5367$$
$$\underline{B^- \quad B^0} \quad 5279 \tag{5.33}$$

> The 0^- and 1^- meson states containing one b quark.

and the 1^- states that decay to these by photon emission

$$\underline{B_s^{*0}} \quad 5415$$
$$\underline{B^{*-} \quad B^{*0}} \quad 5325 \tag{5.34}$$

These states are explained as $b\bar{u}$, $b\bar{d}$, and $b\bar{s}$ bound states. Each state has a corresponding antiparticle. The 4S and higher states of the Υ spectrum decay to pairs of these B mesons. The full picture is again consistent with expectations from the quark model. The connection between the light and heavy quarks gives the electric charge assignments

$$Q_c = Q_u \qquad Q_b = Q_d = Q_s = Q_u - 1 \ . \qquad (5.35)$$

One more, very heavy quark is known. This is the *top quark*, t, with mass about 173 GeV. Because of its large mass, it has a very rapid decay, with a width of 1.2 GeV. The top quark actually decays before it can form hadrons. So there are no hadron states containing t, but the rapid decay allows us to observe the production and decay of the t at the quark level rather than the hadron level.

There are then 6 types of quarks:

$$\begin{matrix} u & c & t \\ d & s & b \end{matrix} \qquad \begin{matrix} Q = Q_u \\ Q = Q_d \end{matrix} \qquad (5.36)$$

From here on, I will refer to the identifying label of a quark (u, d, *etc.*) as its *flavor*.

Definition of *flavor*.

5.5   The baryons

There is another class of strongly interacting particles that includes the proton and neutron. These are *fermions*, called *baryons*. Baryons cannot be created or destroyed singly; rather, they can only be created as particle-antiparticle pairs, or by the conversion of one type of baryon to another. This rule can be described as the presence of a conserved quantum number, called *baryon number B*, with $B = +1$ for baryons and $B = -1$ for antibaryons. Baryon number conservation requires that the lightest baryon, the proton, is absolutely stable. In principle, the proton could decay by reactions such as

$$p \to e^+ \pi^0 \ , \qquad p \to \nu K^+ \ . \qquad (5.37)$$

These modes have been searched for; the lifetime limits are

$$\tau > 8.2 \times 10^{33} \ \text{yr} \qquad \tau > 6.7 \times 10^{32} \ \text{yr} \qquad (5.38)$$

(Note that observing 1 cubic meter of water (3×10^{28} water molecules) for a year and seeing *no* decays places limits at the level of 10^{28} yr.) Other baryons can decay by strong or weak interactions, eventually decaying down to the proton. The neutron is unstable to β decay

$$n \to p e^+ \nu \ . \qquad (5.39)$$

This process requires the weak interaction, which we will study in Part III of this book.

As with the mesons, there are distinctive, relatively light, families of baryons, in which the heavier members have increasing values of the strangeness quantum number. There is a set of 8 spin-$\frac{1}{2}$ states (the *octet*)

The family of light spin-$\frac{1}{2}$ baryons.

$$
\begin{array}{ccc}
\underline{\Xi^-} \quad \underline{\Xi^0} & & 1315 \\[4pt]
\underline{\Sigma^-} \quad \underline{\Sigma^0} \quad \underline{\Sigma^+} & & 1192 \\[2pt]
\overline{\Lambda^0} & & 1116 \\[2pt]
\underline{n} \quad \underline{p} & & 938
\end{array}
\tag{5.40}
$$

and a set of 10 spin-$\frac{3}{2}$ states (the *decuplet*)

The family of light spin $\frac{3}{2}$ baryons.

$$
\begin{array}{c}
\underline{\Omega^-} \qquad 1672 \\[4pt]
\underline{\Xi^{*-}} \quad \underline{\Xi^{*0}} \qquad 1532 \\[4pt]
\underline{\Sigma^{*-}} \quad \underline{\Sigma^{*0}} \quad \underline{\Sigma^{*+}} \qquad 1385 \\[4pt]
\underline{\Delta^-} \quad \underline{\Delta^0} \quad \underline{\Delta^+} \quad \underline{\Delta^{++}} \qquad 1232
\end{array}
\tag{5.41}
$$

All of these states are assigned $P = +1$. The choice $P = +1$ for the proton is a convention. The intrinsic parities of the other baryons are assigned relative to this convention. The relative parity of the heavier baryons can be inferred from their decays. For example, since we know that the pion has $P = -1$, the assignment $P = +1$ for the parity of the Δ requires that the decay $\Delta^0 \to p\pi^-$ have nonzero orbital angular momentum. Indeed, experiments observe that the decay pion is in a state of $L = 1$.

The almost degenerate sets of baryons fall into isospin representations

$$
\begin{array}{lll}
N: \quad I = \dfrac{1}{2} & \Sigma: \quad I = 1 & \Xi: \quad I = \dfrac{1}{2} \\[10pt]
\Delta: \quad I = \dfrac{3}{2} & \Sigma^*: \quad I = 1 & \Xi^*: \quad I = \dfrac{1}{2}
\end{array}
\tag{5.42}
$$

The complete family sizes 8 and 10 are the dimensions of irreducible representations of $SU(3)$.

A few simple facts about the irreducible representations of $SU(3)$.

A clue to the structure of the baryons is the observation that the 10-dimensional representation of $SU(3)$ arises as the set of 3-index symmetric tensors. To understand this statement, we should discuss a few properties of the simplest irreducible representations of $SU(3)$. The fundamental representation of $SU(3)$ is a 3-component vector transformed by a 3×3 unitary matrix,

$$
\xi_a \to U_{ab}\xi_b \ .
\tag{5.43}
$$

The complex conjugate of this representation, called $\bar{3}$, is the a 3-component vector transformated by

$$
\bar{\xi}_b \to \bar{\xi}_a U_{ab}^\dagger \ .
\tag{5.44}
$$

A tensor with two indices, each of which runs over $i = 1, 2, 3$, transforms as

$$
A_{ab} \to U_{ac}U_{bd}A_{cd} \ .
\tag{5.45}
$$

This tensor has 9 components. These can be split into 6 components of a symmetric 3×3 matrix and 3 components of an antisymmetric 3×3 matrix. Each of these objects transforms independently of the other under the transformation (5.45), leading to 6- and 3-dimensional representations of $SU(3)$. This is an example of the idea discussed in Sections 2.3 and 2.4 that a group representation can often be split into a sum of smaller, irreducible representations. For $SU(3)$, the 3-dimensional antisymmetric tensor representation can be shown to be equivalent to the $\bar{3}$ representation presented in (5.44).

In a similar way, a tensor with three indices running over $i = 1, 2, 3$ and totally symmetric in those indices forms an irreducible representation of $SU(3)$. The number of components of this tensor gives the dimension of the representation; this is

$$\frac{3 \cdot 4 \cdot 5}{3!} = 10 . \tag{5.46}$$

For a 3-index symmetric tensor in which the indices take only the values $j = 1, 2$, the number of components is

$$\frac{2 \cdot 3 \cdot 4}{3!} = 4 . \tag{5.47}$$

This describes the symmetric combination of three spin-$\frac{1}{2}$ objects, which is just the spin-$\frac{3}{2}$ representation of $SU(2)$.

From these considerations, it is highly suggestive that the states in the 10 are symmetric combinations of 3 quarks. For example, we can assign

$$\left| \Delta^{++}(S^3 = \frac{3}{2}) \right\rangle = |u \uparrow \ u \uparrow \ u \uparrow\rangle$$

$$\left| \Omega^{-}(S^3 = \frac{3}{2}) \right\rangle = |s \uparrow \ s \uparrow \ s \uparrow\rangle , \tag{5.48}$$

all in a relative S-wave wavefunction. This gives a simple explanation of the spin-$\frac{3}{2}$ nature of the decuplet states and of their flavor quantum numbers.

Actually, if we count both spin and flavor, the light quarks come in 6 states

$$u \uparrow , \quad u \downarrow , \quad d \uparrow , \quad d \downarrow , \quad s \uparrow , \quad s \downarrow \tag{5.49}$$

The number of states that we can build by taking three quarks in a totally symmetric combination is

$$\frac{6 \cdot 7 \cdot 8}{3!} = 56 . \tag{5.50}$$

The decuplet states fill out $10 \cdot 4$ or 40 of these states. What remains is

$$56 - 40 = 16 = 8 \cdot 2 . \tag{5.51}$$

that is, just enough states to fill out the baryon octet.

To see how this works in more detail, I will construct some baryon wavefunctions. Start from the wavefunction of the Δ^{++} with spin $S^3 = \frac{3}{2}$ given in (5.48). The isospin lowering operator $I^- = I^1 - iI^2$ commutes with $I^2 = I(I+1)$ and lowers I^3. This gives the Δ^+ wavefunction

$$\left|\Delta^+(S^3 = \frac{3}{2})\right\rangle = \frac{1}{\sqrt{3}}\left[|u\uparrow\ u\uparrow\ d\uparrow\rangle + |u\uparrow\ d\uparrow\ u\uparrow\rangle + |d\uparrow\ u\uparrow\ u\uparrow\rangle\right].$$
(5.52)

Explicit forms for the Δ and nucleon wave functions in the quark model.

Applying also the spin-lowering operator S^-, we find

$$\left|\Delta^+(S^3 = \frac{1}{2})\right\rangle = \frac{1}{\sqrt{9}}\left[|u\uparrow\ u\uparrow\ d\downarrow\rangle + |u\uparrow\ u\downarrow\ d\uparrow\rangle + |u\downarrow\ u\uparrow\ d\uparrow\rangle\right.$$
$$+ |u\uparrow\ d\uparrow\ u\downarrow\rangle + |u\uparrow\ d\downarrow\ u\uparrow\rangle + |u\downarrow\ d\uparrow\ u\uparrow\rangle$$
$$\left. + |d\uparrow\ u\uparrow\ u\downarrow\rangle + |d\uparrow\ u\downarrow\ u\uparrow\rangle + |d\downarrow\ u\uparrow\ u\uparrow\rangle\right].$$
(5.53)

This is a state with electric charge $+1$, strangeness 0, $S^3 = \frac{1}{2}$, $I = \frac{3}{2}$, and $S = \frac{3}{2}$. Among the octet and decuplet states, there is only one other state that has $Q = +1$, strangeness 0, and $S^3 = \frac{1}{2}$. That state is the spin-up proton state. Since that state has $I = S = \frac{1}{2}$, it must be orthogonal to the state written above. There is only one totally symmetric state with this property, so the proton spin-up state must have the form

$$\left|p(S^3 = \frac{1}{2})\right\rangle = \frac{1}{\sqrt{18}}\left[2|u\uparrow\ u\uparrow\ d\downarrow\rangle - |u\uparrow\ u\downarrow\ d\uparrow\rangle - |u\downarrow\ u\uparrow\ d\uparrow\rangle\right.$$
$$- |u\uparrow\ d\uparrow\ u\downarrow\rangle + 2|u\uparrow\ d\downarrow\ u\uparrow\rangle - |u\uparrow\ d\uparrow\ u\downarrow\rangle$$
$$\left. - |d\uparrow\ u\uparrow\ u\downarrow\rangle - |d\uparrow\ u\downarrow\ u\uparrow\rangle + 2|d\downarrow\ u\uparrow\ u\uparrow\rangle\right].$$
(5.54)

All of the 8 $S^3 = +\frac{1}{2}$ states of the baryon octet can be constructed in this way. The 6 states around the boundary of the octet have forms precisely analogous to that of the proton. The construction of the Λ^0 and Σ^0 states is more subtle. One must first construct the $(I = 1)$ Σ^0 state as

$$|\Sigma^0\rangle = \frac{1}{\sqrt{2}}I^- |\Sigma^+\rangle$$
(5.55)

and then write the Λ^0 as a state orthogonal to this one.

These wavefunctions may seem complicated, but they pay an immediate dividend in explaining the values of the baryon magnetic moments. You can work out the details in Exercise 5.3.

From the quark content of the baryons, we can find the absolute electric charges of the quarks. We find

$$Q_u = +\frac{2}{3} \qquad Q_d = -\frac{1}{3}.$$
(5.56)

This is decidedly odd! Fractional electric charges have never been convincingly observed in nature. After Gell-Mann and Zwieg proposed the quark model, intensive searches were made for fractional charge in rocks, sea water, clam shells, moon rocks, *etc.* No evidence for charge $\frac{1}{3}$ was found (Perl *et al.* 2001). Apparently, the quarks are inside hadrons, but they cannot get out. In Part II of this book, we will see considerable evidence that there are indeed charge $\frac{2}{3}$ and $-\frac{1}{3}$ quarks inside hadrons.

There is one more odd feature of the baryons. Spin-$\frac{1}{2}$ particles are fermions, for which the spin-statistics theorem requires that the quantum states are completely antisymmetric. But the baryon wavefunctions that we constructed, beginning with (5.48), are totally *symmetric*. Han and Nambu proposed that this could be understood if quarks have an additional quantum number, called *color*, taking three values (red, green, blue) (Han and Nambu 1965). If the baryon wavefunction is required to be totally *antisymmetric* in color, it must be totally *symmetric* in all other quantum numbers. The transformation among colors of quark can be described as another $SU(3)$ transformation. If we write indices $i = 1, 2, 3$ in the 3 representations as lowered and indices in the $\bar{3}$ as raised, the basic invariants of $SU(3)$ are

$$\delta_b^a , \qquad \epsilon^{abc} , \qquad \epsilon_{abc} . \tag{5.57}$$

So color-invariant combinations of quarks and antiquarks are

$$\bar{q}^a q_a \qquad \epsilon^{abc} q_a q_b q_c \qquad \epsilon_{abc} \bar{q}^a \bar{q}^b \bar{q}^c . \tag{5.58}$$

These are exactly the mesons, baryons, and antibaryons. These considerations strongly suggest (1) that the color quantum number and color $SU(3)$ symmetry exists, and (2) that physical hadrons are invariant under color $SU(3)$ transformations.

Up to this point, I have only argued that the quark model, with six quarks and color $SU(3)$, gives a plausible explanation for the quantum numbers of the most prominent hadrons. The model suggests that quarks are spin-$\frac{1}{2}$ fermions. Nevertheless, on the basis of the evidence I have offered so far, this is at best still a hypothesis. To find the precise nature of the strong interaction, we will need to look at experiments that are more sensitive to the details of the interactions of quarks with one another and with electromagnetic probes. We will take up this analysis in Part II.

Quarks have fractional electric charges. However, no particles with fractional electric charge have been observed by experiment. To explain this, we must insist that quarks can never be liberated from inside hadrons. A fundamental theory of the strong interaction must address this issue.

The symmetry property of baryon wavefunctions suggest the existence of an additional quark quantum number, *color*.

Exercises

(5.1) This problem will give you a chance to dip into the tables of elementary particle properties produced by the Particle Data Group (Patrignani *et al.* 2016) and to use this information to understand better the systematics of ψ family particle decays.

To work this problem, you should recall that a decay rate in quantum mechanics is given by a *partial width* $\Gamma(A \to f)$, with units of energy, that is (time)$^{-1}$. A partial width gives the rate of a basic quantum mechanical process. The *total width* of a resonance is

$$\Gamma_A = \sum_f \Gamma(A \to f) . \qquad (5.59)$$

That is, it is the sum of the rates for all possible decay processes. The lifetime of the resonance is $\tau = \hbar/\Gamma_A$. The *branching ratio* to the decay channel f, the probability that a particular decay of A gives the final states f, is

$$BR(A \to f) = \Gamma(A \to f)/\Gamma_A . \qquad (5.60)$$

Usually, it is easiest to meaure branching ratios, but the real physics is in the actual rates. To obtain these, we must extract the partial widths from the information that we are given.

(a) The J/ψ can decay in four different ways. (1) decay by $c\bar{c}$ annihilation directly to hadrons; (2) decay by $c\bar{c}$ annihilation to a virtual photon (a short-lived state of electromagnetic fields), which then materializes into an e^+e^- or $\mu^+\mu^-$ pair. The J/ψ is produced in e^+e^- annihilation by e^+e^- annihilation into a virtual photon which then materializes as a J/ψ. This decay is the reverse of that process; (3) decay by $c\bar{c}$ annihilation to a virtual photon, which then materializes into hadrons; (4) decay to 1 photon plus hadrons. There is also a decay to 3 photons with a very small branching ratio (about 10^{-5}).

Look up the listing for the J/ψ at the Particle Data Group web site. The heading "pdgLive" gives the most recently updated information. Look under $c\bar{c}$ to find the information for the J/ψ. The entry $J/\psi \to ggg$ gives the branching ratio for direct decays to hadrons, mode

(1) above. Similarly, the entry $J/\psi \to \gamma gg$ gives the branching ratio for mode (4) above. Write the branching ratio for each of the decay modes (1)–(4). (These should add up to 100%, within the measurement errors.) Using the tabulated total width, find the partial width for each channel.

(b) The $\psi(2S)$ can decay by the 4 modes above and also by 3 additional modes: (5) decay to heavy leptons $\tau^+\tau^-$; (6) decay to J/ψ plus hadrons ($\pi\pi$, π^0, or η); (7) radiative decay to the 1P states χ_c.

Using the information in the entry for the $\psi(2S)$, write the branching ratio for each of the decay modes (1)–(7). (Again, these should add up to 100%, within the measurement errors.) Using the tabulated total width, find the partial width for each channel.

(c) Compute the ratios of the partial widths between the J/ψ and the $\psi(2S)$ for each of the processes (1)–(4). How do these ratios compare? Why would this result be expected?

(5.2) Consider the reaction of pion-nucleon scattering at energies of a few hundred MeV. Two prominent resonances are seen as the center of mass energy is varied. These are the Δ resonances at 1232 MeV and the N^* ("Roper") resonance at 1440 MeV. The Δ has $I = \frac{3}{2}$, $S = \frac{3}{2}$. The Roper has $I = \frac{1}{2}$, $S = \frac{1}{2}$ and can be thought of as a radial excitation of the nucleon. The absolute rates of the reactions that form these resonances need to be computed from a dynamical strong interaction theory. However, the relative rates of different reactions producing the same resonances can be computed using isospin symmetry and Clebsch-Gordan coefficients.

The initial states in the reaction are the π mesons, an $I = 1$ multiplet (π^-, π^0, π^+), and the nucleons, an $I = \frac{1}{2}$ multiplet $N = (p, n)$. The quantum mechanical amplitude to produce a resonance of isospin I from initial states with isospins (I_1, I_1^3) and (I_2, I_2^3) is proportional to the Clebsch-Gordan coefficient

$$\langle I_1 I_2 I_1^3 I_2^3 \mid I_1 I_2 I I^3 \rangle \qquad (5.61)$$

with $I^3 = I_1^3 + I_2^3$. The amplitude for the decay of a resonance to two particles of definite isospin is similarly proportional to the relevant Clebsch-Gordan coefficient. You can find a very readable table of Clebsch-Gordan coefficients for $SU(2)$ at the Particle Data Group web site, under "Mathematical Tools".

(a) There are 4 Δ states: $(\Delta^{++}, \Delta^{+}, \Delta^{0}, \Delta^{-})$. These decay exclusively to 2-particle states πN. Using isospin Clebsch-Gordan coefficients, compute the branching ratios for each state to the 6 possible channels

$$(\pi^{+}\pi^{0}, \pi^{-}) \times (p, n)$$

(b) A crude description of the N^* decays is that 60% of the decays go to πN and 40% go to $\pi\Delta$, Using these values and the Clebsch-Gordan coefficients, compute the branching ratios of the N^* states (N^{*+}, N^{*0}) to the 6 πN states in (a).

(c) The decay of the N^* to $\pi\Delta$ followed by the decay of the Δ leads to the final state $\pi\pi N$. It is easy to compute the branching ratios to the various $\pi\pi N$ states if we assume that there is no quantum mechanical interference between two decay processes. (This would be correct if two pions emitted have significantly different energies, which is actually not so true in this case.) Using this approximation, compute the branching ratios of N^* to the various possible $\pi\pi N$ states.

(5.3) The quark model gives a theory of the magnetic moments of the proton and neutron. If a quark were an elementary Dirac fermion, its magnetic moment would be

$$\vec{\mu} = g\frac{Q_f e}{2m_f}\vec{S} \tag{5.62}$$

with Q_f the quark charge, m_f the quark mass, $\vec{S}$ the quark spin. The Dirac equation predicts the

value of the Landé g-factor $g = 2$. In the proton and neutron, we have only the u and d quarks. By isospin symmetry, the u and d quark masses should have the same value, $m_q \approx 300\,\mathrm{MeV}$. We could then model the baryon magnetic moment as the sum of the three quark magnetic moments,

$$\vec{\mu}_B = \frac{2Q_1 e}{2m_q}\vec{S}_1 + \frac{2Q_2 e}{2m_q}\vec{S}_2 + \frac{2Q_3 e}{2m_q}\vec{S}_3 , \tag{5.63}$$

where Q_1, Q_2, Q_3 = +2/3 or -1/3 depending on whether the quark is u or d.

(a) Using the quark model wavefunction for the proton state with $S^3 = \frac{1}{2}$ written down in (5.54), compute the magnetic moment of the proton in this approximation. This is most easily done by computing the diagonal matrix element of the $\hat{3}$ component of the operator $\vec{\mu}_B$, given by (5.63), in this state. Express the result by computing the proton g factor g_p given by

$$\vec{\mu}_p = g_p\frac{e}{2m_p}\vec{S}_p \tag{5.64}$$

You will find that the result depends on the ratio m_p/m_q.

(b) Using the same method, compute the g factor of the neutron, defined by

$$\vec{\mu}_n = g_n\frac{e}{2m_p}\vec{S}_n . \tag{5.65}$$

(c) The g factors for the proton and neutron are very different from the value 2 predicted by the Dirac equation. The measured values are

$$g_p = +5.586 \qquad g_n = -3.826 .$$

Compare these results to the predictions of the quark model given in parts (a) and (b). What value of the quark mass m_q best accounts for the data?

Detectors of Elementary Particles

Thus far in this book, my discussion of experimental results has only been semi-quantitative. We have looked at symmetry principles and energies of resonances but not yet at the values of rates of elementary particle reactions. To understand measurements of rates, and to compare to experimental data, we need to understand how the momenta and energies of elementary particles are measured. This chapter will give a very brief introduction to that subject. In-depth presentations of the physics and design of particle detectors can be found in (Green 2005) and (Grupen and Shwartz 2008).

Particle detectors are, generally, of one of two types. The first is a detector of ionization or other energy loss. A particle comes in, deposits energy in a sensor, and exits. The position of the energy deposition gives a point on the particle's trajectory, and the amount of energy deposited contains some additional information about the particle's momentum. A detector of this type is called a *tracker*. The second type of detector is one that attempts to convert the entire energy of a particle into a measurable signal. A detector of this type is called a *calorimeter*.

Two types of elementary particle detectors: *trackers* and *calorimeters*.

Particle detectors measure the properties of stable particles, or, at least, of particles that have a macroscopic flight path of length of a millimeter or greater. Unstable particles are measured by observing the particles into which they decay. Trackers are usually sensitive only to charged particles. Calorimeters are sensitive to charged and neutral particles, but only to particles with strong interactions, or to electrons and photons.

All particle detectors have their basis in the effects produced by a relativistic particle moving through matter. Such a particle will knock electrons out of atoms, producing ionization. It will interact electromagnetically with atomic nuclei, transferring momentum and, in some cases, producing additional photons, electrons, and positrons. High energy hadrons will also scatter from nuclei through the strong interaction. A fast charged particle will also interact with the material it passes through in a collective way, through the macroscopic dielectric properties of the medium. All of these effects are used to create different types of parti-

Concepts of Elementary Particle Physics. Michael E. Peskin.
© Michael E. Peskin 2019. Published in 2019 by Oxford University Press.
DOI: 10.1093/oso/9780198812180.001.0001

cle detectors that measure energy-momentum in different ways and also discriminate one type of particle from another.

The theory of particle motion through matter is highly technically developed. In this chapter, I will cover only some simple aspects of this theory. Excellent, more detailed, introductions to this subject can be found in (Bichsel, Groom, and Klein 2016) and other articles from the *Review of Particle Physics* (Patrignani *et al.* 2016), in the relevant chapter of (Jackson 1998), and in the textbooks cited above.

Modern particle detectors are modular systems in which different single-purpose detectors are placed one inside another. These systems attempt to measure all properties of the particles produced in an elementary particle reaction, in a definite sequence. After we review the basic mechanisms used by particle detectors, I will discuss some of these larger-scale detector systems.

6.1 Energy loss by ionization

Scattering of atomic electrons by a relativistic particle.

We begin with the theory of ioniziation. Consider a fast particle of charge Qe that interacts electromagnetically with an electron in an atom. This particle will kick electrons out of the atom and lose energy in the process. I will assume that the fast particle suffers only a small deflection. The geometry of the interaction is

$$\tag{6.1}$$

where b is the *impact parameter*.

If the fast particle is moving relativistically, its field is a pancake in the frame of the electron,

$$(E_{\parallel}, E_{\perp}) = \frac{eQ}{4\pi} \frac{(\gamma vt, \gamma b)}{[b^2 + (\gamma vt)^2]^{3/2}} \;, \tag{6.2}$$

Scattering of a relativistic particle from an electron in an atom.

with $\beta = v/c$, $\gamma = 1/\sqrt{1 - \beta^2}$. The electric field transfers momentum to the electron

$$\Delta p = \int dt \, eE_{\perp} = \frac{2e^2 Q}{4\pi} \frac{1}{bv} \;. \tag{6.3}$$

Note that this is independent of γ. The energy transferred to the electron is

$$\Delta E = \frac{(\Delta p)^2}{2m_e} = 2\left(\frac{e^2}{4\pi}\right)^2 Q^2 \frac{1}{b^2} \frac{1}{m_e v^2} \;. \tag{6.4}$$

The angular deflection of the incoming particle is

$$\theta = 2Q \frac{e^2}{4\pi} \frac{1}{bv} \frac{1}{p} \;. \tag{6.5}$$

These formulae are of course just a crude approximation of the full problem of a particle interacting with an electron bound to an atom.

To calculate the energy loss of the fast particle, we need to average the possible values of the impact parameter b over a plane perpendicular to the trajectory of the particle. The average energy loss is then given by the density of atomic electrons multiplied by the integral

$$\int d^2b \, \Delta E(b) \sim 2\pi \int db \, b \, \frac{1}{b^2} \sim 2\pi \int \frac{db}{b} \, . \tag{6.6}$$

To estimate this logarithmic integral we need an estimate of the maximum and minimum values of b for which the scattering is effective. The maximum value is set by the condition that, if b is large, the particle passes by an atom over a long time, and the position of the electron must be averaged over its orbit around the nucleus. The time for the particle to pass the atom is

$$\Delta t \sim \frac{b}{v\gamma} \, . \tag{6.7}$$

Let ω be a typical atomic frequency or energy difference, of the order of eV. If $\Delta t > \hbar/\omega$, the particle cannot act coherently on the electron; this gives

$$b_{max} \sim \frac{\gamma v}{\omega} \, . \tag{6.8}$$

The minimum value is set by the quantum-mechanical uncertainty in the electron's position, as seen by the incoming particle. This is

$$b_{min} \sim \frac{\hbar}{\gamma m_e v} \, . \tag{6.9}$$

We then find for the energy loss per unit distance x (in cm)

$$
\begin{aligned}
\frac{dE}{dx} &= -nZ \int db \, b \, 2\pi \, \Delta E(b) \\
&= -4\pi\alpha^2 Q^2 \frac{nZ}{m_e v^2} \int_{b_{min}}^{b_{max}} \frac{db}{b} \\
&= -4\pi\alpha^2 Q^2 \frac{nZ}{m_e v^2} \log\left[\frac{\gamma^2 m_e v^2}{\hbar\omega}\right] ,
\end{aligned}
\tag{6.10}
$$

where n is the number density of atoms (atoms/cm^3) and Z is the atomic number or the number of electrons per atom.

This simple derivation captures the main features of Hans Bethe's classic treatment of this problem (Bethe 1930). By a somewhat more sophisticated analysis, Bethe found

$$\frac{dE}{dx} = -4\pi\alpha^2 Q^2 \frac{nZ}{m_e v^2} \left[\log\frac{2\gamma^2 m_e v^2}{\hbar\omega} - \frac{v^2}{c^2}\right] . \tag{6.11}$$

Bethe's formula for the energy loss of a charged particle due to ionization.

A large number of phenomenological improvements to this formula are discussed in (Bichsel, Groom, and Klein 2016).

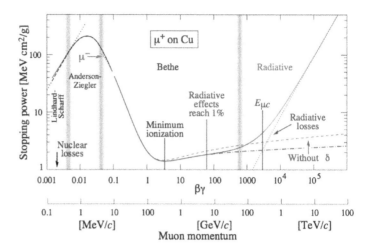

Fig. 6.1 Energy loss in MeV/(g/cm^2) for positive muons in copper as a function of $\beta\gamma$, from (Bichsel *et al.* 2016).

The general form of the energy loss function dE/dx is

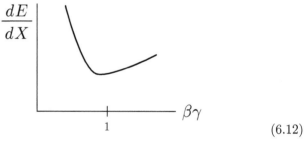

$$(6.12)$$

Qualitative features of Bethe's formula for the energy loss of charged particles include its rapid rise at low velocities, its ionization minimum, and its relativistic rise with $\log E$.

The formula depends on the *velocity* of the particle, but not its *momentum*. (It also depends on the charge Q, but, for most of the particles we consider, $Q = \pm 1$.) The ionization increases rapidly as the particle slows down, as $1/v^2$; it also increases logarithmically as the particle becomes very relativistic. The latter effect is called the *relativistic rise*. Its size depends on the absorbing material. The curve has a minimum for $\beta\gamma \sim 1$, this is called *minimum ionization*. The numerical value of the minimum ionization is a few MeV/cm. More accurately, this value is given by 1.5 MeV·ρ, where ρ is the density in g/cm^3. The minimum of the curve is quite shallow, so single relativistic particles are recognized as contributing an energy deposition of one minimum ionizing particle (1 MIP).

Definition of a *minimum ionizing particle*.

Figure 6.1 shows in more detail the energy loss dE/dx for a muon passing through copper. At the lowest energies, the $1/v^2$ divergence is rounded off by more careful consideration of the atomic physics. At very high energies, another effect comes in, which I will describe in Section 6.2.

The path of a particle in a magnetic field depends on the momentum, but not the velocity; the ionization depends on the velocity but not the momentum. It is thus possible to use dE/dx measurements to measure the particle mass. A heavier particle has a dE/dx curve shifted to higher values of the momentum,

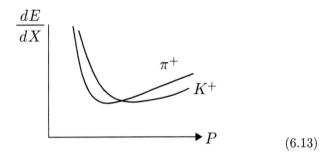

$$(6.13)$$

Measurement of dE/dx requires understanding of one more subtlety. The theory I have given applies to the average value of dE/dx. However, the energy loss in each atomic collision varies strongly with the impact parameter of the scatter. If dE/dx is measured by sampling in slices of an ionizing medium, the sampled values of energy loss will vary according to a probability distribution, first computed by Landau (1944), that includes rare events of very high energy loss. Then it is awkward to average measured values of dE/dx; rather, the particle velocity is estimated from the most probable energy loss, given approximately by

The energy loss by ionization includes the possibility of large positive fluctuations. This must be taken into account in measuring dE/dx.

$$\Delta E = -\xi \left[\log \frac{2\gamma^2 m_e v^2 \xi}{\hbar \omega} - \frac{v^2}{c^2} \right] . \qquad (6.14)$$

where $\xi = (2\pi \alpha^2 Q^2 n Z / m_e v^2) \Delta x$, with Δx the thickness of the sampler. For more details, see (Bichsel 2016).

6.2 Electromagnetic showers

For very relativistic particles, another energy loss mechanism takes over. A very high energy electron can emit a photon, moving roughly collinear with electron, that carries a large fraction of its energy. This effect is called *bremsstrahlung*. Similarly, a very high energy photon easily converts to an electron-position pair, with both members of the pair moving in the same direction as the photon.

Important energy loss mechanisms for relativistic electrons and photons are *bremsstrahlung* and *pair conversion*.

Bremsstrahlung and pair-production are typically interactions between a high energy particle and an atomic nucleus. They occur infrequently along the path of a particle, but they are also significant events that transfer substantial momentum. As such, they should not be described by their average effects but rather by individual collisions occuring with given probability along the path. The probability of scattering in a small interval of the path dx is written as dx/λ. Then the probability $P(x)$ that the particle still has not scattered after a path length x satisfies the equation

$$\frac{d}{dx} P(x) = -\frac{1}{\lambda} P(x) \quad \text{or} \quad P(x) = e^{-x/\lambda} . \qquad (6.15)$$

The parameter λ is called the *mean free path*.

Bremsstrahlung and pair-production are the result of a peculiar property of relativistic kinematics. An electron at rest cannot spontaneously convert to an electron and a photon; this violates energy-momentum conservation. However, if the energies of these particles are much larger than $m_e c^2$, the required nonconservation of energy and momentum is small. The momentum 4-vector of a relativistic electron can be written

$$p^\mu = (E, 0, 0, \sqrt{E^2 - m_e^2}) \approx (E, 0, 0, E - \frac{m_e^2}{2E}) \,. \tag{6.16}$$

Relativistic kinematic origin of bremsstrahlung and e^+e^- pair production.

Now imagine that the electron splits

$$e^-(P) \rightarrow e^-(p') + \gamma(q) \tag{6.17}$$

into a photon carrying a fraction z of the original energy and an electron carrying a fraction $(1 - z)$. The new 4-vectors are

$$q = (zE, p_\perp, 0, zE - \frac{p_\perp^2}{2zE})$$

$$p' = ((1 - z)E, -p_\perp, 0, (1 - z)E - \frac{p_\perp^2 + m_e^2}{2(1 - z)E}) \tag{6.18}$$

For a momentum transfer of order $p_\perp \sim m_e$, the required transfer of energy or longitudinal momentum is of order

$$\frac{m_e^2}{E} \,, \tag{6.19}$$

which, for a GeV electron, is of the order of keV. This is easily supplied by the scattering of an electron from the electrostatic field of a heavy nucleus.

An individual electron-nucleus scatter, then, can split the 4-momentum of a relativistic electron into two pieces, giving an arbitrary fraction z of the energy to a bremsstrahlung photon. In the same way, a photon-nucleus scatter can split the 4-momentum of the photon into the two momenta of an electron-positron pair.

A detailed calculation of the cross section for electron splitting gives a formula for the electron scattering rate of the form of (6.15) but also differential in the energy fraction z taken by the photon. The probability of a scatter at a position x is written as an integral over z, the fraction of the electron's momentum that is transfered to the photon. Then

Probability formula for bremsstrahlung.

$$\frac{d}{dx}P(x) = -\int dz \left\{ \frac{1}{X_0} \frac{1}{z} \left[\frac{4}{3}(1 - z) + z^2 \right] \right\} P(x) \,. \tag{6.20}$$

Probability formula for e^+e^- pair production.

For $\gamma \rightarrow e^+e^-$, the corresponding formula is

$$\frac{d}{dx}P(x) = -\int dz \left\{ \frac{1}{X_0} \left[1 - \frac{4}{3}z(1 - z) \right] \right\} P(x) \,. \tag{6.21}$$

The quantity X_0 is called the *radiation length*; it is given approximately by

$$\frac{1}{X_0} = \frac{4\alpha^3}{m_e^2} nZ^2 \log \frac{m_e}{Q_s} \,, \tag{6.22}$$

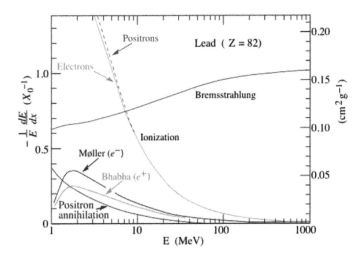

Fig. 6.2 Fractional energy loss per radiation length for electrons and positrons in lead as a function of the electron or positron energy, from (Bichsel *et al.* 2016).

where

$$\frac{1}{Q_s} = 1.4\, Z^{-1/3} a_0 \;, \tag{6.23}$$

where a_0 is the Bohr radius. Note that X_0 depends strongly on the nuclear charge Z. The length $1/Q_s$ is the distance outside the core of a heavy atom at which the nuclear charge is screened by the electrons, computed in the Thomas-Fermi approximation. The appearance of this screening length emphasizes that, while ionization is an interaction with electrons, bremsstrahlung and pair production are interactions with the atomic nuclei.

The formula (6.20) implies that the mean free path for an electron to radiate a hard bremsstrahlung photon is of order X_0, while soft bremsstrahlung photons are emitted more frequently. To be more quantitative, let $\langle E(x) \rangle$ be the average energy of the electron after a distance x. The energy lost in a bremsstrahlung emission is $z \langle E(x) \rangle$. From (6.20), the expected energy obeys

$$\frac{d}{dx} \langle E(x) \rangle = - \int dz \left\{ \frac{1}{X_0} \frac{1}{z} \left[\frac{4}{3}(1-z) + z^2\right] \right\} z \langle E(x) \rangle \;. \tag{6.24}$$

Performing the integral, we find

$$\frac{d}{dx} \langle E \rangle = -\frac{1}{X_0} \langle E \rangle \;, \tag{6.25}$$

that is, X_0 is the mean free path for the energy carried by the initial electron. For photon splitting, the energy sharing is roughly equal between the electron and positron. We can simply integrate the right-hand side of (6.21) over z and find

$$\frac{d}{dx} P(x) = -\frac{7}{9X_0} P(x) \;. \tag{6.26}$$

Mean free paths for bremsstrahlung and pair production.

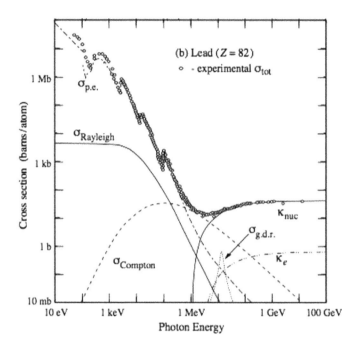

Fig. 6.3 Cross section for photon scattering from lead as a function of the photon energy, from (Bichsel, Groom, and Klein 2016). The various reactions that contribute are shown as separate curves; p.e. denotes the photoelectric effect.

Then the mean free path for a photon to convert to an electron-positron pair is $\lambda = (9/7)X_0$.

An important quantity related to this physics is the *critical energy* E_c. This is the energy below which ionization energy loss dominates over bremsstrahlung. This cross-over is shown, for electrons in lead, in Fig. 6.2. Photons have a similar low-energy cutoff for pair production, just below the e^+e^- threshold. At still lower energies, their energy loss is dominated by the photoelectric effect, as shown in Fig. 6.3.

Here is a table of the radiation length and critical energy for some commonly used materials. I also include the *pion interaction length* λ_I, the mean path for a π^+ to travel in the material before suffering an inelastic collision with a nucleus. The values for many more materials can be found in a useful table in (Patrignani *et al.* 2016).

Radiation lengths and pion interaction lengths in some representative materials.

	X_0 (cm)	E_c (MeV)	λ_I (cm)
Be	35.3	114	59.5
C	18.9	82	38.2
Fe	1.76	22	20.4
W	0.35	8	11.3
Pb	0.56	7	19.9

(6.27)

Once a photon or electron has been created by bremsstrahlung or pair creation, it is free to initiate new processes of these kinds. Roughly,

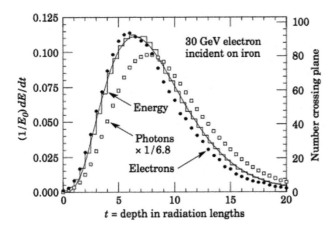

Fig. 6.4 Simulation of an electromagnetic shower in iron, showing the fractional energy loss per radiation length as a function of depth, from (Bichsel, Groom, and Klein 2016).

then, the number of relativistic particles doubles every radiation length. The result is an *electromagnetic shower*. The number of relativistic particles grows exponentially up to 5-8 radiation lengths. Then, the electrons, positrons, and photons drop below the critical energy and dissipate their energy directly without further particle production.

Description of an *electromagnetic shower.*

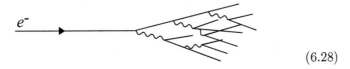

$$(6.28)$$

The energy deposition in a simulated electromagnetic shower in iron is shown in Fig. 6.4. This example is typical, but the details of the particle production will vary from shower to shower. The transverse size of the shower is characterized by the *Molière radius*, given by

$$R_M = X_0 \cdot \frac{21 \text{ MeV}}{E_c} \ . \qquad (6.29)$$

A cylinder with radius R_M contains 90% of the energy deposition of an electromagnetic shower.

6.3   Further effects of nuclear scattering

For heavier charged particles, even when they are relativistic, bremsstrahlung does not contribute significantly to their energy loss except at extremely high energy. For example, we see from Fig. 6.1 that the critical energy for a muon in copper is about 3000 GeV. However, there are two more effects of nuclear scattering that play an important role in particle detectors.

First, a Coulomb scatter from an atomic nucleus can significantly change the direction of the particle's motion. In (6.5), we saw that a fast particle scattering from an electron suffers a small deflection. Scattering from all of the electrons in an atom, this effect is of order Z. But scattering coherently from an atomic nucleus gives a deflection of order Z^2, and also one that is not cut off as strongly for larger momentum transfer. Through the collective action of many such scatterings, the orientation of the particle is smeared in angle, an effect called *multiple scattering*. The increase in the mean square deflection per unit path length travelled is

Deflection of the path of a charged particle by *multiple scattering*.

$$\frac{d\langle\theta^2\rangle}{dx} = \frac{13.6 \text{ MeV}}{\beta p} \cdot Q \cdot \sqrt{\frac{x}{X_0}} . \tag{6.30}$$

where, in this formula, the radiation length X_0 defined in (6.22) again sets the scale of distance.

In designing a tracking detector, it is necessary to compromise between having enough material to see the particle track accurately and having a sufficiently small amount of material that the angle of the track is not smeared by multiple scattering. The balance between these effects is explored in Exercise 6.3.

If the particle traversing the medium is a hadron, it can also interact with atomic nuclei through the strong interaction. For example, a pion moving through detector material will suffer an inelastic collision in the distance called λ_I in (6.27). This collision will take energy from the pion and convert this to the energy of several additional charged and neutral hadrons. After many scatterings, the energy of the pion is converted to the energy carried by many approximately collinear hadrons and their reaction products. This process is called a *hadronic shower*.

Description of a *hadronic shower*.

Hadronic showers are more complex than electromagnetic showers, because they involve a wider variety of processes with different length scales. When a π^+ has an inelastic collision with a nucleus, it creates a large number of relativistic particles, including π^+, π^-, and π^0. The π^0's are very short-lived, decaying almost immediately to 2γ. The characteristic flight distance for a π^0 is $c\tau = 25$ nm. The photons initiate electromagnetic showers, whose depth is set by X_0. If protons are ejected from the nucleus in the collision, these deposit ionization in an even shorter distance. On the other hand, the π^+ and π^- travel as minimum ionizing particles for a distance of order the interaction length λ_I before their next inelastic collision.

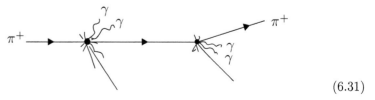

$$\tag{6.31}$$

Hadronic showers thus develop over longer distances than electromagnetic showers, and also contain considerably more fluctuation and irregularity in their components.

6.4 Energy loss through macroscopic properties of the medium

There are two more mechanisms of energy loss that play a role in more specialized detectors. These both exploit macroscopic electromagnetic properties of media. The first is *transition radiation*. When a relativistic particle crosses an interface between vacuum and a medium, there is a mismatch of its electromagnetic fields across the boundary. To repair this, a burst of radiation is emitted. The intensity of this transition radiation, for a conducting film, is estimated as

Description of *transition radiation.*

$$I = \alpha Q^2 \gamma \frac{\hbar \omega_p}{3} ,\qquad (6.32)$$

where ω_p is the plasma energy in the film. Note the dependence on γ. We can discriminate electrons from pions by observing the difference in their transition radiation, at equal momentum, passing through a stack of Mylar foils.

The second of these mechanisms is *Cherenkov radiation*. A relativistic particle can easily move faster than the speed of light in a medium $c_n = c/n$. It is then accompanied by a shock wave of radiation similar to a sonic boom. This is an outwardly moving cone of light, typically peaking in the near ultraviolet. The direction of the radiation is

Description of *Cherenkov radiation.*

$$\cos \theta_C = \frac{1}{n\beta} . \qquad (6.33)$$

Cherenkov light is a sharp discriminator of particle velocity, since it is present only when $\beta > 1/n$. Special materials, called aerogels, are made with index of refraction very close to 1 to discriminate relativistic pions and kaons of equal momentum.

Cherenkov radiation can also be used as a tracking technology, by using an array of photodetectors to measure the position and angle of the cone of Cherekov light emitted by a relativistic particle. This is the tracking technology used by very large water detectors for neutrinos whose results we will discuss in Chapter 20.

6.5 Detector systems for collider physics

Dectector elements based on these mechanisms for particle energy loss can be assembled into detector systems meant to visualize all aspects of an elementary particle collision. Today, experiments at the highest energies are colliding beam experiments that bring together beams with particles of equal energy at a collision point.

$$\longrightarrow \ast \longleftarrow \qquad (6.34)$$

An important concept now used in all colliding beam experiments is the idea of a cylindrical detector surrounding the beams and the collision point. Different types of detectors are placed on concentric cylinders in a definite order, inside to outside, to obtain as much information as possible about the particles produced in the interaction. The first multipurpose cylindrical collider detector was the Mark I detector of the SLAC-LBL collaboration used in the discovery of the J/ψ described in Section 5.2 (Augustin *et al.* 1975).

A cylindrical collider detector must be set up in such a way that the measurements done by detectors in the inner cylinders do not unduly compromise the measurements done by the outer detectors. This means that the inner detectors will contain a low-mass tracker. Calorimeters, which destructively measure total energy, must be placed outside all other important elements. The design must also have a place for a magnet that can provide a solenoidal magnetic field to bend the trajectories of charged particles and allow a momentum measurement. A typical plan is

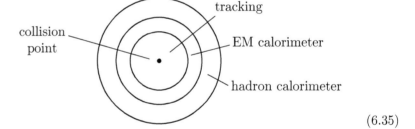

Plan of a typical multipurpose cylindrical collider detector.

$$(6.35)$$

I will now describe the various components of a multipurpose detector from inside to outside. An excellent introduction to the two large LHC detectors has been given by Froidevaux and Sphicas (2006). I take as my primary example the ATLAS detector at the CERN Large Hadron Collider (LHC), which is thoroughly documented in (Aad *et al.* 2008). The passage of particles of different types through the ATLAS detector is illustrated in Fig. 6.5.

We begin with the momentum measurement, which involves the interior tracker and the magnetic field. A relativistic particle moving perpendicular to a magnetic field travels in a circle. Measurement of points on the trajectory can determine the radius of this circle. In practical units,

$$\frac{1}{R} = QB\,\frac{0.3}{p}\;,\qquad(6.36)$$

Charged particle momentum measurement by tracking the motion of the particle through a magnetic field.

where R is given in meters, B is in Tesla, and p is in GeV. More generally, a charged particle in a magnetic field travels in a helix, whose cross section depends on the component of the momentum perpendicular to the magnetic field.

Measurements of the particle trajectory can be made by finely space electrodes in volume of ionizing gas, or by finely etched silicon sensors. These measurements give us the curvature κ of the path, which is equal

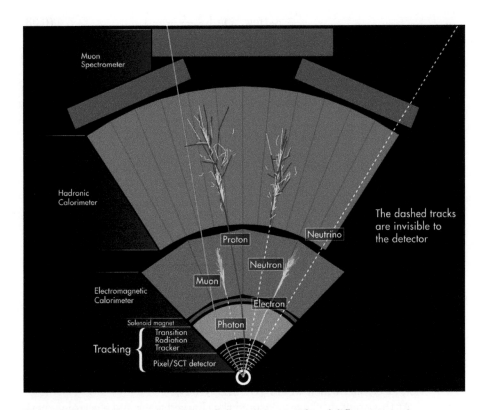

Fig. 6.5 Simulated energy deposition of elementary particles of different types in a slice of the ATLAS detector (figure courtesy of CERN and the ATLAS collaboration). The interaction point at the center of the detector is at the bottom of this figure. Particles produced in collisions move upward from this point. From left to right, we see the signals of a muon, which penetrates the whole detector, a photon, which makes no track but is visible in the electromagnetic calorimeter, a proton and a neutron, which are visible in the hadronic calorimeter, and an electron. Finally, we see the trace of a neutrino, which makes no signal at all.

to $1/R$ and, thus proportional to $1/p_\perp$. An uncertainty in the measurement of κ produces an uncertainty

$$\Delta \frac{1}{p_\perp} = \frac{\Delta p_\perp}{p_\perp^2} \ . \tag{6.37}$$

Thus, the uncertainty in the momentum mesurement rises with momentum. In the detectors for the Large Hadron Collider (LHC),

$$\frac{\delta p_\perp}{p_\perp} \sim (\text{few } \%) \cdot \left(\frac{p_\perp}{100 \text{ GeV}} \right) \ . \tag{6.38}$$

At GeV energies, this uncertainty is sufficiently small that deflections of the trajectory from multiple scattering are also important. Notice that, for tracks of very high momentum, in the range of multi-TeV, even the *sign* of the bending, which gives the sign of the electric charge, is uncertain.

The next element, going outward, is the electromagnetic calorimeter. This device contains and measures electromagnetic showers produced by electrons and photons (with most of the photons from $\pi^0 \to 2\gamma$ decays). In the ATLAS detector at the LHC, the electromagnetic calorimeter is

Electron and photon energy measurement by calorimetry.

a set of lead plates in a bath of liquid argon. Lead (Pb) is chosen as the showering medium because its small value of X_0 gives a relatively compact detector. The depth of the calorimeter is 24 X_0, enough to contain the shower quite well. Charged particles created in the shower leave ionization in the liquid argon; the ionization electrons can be drifted in this inert medium to electrodes, where they are counted to estimate the deposited energy. Only a fraction of the total energy is collected. This uncertainty in the energy measurement of the order of

$$\frac{\Delta E}{E} \sim \frac{10\%}{\sqrt{E}} \ , \tag{6.39}$$

with E in GeV. The uncertainty is dominated by the counting of ionization electrons, which would give $\Delta E \sim \sqrt{N_e}$.

At this point, we have measured the charged track and electromagnetic components of the event. What remains are neutral hadrons such as n, Λ^0, K^0, etc., whose energies must be measured by the creation of hadronic showers. For uniformity, ATLAS measures the total energy of all hadrons by the same calorimetric technique.

We have already noted at the end of Section 6.3 that hadronic showers are more complex than electromagnetic showers, due to the variety of interactions that they contain. In particular, most of the energy deposition comes from ionization in electromagnetic showers, which have a

Hadronic energy measurement by calorimetry.

size set by X_0, while the size of the whole shower is set by the nuclear interaction length λ_I. To measure the energy of a hadronic shower, the calorimeter must compromise between having enough material to provide a depth of many λ_I, while at the same time having sufficient segmentation to minimize the sampling error. ATLAS uses iron as the absorber and scintillating tile as the medium for sampling ionization. The depth of the calorimeter, in the central region, is 11 λ_I.

Hadron calorimeters also have different performance in measuring the energies of π^+ and π^0. For π^0, all of the energy is deposited in electromagnetic showers, while, for π^+, a significant amount of energy goes into nuclear breakup and other mechanisms that are more difficult to sample. Thus, the fluctuations in the fraction of π^0's generated in the first few inelastic collisions increase the uncertainty in the energy measurement. The performance of the ATLAS calorimeter is of the order of

$$\frac{\Delta E}{E} \sim \frac{50\%}{\sqrt{E}} . \tag{6.40}$$

Muons have no strong interactions and only rarely radiate photons to produce electromagnetic showers. Thus, they travel through all of the various layers of the cylindrical detector as simple minimum ionizing particles. To first approximation, any particle that makes it through the whole detector system and is observed as a track in the outer detector layers is a muon. Tracking chambers are placed on the outside of the detector to locate the muon tracks that penetrate through the calorimeters and associate them with tracks measured in the inner detector.

Muons in large collider detectors.

Neutrinos have no strong or electromagnetic interactions, so they do not interact with the detector through any of the mechanism discussed in this chapter. Almost always, neutrinos produced in a particle collision escape the detector without making any signal. The presence of neutrinos (or other possible neutral, weakly interacting particles) can be inferred if the total momentum of observed final-state particles is seen to be unbalanced. Neutrinos do interact through the weak interaction. Such neutrino reactions can be observed, as we will discuss in Chapter 15, using very massive detectors and high neutrino fluxes to compensate for the very small rates of weak interaction processes.

Neutrinos in large collider detectors.

The designs of two other large particle detectors are shown in Fig. 6.6 and Fig. 6.7. Figure 6.6 shows the overall design of the CMS detector at the LHC. In this detector, the solenoidal magnet is placed outside the electromagnetic and hadron calorimeters. Because of this, however, the hadron calorimeter is rather thin and relies on the iron outside the magnet to complete the absorption of the hadronic shower. The iron outside the magnet also returns the magnetic flux from the solenoid, so it is magnetized in the direction opposite to the interior. The reverse bending of the muon in this region, is used to improve the muon momentum measurement at high energies.

Examples of complete detector designs.

Figure 6.7 shows the BaBar detector used at SLAC in the 2000's for studies of the weak interactions of B mesons. The colliding beam system was designed to be asymmetric, colliding 9 GeV electrons and 3 GeV positrons to produce the $\Upsilon(4S)$, for reasons that will be discussed when I review these experiments in Chapter 19. The collisions are then boosted to the right, and this is reflected in the detector layout. Two new detector components are apparent here. The first (also present in CMS and ATLAS, but too small to be visible in Fig. 6.6) is a silicon detector located within cm of the interaction point to locate points on the charge particle trajectories very precisely. This *vertex detector* specifically iden-

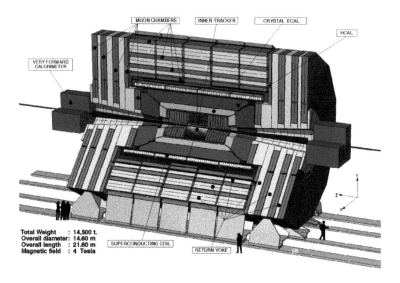

Fig. 6.6 Overall design of the CMS detector at the CERN Large Hadron Collider; figure courtesy of CERN. The figure shows the layered design, with (outward from the center) a silicon tracking detector, an electromagnetic calorimeter, a hadronic calorimeter, the solenoidal magnet, and instrumented iron to return the magnetic flux and identify muons.

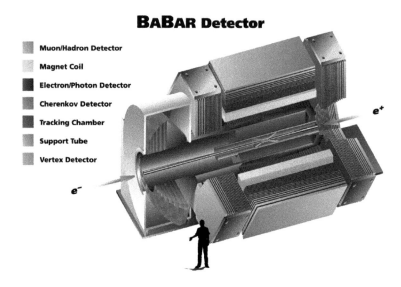

Fig. 6.7 Overall design of the BaBar detector operating at the PEP-2 collider at SLAC (figure courtesy of SLAC and the BaBar collaboration). The figure shows the tracking and calorimetry layers, and also the vertex detector just around the inter-action point and the Cherenkov ring imaging detector used to separately indentify π and K mesons.

tifies B mesons, for which $c\tau = 0.5$ mm. The second is a set of quartz bars, shown in green, that form a Cherenkov ring imaging system. This contributes to the tracking and separates π from K mesons.

We will see other ways of deploying the basic detectors in specific experimental arrangements later in our discussion.

Exercises

(6.1) For a 100 GeV electron moving through iron, estimate the fraction of its energy that it loses to ionization over a distance of 1 X_0. For a 100 GeV charged pion moving through iron, estimate the fraction of its energy that it loses to ionization over a distance of 1 λ_i.

(6.2) An extensive cosmic ray shower is the result of a collision of a very high energy proton from space with the nucleus of a molecule of air in the upper atmosphere. Consider for definiteness a collision that takes place at a height of 5 km. Take air to have a uniform density of 10^{-3} g/cm^3, and rock to have a uniform density of 2.6 g/cm^3. About 1000 charged pions might be produced in a very high energy cosmic ray interaction.

   (a) The probability of a $\pi^+ p$ interaction can be estimated by assigning the p an effective cross sectional area (cross section) of 3 fm^2 = 3×10^{-26} cm^2. Using this quantity, estimate λ_i for this standard air and rock. (For π^+-nucleus scattering, the cross section is should be multiplied by $A^{2/3}$, where A is the nucleon number. Why should this be?)

   (b) If a π^+ of 1 GeV is produced in the original p-nucleus collision, what is the probability that it suffers a nuclear collision before hitting the earth? What is the probability that it decays? The pion lifetime at rest is 2.6×10^{-8} s. The primary decay mode of the π^+ is $\pi^+ \to \mu^+ \nu$, with the μ^+ taking most of its energy.

   (c) The muons from π^+ decay enter the earth. How far do they go before stopping?

   (d) About 10 pions might be produced with energies of 100 GeV. Do they have time to decay to muons? If they decay, how far into the earth do the muons penetrate?

(6.3) This problem illustrates the factors that influence a momentum measurement with a tracking detector.

Consider a charged particle emitted from a high-energy interaction, moving through a cylindrical tracking chamber of radius L under the influence of a solenoidal magnetic field B. For simplicity, assume that the particle moves in a plane perpendicular to the axis of the cylinder and the direction of the magnetic field.

   (a) Since the initial direction of the particle is not known *a priori*, the curvature is measured from the *sagitta s* of its curved trajectory, defined to be the maximum deviation of the curve from a straight line between the point of origin and the point where the particle exits the chamber at radius L.

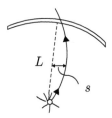

Show that, for small curvature,

$$s = \frac{L^2}{8R} . \tag{6.41}$$

Using (6.36), relate the sagitta to the magnetic field strength and the momentum of the particle. If Δs is the uncertainty in the measurement of the sagitta, obtain a formula for $\delta p/p$ in terms of p, L, B, and Δs.

   (b) It can be shown that, if the tracking detector makes N equally spaced position measurement, each with resolution ϵ, the uncertainty in the measurement of the sagitta is

$$\delta s = \frac{3.4\epsilon}{\sqrt{N+5}} . \tag{6.42}$$

For $N = 50$, $\epsilon = 100$ μm, $L = 1$ m, and $B = 1$ T, estimate the uncertainty in the obtained value of $p_\perp$.

(c) As the particle moves through an ionizing gas, it will multiple scatter. If the cylinder in this exercise is filled with nitrogen gas at atmospheric pressure, compute the expected $\Delta\theta$ from multiple scattering over a distance $L/2$ as a function of p. (The radiation length in N_2 is $X_0 = (38/\rho)$ cm, where ρ is the gas density in g/cm^3.) The error in the sagitta from this source is roughly

$$\delta s = \frac{L}{2}\delta\theta \ . \qquad (6.43)$$

At what value of p is multiple scattering a more important effect than the resolution of the position measurements?

Tools for Calculation

7

To compare the results of elementary particle experiments to proposed theories of the fundamental forces, we must think carefully about what quantities we can compute and measure. We cannot directly measure the force that one elementary particle exerts on another. Most of our information about the subnuclear forces is obtained from scattering experiments or from observations of particle decay.

In scattering experiments, the basic measureable quantity is called the *differential cross section*. In particle decay, the basic measureable quantity is called the *partial width*. In this chapter, I will define these quantities, and I will give formulae that will allow us to predict the values of these quantities from theoretical models. These will provide the calculational tools that we will use in Parts II and III to test possible theoretical ideas for elementary particle interactions against experimental results.

Experiments on elementary paraticles are set up to measure *widths* and *cross sections*, which can then be compared to the predictions of theoretical models.

7.1 Observables in particle experiments

The basic observable quantity associated with a decaying particle is the rate of decay. In quantum mechanics, an unstable particle A decays with the same probability in each unit of time. The probability of survival to time t then obeys the differential equation

$$\frac{dP}{dt} = -\frac{P}{\tau_A} , \tag{7.1}$$

for which the solution is

$$P(t) = e^{-t/\tau_A} . \tag{7.2}$$

The decay rate τ_A^{-1} is also called the *total width* Γ_A. Note that its units 1/sec are equivalent to GeV up to factors of $\hbar$ and c.

If there are numerous decay processes $A \to f$, each process has a rate

$$\Gamma(A \to f) . \tag{7.3}$$

This quantity is called the *partial width*. The total decay rate is

$$\Gamma_A = \sum_f \Gamma(A \to f) . \tag{7.4}$$

Definition of the lifetime of a particle τ_A and the width of the particle Γ_A. In conventional units, $\Gamma_A = \hbar/\tau_A$.

Concepts of Elementary Particle Physics. Michael E. Peskin.
© Michael E. Peskin 2019. Published in 2019 by Oxford University Press.
DOI: 10.1093/oso/9780198812180.001.0001

The probability that A will decay to the final state f is called the *branching ratio*, $BR(A \to f)$. This is given by

$$BR(A \to f) = \Gamma(A \to f)/\Gamma_A . \tag{7.5}$$

The rate of a particle collision process is characterized by a *cross section*. Imagine first that we shoot a beam of A particles of density n_A and velocity v_A at a fixed center B.

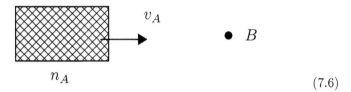

$$\tag{7.6}$$

The rate at which we see scatterings from the beam has the form

$$\text{events/sec} = n_A \, v_A \cdot \sigma \tag{7.7}$$

where σ has units of cm^2. This quantity is called the *cross section* for the reaction. It is the effective area that the target B presents to the beam. An alternative definition is given by the following situation: Imagine two bunches of particles A and B aimed at one another. Let one bunch, for example, B, have a smaller length and area, so that it fits inside the other.

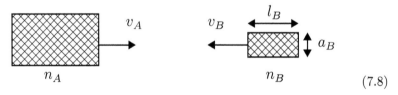

$$\tag{7.8}$$

As the bunch B passes through the bunch A, the rate of scatters of A particles from B particles is

$$\text{events/sec} = n_A \, n_B \, \ell_B \, \mathcal{A}_B \, |v_A - v_B| \cdot \sigma \tag{7.9}$$

A typical scattering process is a reaction with n particles in the final state

$$A + B \to 1 + 2 + 3 + \cdots + n . \tag{7.10}$$

We can represent the probability of finding each given momentum configuration of the final particles by a *differential cross section*

$$\frac{d\sigma}{d^3p_1 d^3p_2 \cdots d^3p_n} . \tag{7.11}$$

The integral over final momenta gives the total rate or the total cross section

$$\sigma(A + B \to 1 + \cdots + n) = \int d^3p_1 \cdots d^3p_n \, \frac{d\sigma}{d^3p_1 d^3p_2 \cdots d^3p_n} . \tag{7.12}$$

7.2   Master formulae for partial width and cross sections

Now I will write the formula for computing partial widths and differential cross sections. This formula is called *Fermi's Golden Rule*. Versions of this formula are derived in standard quantum mechanics textbooks. Here, I will write the formula in the way that is most appropriate for reactions involving relativistic particles.

Begin with the decay rate. For this, we need the quantum mechanical transition matrix element

$$\langle 12\cdots n|\, T\, |A(p_A)\rangle = \mathcal{M}(A \to 1+\cdots+n)\,(2\pi)^4\delta^{(4)}(p_A - \sum_j p_j)\,,\quad (7.13)$$

where T is an appropriate operator representing time evolution. The final state f contains particles $1, 2, \ldots, n$. The matrix element (7.13) must contain an energy-momentum conserving delta function. The factor $\mathcal{M}$ in front of this delta function is called the *invariant matrix element*. If indeed T is time evolution through the process and the states are relativistically normalized, the invariant matrix element must be Lorentz-invariant.

It is useful to work out the dimensions of $\mathcal{M}$. The operator T is dimensionless, and, according to (3.93), the states have total dimension $\mathrm{GeV}^{-(n+1)}$. The delta function has units GeV^{-4}. Then the invariant matrix element has the units

$$\mathcal{M} \sim \mathrm{GeV}^{3-n}\,. \quad (7.14)$$

To find the total rate, we must integrate over all possible values of the final momenta. This integral is called *phase space*. For n final particles, the expression for the phase space integral is

$$\int d\Pi_n = \int \frac{d^3p_1}{(2\pi)^3 2E_1}\cdots\frac{d^3p_n}{(2\pi)^3 2E_n}\,(2\pi)^4\delta^{(4)}(P - \sum_j p_j)\,,\quad (7.15)$$

where P is the total 4-momentum. Notice that I use the Lorentz invariant integral over relativistically normalized momentum states (3.89). The delta function, which is also Lorentz invariant, enforces energy and momentum conservation. Then the whole expression for phase space will be Lorentz invariant and can be used together with the Lorentz invariant matrix element $\mathcal{M}$ defined in (7.13). Similarly, a relativistically normalized initial state $|A\rangle$ will yield the factor $1/2E_A$. Phase space has the dimensions

$$\Pi_n \sim (\mathrm{GeV}^2)^n \cdot \mathrm{GeV}^{-4} = \mathrm{GeV}^{2n-4}\,. \quad (7.16)$$

The Fermi Golden Rule formula for a partial width to an n-particle final state f is

$$\Gamma(A \to f) = \frac{1}{2M_A}\int d\Pi_n\,|\mathcal{M}(A \to f)|^2\,. \quad (7.17)$$

Definition of the invariant matrix element $\mathcal{M}$ for a decay process.

Phase space is the volume of momentum space for n particles, subject to the constraint of fixed total energy and momentum. It is an important ingredient in the calculation of widths and cross sections.

The master formula for the computation of a partial width.

I have not given you a derivation of this equation, but, on the other hand, I have not given you a precise definition of $\mathcal{M}$ or told you how to compute it. A proper definition of the invariant matrix element requires more advanced concepts from quantum field theory. For the computations done in this book, you only need to accept that this formula has the correct structure. And it does. The rate is given by the square of a quantum mechanical matrix element, integrated over the momenta of the possible final state particles. The expression for the decay rate is completely Lorentz-invariant. The expression has total dimension

$$\Gamma \sim \text{GeV}^{-1} \cdot \text{GeV}^{2n-4} \cdot (\text{GeV}^{3-n})^2 \sim \text{GeV} , \qquad (7.18)$$

which is correct.

In computing particle decay rates, we must define the spins of the initial and final particles. Alternatively, we average over the initial spin direction and sum over the spin states of the final particles.

If the final state particles have *spin*, we need to sum over final spin states. The initial state A is in some state of definite spin. If we have not defined the spin of A carefully, an alternative is to *average* over all possible spin states of A. By rotational invariance, the decay rate of A cannot depend on its spin orientation.

The formula for a cross section is constructed in a similar way. We need the matrix element for a transition from the two initial particles to the final particles through the interaction. This is written

Definition of the invariant matrix element $\mathcal{M}$ for a scattering process.

$$\langle 12 \cdots n | \, T \, | A(p_A) B(p_B) \rangle = \mathcal{M}(A + B \to 1 + \cdots + n)$$
$$\cdot (2\pi)^4 \delta^{(4)} \Big(p_A + p_B - \sum_j p_j \Big) . \qquad (7.19)$$

As before, the *invariant matrix element* $\mathcal{M}(A + B \to 1 + \cdots + n)$ is indeed Lorentz invariant if the states are relativistically normalized. The dimension of $\mathcal{M}$ can be computed as we did in the previous case. Here we find

$$\mathcal{M} \sim \text{GeV}^{2-n} . \qquad (7.20)$$

The master formula for the computation of a cross section.

The formula for a cross section is then

$$\sigma(A + B \to f) = \frac{1}{2E_A 2E_B |v_A - v_B|} \int d\Pi_n \, |\mathcal{M}(A + B \to f)|^2 . \quad (7.21)$$

The factor $2E_A 2E_B$ in the prefactor comes from the relativistic normalization of the state $|AB\rangle$. The factor $|v_A - v_B|$ reflects the definition (7.9), in which the cross section is multiplied by a flux factor to obtain the rate of particle reactions. The dimension of the cross section should be cm^2, or GeV^{-2} in natural units. The formula (7.21) gives

$$\sigma \sim \text{GeV}^{-2} \cdot \text{GeV}^{2n-4} \cdot (\text{GeV}^{2-n})^2 \sim \text{GeV}^{-2} , \qquad (7.22)$$

which is correct.

Typically, in computing cross sections, we average over the spin states of the initial particles and sum over the spin states of the final particles.

The formula (7.21) should be summed over final particle spin states. If we do not take care to prepare the initial state in a definite spin state, the formula should be averaged over the initial spins.

The basic formulae for computing widths and cross sections are summarized in Appendix D.

7.3   Phase space

Phase space plays a very important role in particle physics. The default assumption is that final state particles are distributed according to phase space. This assumption is correct unless the transition matrix element has nontrivial structure. So, to look for structure that gives clues about the underlying dynamics, we must compare the results of experiments with the results that would be expected if the matrix element were constant and the process were shaped simply by phase space.

Most of the reactions we will discuss will have two particles in the final state. So, it will be useful if I now simplify the expression for the two-particle phase space once and for all. I will assume that the two particles have arbitrary masses m_1, m_2. Then

$$\int d\Pi_2 = \int \frac{d^3 p_1}{(2\pi)^3 2E_1} \frac{d^3 p_2}{(2\pi)^3 2E_2} (2\pi)^4 \delta^{(4)}(P - p_1 - p_2) \ . \qquad (7.23)$$

Work in the center of mass (CM) frame, where $\vec{p}_1 + \vec{p}_2 = 0$. The integral over the 3-momentum delta function enforces

$$\vec{p}_1 = -\vec{p}_2 \ . \qquad (7.24)$$

Then

$$P = (E_{CM}, \vec{0}) \ , \quad p_1 = (E_1, \vec{p}) \ , \quad p_2 = (E_2, -\vec{p}) \ , \qquad (7.25)$$

with

$$E_1 = \sqrt{p^2 + m_1^2} \ , \qquad E_2 = \sqrt{p^2 + m_2^2} \ . \qquad (7.26)$$

and (7.23) becomes

$$\int d\Pi_2 = \int \frac{d^3 p}{(2\pi)^3} \frac{1}{2E_1 2E_2} (2\pi)\delta(E_{CM} - E_1 - E_2) \ . \qquad (7.27)$$

It is most convenient to view the remaining momentum integral in spherical coordinates,

$$d^3 p = dp\, p^2\, d\theta\, \sin\theta\, d\phi = dp\, p^2\, d\Omega \ . \qquad (7.28)$$

Then the integral over the remaining delta function becomes

$$\int dp\, \delta(E_{CM} - E_1(p) - E_2(p)) = \frac{1}{|dE_1/dp + dE_2/dp|}$$

$$= \frac{1}{|p/E_1 + p/E_2|} = \frac{E_1 E_2}{(E_1 + E_2)p} \ . \qquad (7.29)$$

Since $E_1 + E_2 = E_{CM}$, we find

$$\int d\Pi_2 = \int \frac{p^2 d\Omega}{16\pi^2 E_1 E_2} \frac{E_1 E_2}{p E_{CM}} \ , \qquad (7.30)$$

Reduction of the expression for two-body phase space to a simple integral.

or, finally,

$$\int d\Pi_2 = \frac{1}{8\pi}\left(\frac{2p}{E_{cm}}\right)\int\frac{d\Omega}{4\pi} .\qquad (7.31)$$

Oddly, two-body phase space is dimensionless; we could have seen this already from (7.23). In the extreme relativistic limit $E \gg m_1, m_2$, the expression (7.31) reduces to

$$\int d\Pi_2 = \frac{1}{8\pi}\int\frac{d\Omega}{4\pi} .\qquad (7.32)$$

Please note this subtlety: When there are two identical particles in the final state, we must integrate over only half of phase space, so as not to count identical quantum states twice.

There is an important subtlety in integration over 2-body phase space for identical particles. Consider, for example, the possible final state $\pi^0\pi^0$. Bose statistics implies that the two states

$$\left|\pi^0(\vec{p})\pi^0(-\vec{p})\right\rangle \quad\text{and}\quad \left|\pi^0(-\vec{p})\pi^0(\vec{p})\right\rangle\qquad (7.33)$$

are *identical*. In the sum over states, we must sum over this state once and not twice. Thus, for $\pi^0\pi^0$ and other systems of identical particles, the integral over phase space should be taken over *half* of $\int d\Omega$. The same principle applies to multi-particle phase space when two final particles are identical.

It is also possible to reduce the expression for three-body space to a relatively simple formula. Work in the center of mass frame where $\vec{p}_1 + \vec{p}_2 + \vec{p}_3 = 0$. Let the total energy-momentum in this frame be Q, with $Q^0 = E_{CM}$. The three vectors $\vec{p}_1$, $\vec{p}_2$, $\vec{p}_3$ lie in a common plane, called the *event plane*. The integral (7.15) can be written as an integral over the orientation of this plane and over the variables

Reduction of the expression for three-body phase space to a relatively simple integral.

$$x_1 = \frac{2E_1}{E_{CM}} , \quad x_2 = \frac{2E_2}{E_{CM}} , \quad x_3 = \frac{2E_3}{E_{CM}} ,\qquad (7.34)$$

which obey the constraint

$$x_1 + x_2 + x_3 = 2 .\qquad (7.35)$$

I strongly encourage you to work through Exercise 7.2, which derives this formula and gives some applications.

It can be shown that, after integrating over the orientation of the event plane, the integral over three-body phase space can be written as

$$\int d\Pi_3 = \frac{E_{CM}^2}{128\pi^3}\int dx_1 dx_2 .\qquad (7.36)$$

The derivation of this formula is given in Exercise 7.2.

The variables x_1 and x_2 are to be integrated over all kinematically allowed values, but it is often not easy to write the boundary of the region of integration explicitly. When all three particles are massless, the maximum energy of any particle is $E_i = E_{CM}/2$, since a particle achieves its maximum energy when the two other particles are collinear in the opposite direction. Then the integral in (7.36) would be taken over the region

$$\int_0^1 dx_1 \int_{1-x_1}^1 dx_2 .\qquad (7.37)$$

When some particles are massive, the integration domain does not have such a simple form. Its boundaries can be found implicitly by imposing the three constraints $-1 \leq \cos\theta_{ij} \leq 1$, where θ_{ij} is the angle between $\vec{p}_i$ and $\vec{p}_j$ in the event plane.

It can be shown, further, that the integral (7.36) can alternatively be written in terms of the invariant masses of pairs of the three vectors. For example, if $m_{12}^2 = (p_1 + p_2)^2$ and $m_{23}^2 = (p_2 + p_3)^2$, then

$$\int d\Pi_3 = \frac{1}{128\pi^3 E_{CM}^2} \int dm_{12}^2 dm_{23}^2 \ . \tag{7.38}$$

This formula leads to an important construction in hadron physics called the *Dalitz plot*. This is also described in Exercise 7.2.

7.4 Example: $\pi^+\pi^-$ scattering at the ρ resonance

One important type of structure that one finds in scattering amplitudes is a *resonance*. In ordinary quantum mechanics, a resonance is described by the *Breit-Wigner formula*

$$\mathcal{M} \sim \frac{1}{E - E_R + i\Gamma/2} \ . \tag{7.39}$$

where E_R is the energy of the resonant state and$\Gamma$is its decay rate. The Fourier transform of this expression is

$$\psi(t) = \int \frac{dE}{2\pi} \frac{e^{-iEt}}{E - E_R + i\Gamma/2}$$
$$= -ie^{-iE_R t} e^{-\Gamma t/2} \ . \tag{7.40}$$

Evaluate (7.40) by integrating around the contour in the complex E plane

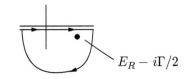

$E_R - i\Gamma/2$

Then the probability of maintaining the resonance decays exponentially

$$|\psi(t)|^2 = e^{-\Gamma t} \ , \tag{7.41}$$

corresponding to the lifetime

$$\tau_R = 1/\Gamma \ . \tag{7.42}$$

For the description of elementary particle reactions, we need a relativistic version of the Breit-Wigner formula. I will write this in a moment.

It is useful to consider a specific example of a resonance in an elementary particle reaction. The ρ^0 meson decays to $\pi^+\pi^-$ and, conversely, it can be produced in $\pi^+\pi^-$ collisions. The ρ^0 is then found in the reaction

$$\pi^+\pi^- \to \rho^0 \to \pi^+\pi^- \tag{7.43}$$

as a resonance at the ρ^0 mass of 770 MeV. We can represent this process by a diagram of the evolution of the process in space-time

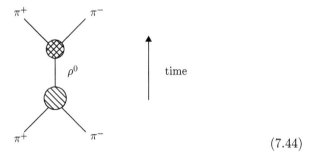

$$(7.44)$$

Using quantum field theory, Richard Feynman introduced a method for computing $\mathcal{M}$ using space-time diagrams of the form shown in (7.44). In that context, these diagrams are called *Feynman diagrams*. Whether or not we use them for computation, I will use such diagrams to visualize the elementary particle processes that we will discuss in this book.

Let us first consider the production of the ρ^0 resonance through the reaction

$$\pi^+(p_A)\pi^-(p_B) \to \rho^0(p_C) , \qquad (7.45)$$

with $p_A + p_B = p_C$. For the moment, I will consider the ρ^0 as a stable particle with a definite mass m_ρ. The production is given by an invariant matrix element

$$\mathcal{M}(\pi^+\pi^- \to \rho^0) \qquad (7.46)$$

We can guess the structure of this matrix element based on the known properties of the π and ρ mesons and Lorentz invariance. The ρ^0 has spin 1, so it has an associated polarization vector ϵ^μ. In the rest frame of the ρ^0, ϵ^μ should point in one of the three spatial directions. These three directions are characterized by the condition

$$p_C \cdot \epsilon = 0 , \qquad (7.47)$$

which is a Lorentz-invariant condition that can be applied in any frame.

Since I am normalizing all states relativistically, the matrix element must be Lorentz invariant. It must also be proportional to $\epsilon^{*\mu}$. The polarization vector is complex-conjugated because the ρ^0 appears in a bra vector. The only possible structure is

In this example, Lorentz invariance and momentum conservation completely fix the form of the invariant matrix element.

$$\mathcal{M}(\pi^+\pi^- \to \rho^0) = g_\rho \epsilon^* \cdot (p_A - p_B) , \qquad (7.48)$$

where g_ρ is a constant. The alternative structure

$$\mathcal{M}(\pi^+\pi^- \to \rho^0) = g_\rho \epsilon^* \cdot (p_A + p_B) , \qquad (7.49)$$

is zero by (7.47), since $p_A + p_B = p_C$. The constant g_ρ can be seen to be dimensionless: According to (7.20), the invariant matrix element has dimensions of GeV, while the right-hand side has dimensions

$$g_\rho \cdot (\text{GeV}) . \qquad (7.50)$$

Now we can write the cross section for formation of the ρ^0 as

$$\sigma(\pi^+\pi^- \to \rho^0) = \frac{1}{2E_A 2E_B |v_A - v_B|} \int \frac{d^3 p_C}{(2\pi)^3} \frac{1}{2E_C}$$

This analysis gives a model for the computation of a cross section.

$$|\mathcal{M}|^2 (2\pi)^4 \delta^{(4)}(p_C - p_A - p_B) \ . \quad (7.51)$$

In the center of mass (CM) frame

$$p_A = (E, p) \qquad p_B = (E, -p) \qquad E = m_\rho/2 \qquad p = [m_\rho^2/4 - m_\pi^2]^{1/2} \ . \quad (7.52)$$

In this frame, the ρ polarization vector ϵ points in a space direction, and so

$$\epsilon^* \cdot (p_A - p_B) = -\vec{\epsilon} \cdot (\vec{p}_A - \vec{p}_B) = -2\vec{\epsilon} \cdot \vec{p}. \quad (7.53)$$

This expression has one power of momentum, so this is a P-wave scattering process, as required for angular momentum conservation. We can rewrite

$$\int \frac{d^3 p_C}{(2\pi)^3} \frac{1}{2E_C} = \int \frac{d^4 p_C}{(2\pi)^4} 2\pi \delta(p_C^2 - m_\rho^2) \quad (7.54)$$

and then integrate $d^4 p_C$ over the energy-momentum conserving delta function. The expression for the cross section reduces to

$$\sigma(\pi^+\pi^- \to \rho^0) = \frac{1}{4(m_\rho/2)^2(4p/m_\rho)} 2\pi\delta((p_A + p_B)^2 - m_\rho^2) g_\rho^2 \cdot 4|\vec{\epsilon} \cdot \vec{p}|^2 \ . \quad (7.55)$$

Summing over ρ^0 polarizations, we find

$$\sigma(\pi^+\pi^- \to \rho^0) = g_\rho^2 \frac{p}{m_\rho}(2\pi)\delta((p_A + p_B)^2 - m_\rho^2) \ . \quad (7.56)$$

Counting the dimension of the delta function as $(\text{GeV})^{-2}$, this has dimensions

$$\text{GeV}^{-2} \sim \text{cm}^2 \ , \quad (7.57)$$

which is the correct result.

Conversely, we can compute the decay rate for $\rho^0 \to \pi^+\pi^-$. The invariant matrix element that we need here is $\mathcal{M}(\rho^0 \to \pi^+\pi^-)$, which is the complex conjugate of (7.48). The Fermi's Golden Rule formula gives

$$\Gamma_\rho = \frac{1}{2m_\rho} \int d\Pi_2 |\mathcal{M}|^2 \ . \quad (7.58)$$

Using the evaluation of 2-body phase space in (7.31), this reduces to

$$\Gamma_\rho = \frac{1}{2m_\rho} \frac{1}{8\pi} \frac{2p}{m_\rho} g_\rho^2 \left\langle 4|\vec{\epsilon} \cdot \vec{p}|^2 \right\rangle \ . \quad (7.59)$$

The average over orientations of the outgoing pions gives

$$\left\langle 4|\vec{\epsilon} \cdot \vec{p}|^2 \right\rangle = 4\frac{p^2}{3} \ . \quad (7.60)$$

Alternatively, we would obtain the same result by averaging over the polarization of the ρ^0. The final result is

$$\Gamma_\rho = \frac{g_\rho^2}{6\pi} \frac{p^3}{m_\rho^2} . \tag{7.61}$$

This, correctly, has units of GeV. The measured width of the ρ^0 is 150 MeV. This requires

$$\frac{g_\rho^2}{4\pi} = 2.9 , \tag{7.62}$$

a rather strong coupling.

Now we can put the pieces together, modelling the ρ^0 as a relativistic Breit-Wigner resonance. I propose the form

Relativistic form of the Breit-Wigner resonance formula.

$$\mathcal{M} \sim \frac{1}{P^2 - m_R^2 + i m_R \Gamma_R} , \tag{7.63}$$

where P is the total momentum 4-vector creating the resonance. If we go to the CM frame and expand

$$P = (m_R + \Delta E, \vec{0}) \qquad P^2 = m_R^2 + 2 m_R \Delta E + \cdots , \tag{7.64}$$

this gives back the earlier Breit-Wigner expression (7.39), with the denominator multiplied by $2m_R$.

With this relativistic formula for the resonance, the formula for the cross section for $\pi^+ \pi^- \to \rho^0 \to \pi^+ \pi^-$ is

$$\sigma(\pi^+(p_A)\pi^-(p_B) \to \rho^0 \to \pi^+(p_A')\pi^-(p_B'))$$
$$= \frac{1}{2E_A 2E_B |v_A - v_B|} \int d\Pi_2$$
$$\left| \sum_\epsilon \frac{\mathcal{M}(\pi^+\pi^- \to \rho^0(\epsilon))\mathcal{M}(\rho^0(\epsilon) \to \pi^+\pi^-)}{(p_A + p_B)^2 - m_\rho^2 + i m_\rho \Gamma_\rho} \right|^2 . \tag{7.65}$$

Inserting the explicit forms for the matrix elements, with the CM momentum of the initial π^+ equal to $\vec{p}$ and that of the final π^+ equal to $\vec{p}\,'$, we find

$$\sigma(\pi^+\pi^- \to \rho^0 \to \pi^+\pi^-)$$
$$= \frac{1}{4m_\rho p} \frac{1}{8\pi} \frac{2p}{m_\rho} \int \frac{d\Omega}{4\pi} \frac{1}{(E_{CM}^2 - m_\rho^2)^2 + m_\rho^2 \Gamma_\rho^2}$$
$$\left| \sum_\epsilon 2g_\rho \vec{\epsilon} \cdot \vec{p} \, 2g_\rho \vec{\epsilon} \cdot \vec{p}\,' \right|^2 . \tag{7.66}$$

Evaluating the sum over ρ polarizations, we find

$$\sigma(\pi^+\pi^- \to \pi^+\pi^-) = \frac{1}{\pi m_\rho^2} \frac{g_\rho^4}{(E_{CM}^2 - m_\rho^2)^2 + m_\rho^2 \Gamma_\rho^2} \int \frac{d\Omega}{4\pi} |\vec{p} \cdot \vec{p}\,'|^2 \tag{7.67}$$

This is a very concrete formula, for which we have determined all of the parameters. It can be compared directly to experimental data on the final energies and angles of the pions. Note that the factor

$$\frac{d\sigma}{d\Omega} \sim |\vec{p} \cdot \vec{p}\,'|^2 \tag{7.68}$$

is characteristic of a resonance in the $L = 1$ partial wave. The shape of the resonance as a function of the CM energy is

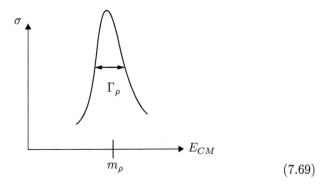

$$(7.69)$$

As a check on the formalism, I will now perform the integral over the final state momenta in the limit of a very long-lived or narrow resonance. To illustrate the generality of the result, I will consider a general decay of the ρ^0 to a final state f. The argument is a bit long, but in the end it will connect nicely to the easier formulae above.

The cross section formula for $\pi^+\pi^- \to \rho \to f$ is

$$\sigma(\pi^+\pi^- \to \rho \to f) = \frac{1}{4m_\rho p} \int d\Pi_f \left| \sum_\epsilon \frac{[2g_\rho \vec{\epsilon}^{\,*} \cdot \vec{p}]\mathcal{M}(\rho^0_\epsilon \to f)}{(p_A + p_B)^2 - m_\rho^2 + im_\rho\Gamma_\rho} \right|^2 .$$

$$(7.70)$$

Rewrite the delta function in the phase space as

$$(2\pi)^4\delta^{(4)}(p_A + p_B - \sum p_j) = \int \frac{d^4 p_C}{(2\pi)^4} (2\pi)^4\delta^{(4)}(p_A + p_B - p_C)$$

$$(2\pi)^4\delta^{(4)}(p_C - \sum p_j) \qquad (7.71)$$

Now we can integrate over the final state phase space to find the total decay rate. There is a subtlety here involving the spins. First, when we sum over all final states, the decay rate is independent of ϵ. Second, since the different ϵ correspond to different angular momentum states, when we integrate over all final state configurations there is no interference between the contributions from different ϵ. With this insight, we can combine the phase space integral in (7.70) with square of the matrix element $\mathcal{M}(\rho^0 \to f)$ to form the partial width for ρ^0 decay to f. Then (7.70) reduces to

Notice how nicely the cross sections to produce a given final state, summed over all final states, reproduces the formula (7.56) for the total rate to produce the resonance.

$$\sigma(\pi^+\pi^- \to \rho \to f) = \frac{1}{4m_\rho p} \int \frac{d^4 p_C}{(2\pi)^4} (2\pi)^4\delta^{(4)}(p_A + p_B - p_C)$$

$$\cdot \sum_\epsilon |2g_\rho \vec{\epsilon}^{\,*} \cdot \vec{p}_A|^2 \frac{1}{(p_C^2 - m_\rho^2)^2 + (m_\rho\Gamma_\rho)^2} 2m_\rho\Gamma(\rho \to f) \quad (7.72)$$

or

$$\sigma(\pi^+\pi^- \to \rho \to f) = \frac{1}{4m_\rho p} 4g_\rho^2 p^2 \frac{2m_\rho \cdot \Gamma(\rho \to f)}{(p_C^2 - m_\rho^2)^2 + (m_\rho\Gamma_\rho)^2} . \quad (7.73)$$

Summing over all possible final states, we find

$$\sum_f \sigma(\pi^+\pi^- \to \rho \to f) = \frac{1}{m_\rho p} 2\pi g_\rho^2 p^2 \cdot \frac{(m_\rho \Gamma_\rho/\pi)}{(p_C^2 - m_\rho^2)^2 + (m_\rho \Gamma_\rho)^2} \cdot \quad (7.74)$$

Finally, notice that, when Γ_ρ is very small, the last factor in this expression approximates a delta function. The normalization of this delta function is given by the integral

$$\int dp_C^2 \frac{m_\rho \Gamma_\rho/\pi}{(p_C^2 - m_\rho^2)^2 + (m_\rho \Gamma_\rho)^2} = \frac{m_\rho \Gamma_\rho}{\pi} \frac{1}{m_\rho \Gamma_\rho} \tan^{-1}\left(\frac{s - m_\rho^2}{m_\rho \Gamma_\rho}\right). \quad (7.75)$$

When the resonance is narrow, we can extend the integral from $-\infty$ to ∞, to find

$$\int dp_C^2 \frac{m_\rho \Gamma_\rho/\pi}{(p_C^2 - m_\rho^2)^2 + (m_\rho \Gamma_\rho)^2} = 1 . \quad (7.76)$$

In this limit, our expression collapses to

$$\sigma(\pi^+\pi^- \to \rho \to f) = \frac{g_\rho^2 p}{m_\rho} \cdot 2\pi\delta((p_A + p_B)^2 - m_\rho^2) \quad (7.77)$$

This agrees precisely with our earlier calculation of the production rate of the resonance.

We will put these formulae to work already in the next chapter.

Exercises

(7.1) At a mass of about 500 MeV, there is a very broad resonance called the σ with spin 0 and isospin 0. It is broad because it decays very rapidly into two pions. (The Particle Data Group calls this the $f_0(500)$.) If we imagine that the σ were, instead, a narrow resonance, we could study it using the methods of Section 7.4.

(a) Write the matrix elements for σ formation as

$$\mathcal{M}(\pi^i \pi^j \to \sigma) = G\delta^{ij} , \quad (7.78)$$

where G is a constant and $i, j = 1, 2, 3$ are isospin indices. The usual pion states are: $\pi^\pm = (\pi^1 \pm i\pi^2)/\sqrt{2}$, $\pi^0 = \pi^3$. Show that the form of (7.78) is consistent with angular momentum, and isospin symmetry. What are the P and G-parity quantum numbers of the σ?

(b) Compute the matrix elements $\mathcal{M}(\pi^+\pi^- \to \sigma)$ and $\mathcal{M}(\pi^0\pi^0 \to \sigma)$ in terms of G.

(c) Compute the decay rate of the σ to $\pi^+\pi^-$. You should find

$$\Gamma(\sigma \to \pi^+\pi^-) = \frac{G^2}{16\pi m_\sigma} \frac{2p}{m_\sigma} , \quad (7.79)$$

where $p = (1/2)(m_\sigma^2 - 4m_\pi^2)^{1/2}$.

(d) Compute the total width of the σ. The answer should be $(3/2)$ of the result in (c). Why? What is the branching ratio $BR(\sigma \to \pi^0\pi^0)$?

(e) Compute the cross sections for the reactions $\pi^+\pi^- \to \pi^+\pi^-$ and $\pi^+\pi^- \to \pi^0\pi^0$, assuming these are dominated by the σ resonance. How do these cross sections reflect the spin 0 and isospin 0 nature of the σ?

(7.2) This problem derives the formula (7.36) for 3-body phase space and demonstrates an important application of that expression. The problem is very long, but it will be worth your time. The very last parts of this problem make direct contact with experimental data.

In this problem, 1, 2, 3 represent three particles with nonzero masses m_1, m_2, m_3, and $Q = p_1 + p_2 + p_3$. In the center of mass (CM) frame, $Q = (E_{CM}, 0, 0, 0)$. Let E_1, E_2, E_3 be the energies of the three particles in this frame.

(a) Define

$$x_1 = \frac{2Q \cdot p_1}{Q^2}, \quad x_2 = \frac{2Q \cdot p_2}{Q^2}, \quad x_3 = \frac{2Q \cdot p_3}{Q^2}, \tag{7.80}$$

Evaluate these quantities in the CM frame and show that

$$x_1 + x_2 + x_3 = 2 \tag{7.81}$$

(b) Write expressions for the CM energies E_i and the CM momentum values p_i in terms of the x_i, $i = 1, 2, 3$.

(c) Show that the invariant mass of the system of particles 1 and 2 is related to x_3 by

$$m_{12}^2 = (p_1 + p_2)^2 = (1 - x_3)Q^2 + m_3^2 . \tag{7.82}$$

There is a similar relation for m_{23}^2 and m_{31}^2.

(d) Let θ_{12} be the angle between the momenta of 1 and 2 in the CM frame. Show that the formula (7.82) determines θ_{12} as a function of the x_i. In fact, the whole configuration of final state momenta is specified, up to an overall rotation, when the x_i are fixed.

(e) Write out the integral over 3-body phase space in the CM frame. There are 9 integrals and 4 delta functions. Three of these delta functions can be removed by integrating out $\vec{p}_3$. Write the resulting expression as an integral over p_1, p_2 and 4 angles, constrained by 1 remaining delta function.

(f) Because we have eliminated $\vec{p}_3$ in terms of $\vec{p}_1$ and $\vec{p}_2$, the quantity E_3 in the delta function depends on $|\vec{p}_1 + \vec{p}_2|$ and therefore on $\cos \theta_{12}$. Do the integral over $\cos \theta_{12}$, eliminating the last delta function.

(g) The remaining three angles simply rotate the overall configuration of momenta. Integrate over these variables.

(h) All that remains are integrals over p_1 and p_2. Using (a), convert these to integrals over x_1 and x_2. Then, using (c), convert these to integrals over m_{23}^2 and m_{13}^2. You should find

$$\int d\Pi_3 = \frac{Q^2}{128\pi^3} \int dx_1 dx_2$$
$$= \frac{1}{128\pi^3 Q^2} \int dm_{23}^2 dm_{13}^2 . \tag{7.83}$$

(i) It is amazing that the integrand has no dependence on x_1, x_2, x_3! Dalitz suggested that, for a 3-body decay $A \to 1+2+3$, we should make a scatter plot of events in the plane of m_{23}^2 vs. m_{13}^2. If the matrix element is constant, the data points will scatter evenly over this plane. Write a formula for $\Gamma(A \to 1+2+3)$ and justify this statement. If there is a resonance, that will be apparent as a clustering of points in some region. The plot of m_{23}^2 vs. m_{13}^2 is called the *Dalitz plot*.

(j) The integral in (7.83) should be taken over all kinematically allowed values. It takes a little work to find the boundary of the integration region. Study this first for the case in which a particle of mass M decays to three particles, all of which are massless. In this case, there are allowed configurations all the way out to the boundaries $m_{13}^2 = 0$, $m_{23}^2 = 0$, $m_{12}^2 = 0$. Draw the region of integration on the (m_{13}^2, m_{23}^2) plane. For each segment of the the boundary, draw a typical momentum configuration. You should find that the boundaries of the Dalitz plot are given by configurations in which two momentum vectors are collinear and the third is directly opposite, balancing the momentum.

(k) Now consider the case of the decay of a particle of mass M to three particles with $m_1 = m_2 = 0$, $m_3 = m > 0$. Again, the boundaries of the Dalitz plot are given by configurations in which two momentum vectors are collinear and the third is directly opposite. Work out the positions of the boundaries in the (m_{13}^2, m_{23}^2) plane. The kinematic formulae (2.19) will be helpful, as will the result in part (c) above.

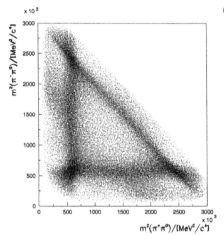

(l) The figure above shows the Dalitz plot for a process in $p\bar{p}$ annihilation at rest,

$$p\bar{p} \to \pi^+ \pi^- \pi^0 , \qquad (7.84)$$

from (Abele 1999). Resonances are apparent. From their masses and quantum numbers, Identify the resonances with specific hadrons.

(m) The following figure shows the Dalitz plot for the decay

$$D^0 \to K^- \pi^+ \pi^0 , \qquad (7.85)$$

from (Kopp 2001). I hope you can make out a heavy horizontal band across the lower part of the plot, a vertical band on the left, and a diagonal band on the right. These bands are obscured by the fact that interference effects cause the bands to be dark in some places but light (zero) in others. Identify these bands with specific hadrons.

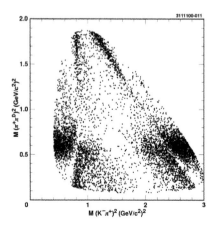

(7.3) In Section 7.4, I wrote a formula for a reaction $A + B \to 1 + 2$ mediated by a narrow resonance R that suggests the following general form:

$$\sigma(A + B \to 1 + 2)$$
$$= \frac{1}{2E_A 2E_B |v_A - v_B|} \int d\Pi_2$$
$$\cdot \left| \sum_\epsilon \frac{\mathcal{M}(A + B \to R(\epsilon)) \cdot \mathcal{M}(R(\epsilon) \to 1 + 2)}{P^2 - m_R^2 + im_R \Gamma_R} \right|^2 , \qquad (7.86)$$

where $P = (p_A + p_B)$. Quite generally,

$$\mathcal{M}(A + B \to R(\epsilon)) = (\mathcal{M}(R(\epsilon) \to A + B))^* . \qquad (7.87)$$

(Technically, this follows from time-reversal invariance.) Assume this statement for the purpose of this problem. In the equation above, ϵ represents the spin state of R. The particles A, B, etc. might also have spin. Let p_A, p_1 be the momenta of A and 1 in the rest frame of R.

(a) Show that

$$\Gamma(R \to 1 + 2) = \frac{1}{16\pi m_R} \left(\frac{2p_1}{m_R} \right)$$
$$\cdot \left\langle |\mathcal{M}(R(\epsilon) \to 1 + 2)|^2 \right\rangle , \qquad (7.88)$$

where the right-hand side is averaged over the directions of the momenta of 1 and 2 and summed over the spins of these particles. Note that the result is independent of ϵ, the spin state of R.

(b) Use this observation to write the expression (7.86) in terms of partial widths. Assume for simplicity that that A, B, 1, 2 are spin 0 particles. Let R have spin J. Show that

$$\sigma(A + B \to 1 + 2)$$
$$= \frac{4\pi m_R^2}{p_A^2} (2J + 1)$$
$$\cdot \frac{\Gamma(R \to A + B)\Gamma(R \to 1 + 2)}{(E_{CM}^2 - m_R^2)^2 + m_R^2 \Gamma_R^2} . \qquad (7.89)$$

(c) Let A, B, 1, 2 have spins J_A, J_B, J_1, J_2. Assuming an unpolarized initial state, generalize the formula in (b) appropriately. Show that the factor $(2J + 1)$ in the expression in (b) is replaced by $(2J + 1)/(2J_A + 1)(2J_B + 1)$, and that this factor does not depend on J_1 or J_2.

(d) Now consider the case of the J/ψ. In Exercise 5.1, we learned that the branching ratios of the J/ψ to e^+e^- and $\mu^+\mu^-$ are equal to $B \approx 6\%$, and that almost all other decays of the J/ψ are to hadrons. Assume for simplicity that these three modes are the only modes of J/ψ decay, with branching ratios B, B, and $(1-2B)$, respectively. Write expressions for the cross sections for $e^+e^- \to e^+e^-$, $e^+e^- \to \mu^+\mu^-$, and $e^+e^- \to hadrons$ in the vicinity of the resonance, in which the only free parameters are $m_{J/\psi}$, $\Gamma_{J\psi}$ and B. You may ignore non-resonant contributions to the scattering amplitude, and consider the electron and the muon to have zero mass.

(e) Evaluate the expressions for the cross sections at the peak of the resonance. Show that the peak value of the cross section for $e^+e^- \to$ hadrons essentially measures B or $\Gamma(J/\psi \to e^+e^-)$. This is somewhat counterintuitive. Write an explicit formula for B in terms of the peak cross section for $e^+e^- \to$ hadrons.

(f) Show that, with B determined, the integral over the cross section through the resonance

$$\int dE_{CM}\ \sigma(e^+e^- \to \text{hadrons}) \qquad (7.90)$$

determines the width $\Gamma_{J/\psi}$. Notice that we can measure $\Gamma_{J/\psi}$ without having to make a detailed measurement of the shape of the resonance.

Part II

The Strong Interaction

Electron-Positron Annihilation

8

We now begin our search for a fundamental theory describing the strong interaction. The first approaches to this problem attempted to build the theory from the properties of the cross sections for the scattering of pions and other mesons at low energies. Much later, it was realized that one could gain much more insight from the study of meson production by the photon and other electromagnetic probes. Most remarkably, the study of electron-positron annihilation to mesons showed simple and remarkable properties that are readily interpreted in relation to the quark model of hadrons discussed in Chapter 5. In this chapter, I will describe the important features of this reaction.

For reference in interpreting the results on hadrons, it will be useful first to understand the purely electromagnetic process of e^+e^- annihilation to a pair of muons. Using this process, I will also introduce the *current-current interaction*, a basic coupling of spin $\frac{1}{2}$ particles. This interaction provides the basis for fermion scattering by electromagnetic forces. Its properties are especially simple at very high energies, where the fermion masses can be neglected. By performing experiments at high energies, we can see whether the characteristic features of this coupling appear in reactions involving the other fundamental forces. Indeed, we will see that it plays a central role in the dynamics for both the strong and the weak interaction.

8.1 The reaction $e^+e^- \to \mu^+\mu^-$

We begin our study with the reaction $e^+e^- \to \mu^+\mu^-$. The matrix elements for this process can be constructed by breaking the process down into components. First, the e^+e^- state is annihilated by an electromagnetic current. This current couples to a quantum state of electromagnetic excitation. Finally, this state couples to another current matrix element describing the creation of the muon pair. These elements are

Concepts of Elementary Particle Physics. Michael E. Peskin.
© Michael E. Peskin 2019. Published in 2019 by Oxford University Press.
DOI: 10.1093/oso/9780198812180.001.0001

visualized in the Feynman diagram

$$(8.1)$$

In drawing this diagram, I label each line with the momentum carried by the particle or resonance. We will make use of these momentum labels later in this chapter.

The evolution of a quantum mechanical process in space-time is described by a Feynman diagram.

I will describe the intermediate photon state as a Breit-Wigner resonance at zero mass. Taking the limit of zero resonance mass in (7.63), it would then contribute to the scattering amplitude a factor

$$\frac{1}{q^2} \, , \tag{8.2}$$

where q is the momentum carried by the photon from the initial to the final state. We consider the reaction at energies large compared to the muon mass and, certainly, very far from the mass shell condition $q^2 = 0$ for a photon. A resonance contributing to an elementary particle reaction very far from its mass shell is called a *virtual particle*. In this case, we say that the reaction is mediated by a *virtual photon*.

A virtual particle is a particle that appears in a process as a resonance off its mass shell.

The remainder of the matrix element is formed from the product of two electromagnetic current operators, one of which annihilates the e^+e^- pair, the other of which creates the muon pair. Explicitly

$$\mathcal{M}(e^+e^- \to \mu^+\mu^-) = (-e)\,\langle \mu^+\mu^- |\, j^\mu\, |0\rangle\, \frac{1}{q^2}(-e)\,\langle 0|\, j_\mu\, |e^+e^-\rangle \; . \tag{8.3}$$

Note that the electric charges of the electron and the muon, $(-e)$ in both cases, appear as the strengths of the couplings of these states to the electromagnetic current.

The basic operator structure

The current-current interaction.

$$j^\mu\, j_\mu \tag{8.4}$$

is called the *current-current interaction*. We have seen in Section 3.3 that the photon interacts with other fields, in the Lagrangian description of electromagnetism, by direct coupling to the current. The current-current structure then arises naturally in electromagnetism. In fact, it will give the basic form of the scattering amplitude in any model in which the interaction is mediated by vector fields.

Our next task is to turn the expression (8.3) into an explicit formula that we can compare to the measured cross section for $e^+e^- \to \mu^+\mu^-$.

8.2 Properties of massless spin-$\frac{1}{2}$ fermions

To evaluate (8.3), we need to compute the matrix elements of the currents between fermion states. For our present purposes, I consider energies so large that both the electrons and muons are moving relativistically and their masses can be neglected. I will now show that the dynamics of fermions and the calculation of matrix elements is dramatically simplified in that limit.

Consider, then, the properties of the Dirac equation when we take the mass of the fermion to zero. In this approximation, the Dirac equation takes the form

> When the fermion mass is set equal to zero, the Dirac equation takes an especially simple form.

$$i\gamma^\mu \partial_\mu \Psi = 0 . \tag{8.5}$$

To analyze this equation, it is convenient to choose the representation (3.42) of the Dirac matrices

$$\gamma^0 = \begin{pmatrix} 0 & 1 \\ 1 & 0 \end{pmatrix} , \qquad \gamma^i = \begin{pmatrix} 0 & \sigma^i \\ -\sigma^i & 0 \end{pmatrix} . \tag{8.6}$$

It is convenient to write this representation as

$$\gamma^\mu = \begin{pmatrix} 0 & \sigma^\mu \\ \bar{\sigma}^\mu & 0 \end{pmatrix} , \tag{8.7}$$

defining the matrices

$$\sigma^\mu = (1, \vec{\sigma})^\mu , \qquad \bar{\sigma}^\mu = (1, -\vec{\sigma})^\mu . \tag{8.8}$$

Using this representation, and writing

$$\Psi = \begin{pmatrix} \psi_L \\ \psi_R \end{pmatrix} \tag{8.9}$$

the Dirac equation splits into two 2-component equations,

$$i\bar{\sigma} \cdot \partial \psi_L = 0 \qquad i\sigma \cdot \partial \psi_R = 0 . \tag{8.10}$$

We will see in a moment that the fields ψ_L and ψ_R annihilate different electron states and create different positron states. These states are not connected by the Dirac equation in this massless limit. When we couple the Dirac equation to electromagnetism, we modify the derivative to include the A_μ field,

$$\partial_\mu \rightarrow D_\mu = (\partial_\mu + ieA_\mu) . \tag{8.11}$$

This preserves the separation of the fields ψ_L and ψ_R and of the associated electrons and positrons. The two pieces of the Dirac field communicate only through the mass term. Thus, for zero electron mass or for very high energy where the mass can be neglected, there are essentially two different species of electrons, e_L^- and e_R^-. Electromagnetic interactions cannot turn electrons of one kind into the other.

We now find the plane wave solutions of the 2-component equations. Look first at the ψ_R equation.

$$(i\partial_t + i\vec{\sigma}\cdot\vec{\nabla})\psi_R = 0 . \tag{8.12}$$

A plane wave solution has the form

$$\psi_R(x) = u_R(p)e^{-iEt+i\vec{p}\cdot\vec{x}} , \tag{8.13}$$

where $u_R(p)$ is a 2-component spinor. For simplicity, look for a plane wave moving in the $\hat{3}$ direction: $\vec{p} = p\hat{3}$. Then

$$(E - p\sigma^3)\, u_R = \begin{pmatrix} E - p & 0 \\ 0 & E + p \end{pmatrix} u_R = 0 . \tag{8.14}$$

There are two solutions. The first has $E = p > 0$; this is

$$\psi_R = \begin{pmatrix} 1 \\ 0 \end{pmatrix} e^{-iEt+iEx^3} . \tag{8.15}$$

This state carries a spinor with spin $S^3 = +\frac{1}{2}$, that is, spin up along the direction of motion. The corresponding electron moves at the speed of light and spins in the right-handed sense,

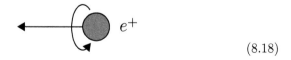

$$(8.16)$$

The field operator $\psi_R(x)$ destroys an electron in this state. The second solution has negative energy $E = -p < 0$.

$$\psi_R = \begin{pmatrix} 0 \\ 1 \end{pmatrix} e^{+i|E|t+i|E|x^3} . \tag{8.17}$$

This solution corresponds to the creation of a positron by the Dirac field. The positron will be moving in the $-\hat{3}$ direction and will have a spin opposite to the spinor shown. This spinor has $S^3 = -\frac{1}{2}$, so the positron has $S^3 = +\frac{1}{2}$, which is spin *down* with respect to the direction of motion. This is a positron moving at the speed of light and spinning in the left-handed sense,

$$(8.18)$$

Definition of *helicity*.

To describe these states, it is convenient to define the *helicity* of a particle, equal to the spin projected along the direction of motion,

$$h = \hat{p}\cdot\vec{S} . \tag{8.19}$$

The solutions of the ψ_R equation correspond to an $h = +\frac{1}{2}$ electron and an $h = -\frac{1}{2}$ positron. These states are particle and antiparticle.

The ψ_L equation

$$(i\partial_t - i\vec{\sigma} \cdot \vec{\nabla})\psi_L = 0 . \tag{8.20}$$

is solved in a similar way. Look for plane waves with $\vec{p} = p\hat{3}$. These have the form

$$\psi_L = \begin{pmatrix} 0 \\ 1 \end{pmatrix} e^{-iEt+iEx^3} . \tag{8.21}$$

for positive energy, and

$$\psi_L = \begin{pmatrix} 1 \\ 0 \end{pmatrix} e^{+i|E|t+i|E|x^3} . \tag{8.22}$$

for negative energy. The first of these describes the destruction of a massless left-handed electron moving in the $+\hat{3}$ direction; the second describes the creation of a massless right-handed positron moving in the $-\hat{3}$ direction.

We can find the solutions for electrons and positrons moving in other directions by rotating the expressions above. These plane wave solutions appear in the matrix elements through which massless Dirac fields create and destroy particles. For example,

$$\langle 0| \psi_R(x) \left|e_R^-(p)\right\rangle = u_R(p)\, e^{-ip\cdot x} ,$$
$$\left\langle e_L^+(p)\right| \psi_R(x) |0\rangle = v_L(p)\, e^{+ip\cdot x} . \tag{8.23}$$

Note that, by convention, the 2-component spinor is called $u(p)$ in the destruction of electrons and $v(p)$ in the creation of positrons. The momentum p here is always a vector with positive energy. The full theory of the quantum Dirac equation is needed to give the correct normalizations of the u and v spinors. Using these normalizations, the complete expressions for the spinors are

In the zero mass or high energy limit, we treat electrons as belonging to two distinct species of particles: e_R^- and its antiparticle e_L^+, and e_L^- and its antiparticle e_R^+.

	destruction		creation
e_R^-:	$u_R(p) = \sqrt{2E}\,\xi_+$	e_L^+:	$v_L(p) = \sqrt{2E}\,\xi_+$
e_L^-:	$u_L(p) = \sqrt{2E}\,\xi_-$	e_R^+:	$v_R(p) = \sqrt{2E}\,\xi_-$

where, in these formulae, ξ_+, ξ_- are the spinors with spin up and spin down, respectively, along the direction of motion. The normalization factor of $\sqrt{2E}$ will give the correct mass dimensions when we use these expressions to evaluate matrix elements. We will see examples of this in the next section.

In the basis (8.9), the full Dirac Lagrangian, including the fermion mass term, takes the form

$$\mathcal{L} = \psi_R^\dagger(i\sigma \cdot \partial)\psi_R + \psi_L^\dagger(i\bar{\sigma} \cdot \partial)\psi_L - m(\psi_R^\dagger\psi_L + \psi_L^\dagger\psi_R) . \tag{8.24}$$

For a fermion at high energy, any flip of helicity from e_R^- to e_L^- or vice versa brings in a factor m/E.

The components ψ_R and ψ_L are mixed by the mass term. Equivalently, any helicity flip from e_R^- to e_L^- or vice versa requires a factor of m and so is suppressed at high energy by a factor m/E.

8.3 Evaluation of the matrix elements for $e^+e^- \to \mu^+\mu^-$

With these ingredients, we can construct the expectation values of j^μ in the expression for the $e^+e^- \to \mu^+\mu^-$ matrix element above. The matrix elements will depend on the spin states of the electron, positron, and muons. That analysis will be important to get an explicit theoretical prediction for this reaction. But also, at the same time, this analysis will illustrate how Feynman diagrams such as (8.1) and the corresponding matrix element formulae such as (8.3) encode the physics of elementary particle interactions. Please follow, in particular, the flow of angular momentum from the e^+e^- system to the virtual photon and then to the $\mu^+\mu^-$ system. This will determine the observable form of the final answer for the cross section.

The evaluation of (8.3), described in this section, is lengthy and somewhat technical. Please work through this derivation carefully, step by step. It is the model for many other calcuations done later in this book.

Begin with the matrix element to annihilate a right-handed electron and a left-handed positron,

$$\langle 0 |\, j^\mu \, | e_R^-(p_-) e_L^+(p_+) \rangle \ . \tag{8.25}$$

We saw in (3.65) that, for a Dirac field, the conserved electromagnetic current is

$$j^\mu = \overline{\Psi}\gamma^\mu\Psi \tag{8.26}$$

Inserting the representation of the Dirac matrices

$$\gamma^0\gamma^\mu = \begin{pmatrix} 0 & 1 \\ 1 & 0 \end{pmatrix} \begin{pmatrix} 0 & \sigma^\mu \\ \overline{\sigma}^\mu & 0 \end{pmatrix} = \begin{pmatrix} \overline{\sigma}^\mu & 0 \\ 0 & \sigma^\mu \end{pmatrix} , \tag{8.27}$$

we find

$$j^\mu = \psi_L^\dagger \overline{\sigma}^\mu \psi_L + \psi_R^\dagger \sigma^\mu \psi_R \ . \tag{8.28}$$

Definition of *helicity conservation*. This special simplification appears specifically for massless fermions interacting through current-current interactions.

Then, also, the current splits into pieces for left- and right-handed electrons. The e_R^- can scatter into an e_R^- or annihilate an e_L^+, but—in the limit of zero electron mass—it cannot turn into an e_L^- or annihilate an e_R^+. These selection rules are called *helicity conservation*. Helicity conservation applies only to the massless limit of the Dirac equation; for a massive fermion with energy E, the amplitude to flip from L to R is proportional to m/E.

We can now evaluate the matrix element in the CM frame

$$e^- \longrightarrow \qquad \longleftarrow e^+ \qquad\qquad \longmapsto \hat{3} \tag{8.29}$$

with electron and positron momenta

$$p_- = (E, 0, 0, E) \qquad p_+ = (E, 0, 0, -E) \tag{8.30}$$

The expression (8.25) becomes

$$\langle 0 |\, \psi_R^\dagger \sigma^\mu \psi_R \, | e_R^-(p_-) e_L^+(p_+) \rangle \ . \tag{8.31}$$

The field ψ_R annihilates the e_R^-, giving a factor $u_R(p_-)$. The field $\psi_R^\dagger$ annihilates the e_L^+, giving a factor $v_L^\dagger(p_+)$. Putting in the explicit values, we find that (8.31) becomes

$$v_L^\dagger(p_+)\sigma^\mu u_R(p_-) = \sqrt{2E}\,(0 \quad 1)\,(1,\vec\sigma)\,\sqrt{2E}\begin{pmatrix}1\\0\end{pmatrix}$$

$$= 2E\,(0,1,+i,0)^\mu\ . \tag{8.32}$$

The result is very attractive. The vector

$$\vec\epsilon_+ = \frac{1}{\sqrt{2}}(\hat 1 + i\hat 2) \tag{8.33}$$

represents angular momentum $J^3 = +1$ along the $\hat 3$ axis. This is the total angular momentum—from the electron and positron spins—entering the reaction.

$$\tag{8.34}$$

The angular momentum is transferred from the e^+e^- system to the virtual photon. Finally, we find

$$\langle 0|\, j^\mu \left| e_R^-(p_-)e_L^+(p_+) \right\rangle = 2E\cdot\sqrt{2}\cdot(0,\vec\epsilon_+)^\mu\ . \tag{8.35}$$

A current has the units of $1/\mathrm{cm}^3$ (for ρ) or $1/\mathrm{cm}^2$ sec (for $\vec j$), both of which are GeV^3 in natural units. The two-particle state with relativistic normalization has the units GeV^{-2}, according to (3.93). Then the matrix element should have units of GeV, and it does.

A similar calculation gives the matrix element for annihilation of $e_L^-e_R^+$. We find

$$\langle 0|\, j^\mu \left| e_L^-(p_-)e_R^+(p_+) \right\rangle = \langle 0|\, \psi_L^\dagger\overline\sigma^\mu\psi_L \left| e_L^-(p_-)e_R^+(p_+) \right\rangle$$

$$= v_R^\dagger(p_+)\overline\sigma^\mu u_L(p_-)$$

$$= \sqrt{2E}\,(1 \quad 0)\,(1,-\vec\sigma)\,\sqrt{2E}\begin{pmatrix}0\\1\end{pmatrix}$$

$$= -2E\,(0,1,-i,0)^\mu\ . \tag{8.36}$$

This gives

$$\langle 0|\, j^\mu \left| e_L^-(p_-)e_R^+(p_+) \right\rangle = -2E\cdot\sqrt{2}\cdot(0,\vec\epsilon_-)^\mu\ , \tag{8.37}$$

where

$$\vec\epsilon_- = \frac{1}{\sqrt{2}}(\hat 1 - i\hat 2) \tag{8.38}$$

is the vector representing the $J^3 = -1$ angular momentum state.

The other two electron helicity combinations, $e_R^-e_R^+$ and $e_L^-e_L^+$, make no contribution to the annihilation rate in the limit of zero electron mass. Similarly, in the muon pair production, only the helicity states $\mu_R^-\mu_L^+$ and $\mu_L^-\mu_R^+$ can appear. The matrix elements for the production of these states by a current are the complex conjugates of the matrix elements computed above, oriented appropriately along the direction of the muon momenta. Let $\vec{\epsilon}\,'_+$ and $\vec{\epsilon}\,'_-$ be the vectors representing angular momentum $+1$ and -1 along this axis. Then

$$\langle \mu_R^-(p'_-)\mu_L^+(p'_+)| \, j^\mu \, |0\rangle = 2E \cdot \sqrt{2} \cdot (0, \vec{\epsilon}\,'^*_+)^\mu$$
$$\langle \mu_L^-(p'_-)\mu_R^+(p'_+)| \, j^\mu \, |0\rangle = -2E \cdot \sqrt{2} \cdot (0, \vec{\epsilon}\,'^*_-)^\mu \, . \qquad (8.39)$$

Now we can assemble the pieces. For the reaction $e_R^- e_L^+ \to \mu_R^- \mu_L^+$, the matrix element (8.3) evaluates to

$$\mathcal{M} = -\frac{e^2}{q^2} \, 2(2E)^2 \, \vec{\epsilon}\,'^*_+ \cdot \vec{\epsilon}_+$$
$$= -2e^2 \vec{\epsilon}\,'^*_+ \cdot \vec{\epsilon}_+ \, . \qquad (8.40)$$

I have used the fact that, in the CM frame, $q = (E_{CM}, \vec{0})$, with $E_{CM} = 2E$. The final result should be dimensionless, as we expect from (7.20). This follows from the observation that we made in (2.34) that the electric charge e is dimensionless in natural units.

So far, we only have abstract expressions for the matrix elements. To make this more concrete, we need to evaluate these expressions for the particular kinematics seen by an experiment. To do this, it is useful first to write out the various momentum vectors and polarization vectors and spinors explicitly in the CM frame.

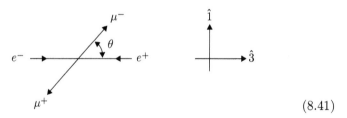

$$(8.41)$$

$$\begin{aligned} p_- &= (E,0,0,E) & p'_- &= (E, E\sin\theta, 0, E\cos\theta) \\ p_+ &= (E,0,0,-E) & p'_+ &= (E, -E\sin\theta, 0, -E\cos\theta) \, , \\ \vec{\epsilon}_\pm &= (1, \pm i, 0)/\sqrt{2} & \vec{\epsilon}\,'_\pm &= (\cos\theta, \pm i, -\sin\theta)/\sqrt{2} \, . \end{aligned} \qquad (8.42)$$

The expressions for $\vec{\epsilon}_\pm$ respect the condition that $\vec{\epsilon}_\pm$ should be orthogonal to the corresponding momentum vector. From these values,

$$\vec{\epsilon}\,'^*_+ \cdot \vec{\epsilon}_+ = \vec{\epsilon}\,'^*_- \cdot \vec{\epsilon}_- = \frac{1}{2}(\cos\theta + 1) \, ,$$

$$\vec{\epsilon}\,'^*_+ \cdot \vec{\epsilon}_- = \vec{\epsilon}\,'^*_- \cdot \vec{\epsilon}_+ = \frac{1}{2}(\cos\theta - 1) \, . \qquad (8.43)$$

Putting the explicit polarization vectors from (8.42) into this equation, we find the four nonzero matrix elements

$$|\mathcal{M}(e_R^- e_L^+ \to \mu_R^- \mu_L^+)|^2 = |\mathcal{M}(e_L^- e_R^+ \to \mu_L^- \mu_R^+)|^2 = e^4(1 + \cos\theta)^2 \, ,$$
$$|\mathcal{M}(e_R^- e_L^+ \to \mu_L^- \mu_R^+)|^2 = |\mathcal{M}(e_L^- e_R^+ \to \mu_R^- \mu_L^+)|^2 = e^4(1 - \cos\theta)^2 \, .$$

$$(8.44)$$

8.4   Evaluation of the cross section for $e^+e^- \to \mu^+\mu^-$

Once we have derived the result (8.44), we can put the expressions for the matrix elements into (7.21) and find the predictions for the $e^+e^- \to \mu^+\mu^-$ cross sections. We can use (7.32) to evaluate the phase space integral.

I will work out the cross sections first for processes in which the leptons have definite polarization. For $e_R^- e_L^+ \to \mu_R^- \mu_L^+$, we find

$$
\begin{aligned}
\sigma &= \frac{1}{2E \cdot 2E \cdot 2} \int d\Pi_2 \, |\mathcal{M}|^2 \\
&= \frac{1}{2E_{CM}^2} \frac{1}{8\pi} \int \frac{d\cos\theta}{2} \, e^4 (1+\cos\theta)^2 .
\end{aligned}
\tag{8.45}
$$

This gives the different cross section for the reaction,

$$
\frac{d\sigma}{d\cos\theta}(e_R^- e_L^+ \to \mu_R^- \mu_L^+) = \frac{\pi\alpha^2}{2E_{CM}^2}(1+\cos\theta)^2 .
\tag{8.46}
$$

The quantity $d\sigma/d\cos\theta$ is a *differential cross section*. It predicts the distribution of events as a function of $\cos\theta$.

Notice that the angular distribution is peaked in the forward direction. The e^+e^- system, which has $J^3 = +1$, transfers its angular momentum to the final state most effectively when the μ_R^- is going forward.

In all, the process $e^+e^- \to \mu^+\mu^-$ has four amplitudes for the various spin states that are permitted by helicity conservation. All of the differential cross sections have the same structure. For $e_R^- e_L^+ \to \mu_R^- \mu_L^+$ and $e_L^- e_R^+ \to \mu_L^- \mu_R^+$,

$$
\frac{d\sigma}{d\cos\theta} = \frac{\pi\alpha^2}{2E_{CM}^2}(1+\cos\theta)^2 ,
\tag{8.47}
$$

and, for $e_R^- e_L^+ \to \mu_L^- \mu_R^+$ and $e_L^- e_R^+ \to \mu_R^- \mu_L^+$,

$$
\frac{d\sigma}{d\cos\theta} = \frac{\pi\alpha^2}{2E_{CM}^2}(1-\cos\theta)^2 .
\tag{8.48}
$$

Typically, to compare predictions from Feynman diagrams to measured cross sections, we must sum over final spin states and average over initial spin states.

It is possible to carefully prepare beams in polarized initial states and to gain information about the the muon polarization by stopping the muons and analyzing their decays. But, typically, high energy beams contain particles with random spin orientations, and the muon polarization is also not observed. To represent this, we average over all possible initial spin states and sum over all possible final states. This gives the final result for the differential cross section

$$
\frac{d\sigma}{d\cos\theta} = \frac{\pi\alpha^2}{2E_{CM}^2}(1+\cos^2\theta) .
\tag{8.49}
$$

The shape of the spin-averaged differential cross section is

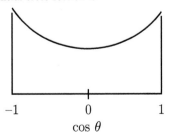

This angular distribution is characteristic of the fact that the muons in the final state have spin $\frac{1}{2}$. In Exercise 8.1, you can compute the angular distribution for e^+e^- to particles of spin 0 and show that the result is qualitatively different from this one.

The integral of the differential cross section over $\cos\theta$ gives the total cross section, which in turn predicts the total rate for muon pair production in e^+e^- annihilation. The result is

$$\sigma = \frac{4\pi\alpha^2}{3E_{CM}^2} \ . \tag{8.50}$$

This result has the units of GeV^{-2}, as expected. No dimensionful parameter appears in the formula except for the center of mass energy, so the cross section must decrease as $1/E_{CM}^2$.

The standard unit used in nuclear and particle physics for expressing cross sections is the *barn*,

$$1 \text{ barn} = 10^{-24} \text{ cm}^2 = 100 \text{ fm}^2 \ . \tag{8.51}$$

This is an area somewhat larger than the cross-sectional area of a large nucleus. The proton-proton scattering cross section at high energies is about 0.1 barn = 100 millibarn (mb). As we will see in Chapter 13, important cross sections at the Large Hadron Collider have the size of nb = 10^{-9} barn, or smaller. The conversion factor from GeV^{-2} to barns is

$$(\hbar c)^2 = 0.389 \text{ GeV}^2 \text{ mb} \ . \tag{8.52}$$

Using this conversion factor, we can write the spin-averaged cross section

The final result for the $e^+e^- \to \mu^+\mu^-$ cross section, in physical units.

for $e^+e^- \to \mu^+\mu^-$ as

$$\sigma(e^+e^- \to \mu^+\mu^-) = \frac{87 \text{ nb}}{E_{CM}^2} \ , \tag{8.53}$$

with E_{CM} given in GeV.

The formulae we have just derived are the leading-order predictions of QED. They do indeed give an accurate description of the rate and angular distribution of the process $e^+e^- \to \mu^+\mu^-$ for energies up to about 30 GeV. Above this energy, effects of the weak interaction must also be included. We will discuss this in Chapter 17.

8.5 e^+e^- annihilation to hadrons

With this well-understood QED process as a reference point, we can now discuss the process of e^+e^- annihilation to hadrons. The main products of this reaction are observed to be π and K mesons. I will consider this process at multi-GeV center of mass energies, energies much higher than the masses of these mesons, and of the related spin 1 mesons ρ, ω, K^*.

The "naive quark model" theory of the cross section for $e^+e^- \to$ hadrons.

The quark model makes a prediction for the cross section for e^+e^- annihilation to hadrons, but it is such a simple one that we are tempted to reject it out of hand. Imagine that we are at center of mass energies at which we can ignore the quark masses. Then, since quarks are spin-$\frac{1}{2}$ particles, the structure of the QED cross section for quark pair production is exactly the same as that in the process of muon pair production that we have just analyzed. If we stop here, we will be ignoring the effects of the strong interaction, which play an essential role in forming the mesons that appear in the final state. However, perhaps this model would be useful as an estimate of the order of magnitude of the cross section or as a reference value.

This model is so simple that we can write the cross section by making just three changes in the calculation of muon pair production. The basic elements of this calculation are unchanged, because we are assuming that quarks are spin $\frac{1}{2}$ particles, and that the energy is high enough that we can ignore their masses. The changes are the following: First, we must sum over the relevant quark species for which we can plausibly ignore the masses at the energy we consider. Second, we need to change the value of the electric charge of the produced particles, from -1 for the muon to $Q_f = +\frac{2}{3}$ for u, c and $Q_f = -\frac{1}{3}$ for d, s, b. In (8.3), the matrix element $\mathcal{M}$ contains one power of the final electric charge, so the cross section is proportional to Q_f^2. Finally, we learned from the structure of baryons that quarks carry a hidden quantum number called *color*, which takes three values. We need to sum over the final color states in computing the total cross section. Thus, our simple model predicts the same angular distribution as before

$$\frac{d\sigma}{d\cos\theta}(e^+e^- \to \text{hadrons}) \sim (1 + \cos^2\theta) , \qquad (8.54)$$

while the total cross section is modified to

$$\sigma(e^+e^- \to \text{hadrons}) = \sum_f 3Q_f^2 \cdot \frac{4\pi\alpha^2}{3E_{CM}^2} , \qquad (8.55)$$

where the sum is taken over $f = u, d, s$ and also c, b if the CM energy is high enough that those quarks can be produced. This expression can also be written in terms of the ratio of the production rates for hadrons and muons, which can be directly measured in the same experiment at any center of mass energy. For the contributions of different sets of quarks, the model gives

$$\frac{\sigma(e^+e^- \to \text{hadrons})}{\sigma(e^+e^- \to \mu^+\mu^-)} = \sum_f 3Q_f^2 = \begin{cases} 2 & u, d, s \\ 3\frac{1}{3} & u, d, s, c \\ 3\frac{2}{3} & u, d, s, c, b \end{cases} \qquad (8.56)$$

The naive quark model gives this very simple prediction.

How well does this oversimplified model work? Figure 8.1 shows the experimental data on the total cross section for the process $e^+e^- \to$ hadrons. The top plot gives the absolute cross section, showing clearly the E_{CM}^{-2} dependence. The bottom plot shows the ratio of the hadronic and $\mu^+\mu^-$ cross sections. The solid green line shows the prediction of the lowest-order theory given above. The horizontal red curves show a more sophisticated theory, to be explained in Chapter 11. The predictions break down in a big way at the energies of the lightest mesons of each new type. In particular, the J/ψ and Υ resonances appear as huge delta functions on this plot. Away from quark thresholds, however, the formula that we have derived works amazingly well. The feature called

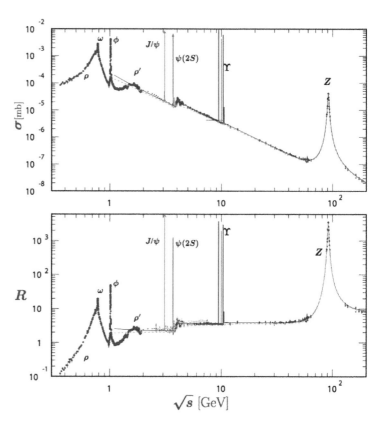

Fig. 8.1 Measurements of the total cross section for e^+e^- annihilation to hadrons as a function of energy, compiled in (Patrignani 2016). The lower figure shows the ratio $R = \sigma(e^+e^- \rightarrow \text{hadrons})/\sigma(e^+e^- \rightarrow \mu^+\mu^-)$. The green dotted curve is the prediction (8.56). The vertical red lines show the ψ and Υ resonances. The horizontal red curve is the prediction (11.72).

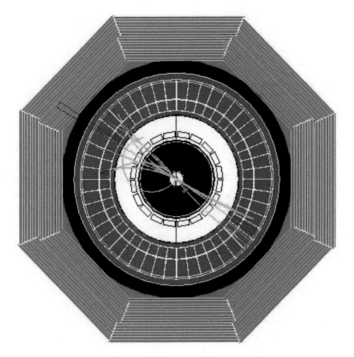

Fig. 8.2 Event display from the SLD experiment showing a typical e^+e^- annihilation to hadrons event at a center of mass energy of 91 GeV (figure courtesy of SLAC and the SLD collaboration).

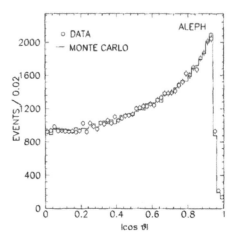

Fig. 8.3 Distribution of the orientations of jet axes in e^+e^- annihilation to hadrons as a function of $|\cos\theta|$, as measured by the ALEPH experiment at the LEP e^+e^- collider, from (Decamp *et al.* (1990)).

In high energy reactions, quarks and antiquarks are seen as *jets*, narrow streams of high-energy hadrons.

Z is due to the weak interaction; we will study this resonance in Chapter 17.

The prediction for the angular distributions can also be tested experimentally. Before considering any method of detailed comparison, we need to ask what $e^+e^- \rightarrow hadrons$ events actually look like at high energies. Figure 8.2 shows a typical event at $E_{CM} = 91$ GeV. The tracks are mostly charged pions and kaons. The tracks clearly form two bundles, with π and K mesons moving in opposite directions. We call such a bundle of hadronic tracks a *jet*. The final states of e^+e^- annihilation to hadrons at high energy typically consist of two back-to-back jets.

It is very tempting to interpret the jets as the observable manifestation of quarks and antiquarks. Quarks are not observed in isolation, only as constituents of hadrons. However, it is not hard to imagine that a high-energy quark might induce the creation of more quark-antiquark pairs and that all of these might reform into pions and other hadrons. In this understanding, the central axes of the jets would be proxies for the original directions of the quarks.

Figure 8.3 shows the orientation of the jet axes in e^+e^- annihilation to hadrons at 91 GeV. It is not easy to tell quark jets from antiquark jets, so the distribution is shown for $|\cos\theta|$. However, the functional form is very close to $(1 + \cos^2\theta)$! Apparently, the overall momentum flow in e^+e^- annihilation events directly reflects the electrodynamic interactions of quarks, and the identification of quarks as spin-$\frac{1}{2}$ particles. There is almost no effect of the strong interactions of quarks on this property of the final state.

How is it possible that the strong interactions can be strong and yet these predictions for hadronic processes can be so accurate? More surprises lie ahead.

Exercises

(8.1) The spectroscopy of mesons and baryons tells us that quarks are spin-$\frac{1}{2}$ particles, but we can also check this from the angular distribution in $e^+e^- \to$ hadrons. To analyze this, consider the alternative hypothesis that quarks are spin 0 particles. Consider their electrodynamic interactions at very high momentum where masses can be neglected.

   (a) The matrix element for the creation of a spin 0 particle of charge 1 and mass m and its antiparticle by the electromagnetic current has the form

$$\langle \phi^-(p_-)\phi^+(p_+) | \, ej^\mu(x) \, | 0 \rangle$$
$$= e \, (p_- - p_+)^\mu e^{+ip_- \cdot x + ip_+ \cdot x} \; . \; (8.57)$$

   To justifiy this, note that the right-hand side of this equation must be a 4-vector built from the boson momenta p_-^μ and p_+^μ. Show, using current conservation ($\partial_\mu j_{EM}^\mu = 0$), that the structure $(p_- + p_+)^\mu$ cannot appear. Note that p_- and p_+ are on shell, i.e., $p_-^2 = p_+^2 = m^2$.

   (b) Draw the Feynman diagram for $e^+e^- \to \phi^+\phi^-$. Write the expression for the matrix element for this process the cases $e_R^- e_L^+$ and $e_L^- e_R^+$. You will need to use (8.57) and the matrix elements of the electromagnetic current between electron states, (8.35) and (8.37).

   (c) Draw a diagram showing the kinematics of the process. Work in the center of mass frame, with the electron and the spin 0 boson having initial energy E. Take the initial electron and positron directions to be along the $\hat{3}$ axis and the final boson directions to be along the vector $\hat{n} = cos\theta \, \hat{3} + \sin \theta \hat{1}$. Write out the four momentum 4-vectors.

   (d) Evaluate the matrix elements from part (b), square them, and compute the differential cross section for $e^+e^- \to \phi^-\phi^+$, averaged over initial spins. Compare to the result (8.49) for production of spin-$\frac{1}{2}$ particles.

   (e) Compute the total cross section for $e^+e^- \to \phi^-\phi^+$. Show that, for $m = 0$, this is $1/4$ of the corresponding result for e^+e^- annihilation to spin-$\frac{1}{2}$ particles.

(8.2) The vector mesons ρ^0, ω, and ϕ can decay to e^+e^- or to $\mu^+\mu^-$. The decay rates to e^+e^- are better known, since these can be measured from the inverse processes $e^+e^- \to \rho^0, \omega, \phi$.

   (a) At the Particle Data Group website (Patrignani *et al.* 2016), look up the total widths of the vector mesons and their branching ratios to e^+e^-. Compute the partial decays widths of the three vector mesons to e^+e^-.

   (b) To understand the relative sizes of these widths, we will need to construct the quark model wavefunctions of the three vector mesons. Here is a mathematical warm-up exercise: For any group G, let $\{|a\rangle\}$ be basis states for a representation, and let $\{|\bar{a}\rangle\}$ be basis states for the complex conjugate representation. The generators of the group act on the states of the representation by

$$T^i \, |a\rangle = (t^i)_{ab} \, |b\rangle \; . \quad (8.58)$$

For example, $SU(2)$ acts on spinors by

$$J^i \, |a\rangle = (\frac{\sigma^i}{2})_{ab} \, |b\rangle \; . \quad (8.59)$$

Then the action on the complex conjugate representation is

$$T^i \, |\bar{a}\rangle = (-t^{iT})_{ab} \, |\bar{b}\rangle \; , \quad (8.60)$$

where T denotes the matrix transpose: $(t^{iT})_{ab} = (t^i)_{ba}$. Verify this by showing that

$$T^i(\sum_a |a\rangle \, |\bar{a}\rangle) = \sum_a (T^i \, |a\rangle) \, |\bar{a}\rangle + |a\rangle \, (T^i \, |\bar{a}\rangle) = 0$$
$$(8.61)$$

Then the state

$$\sum_a |a\bar{a}\rangle \quad (8.62)$$

is invariant under G, as it should be.

   (c) Now write the quark model flavor wavefunctions ρ^0, ω^0, ϕ^0, analogous to the wavefunction

$$|\rho^+\rangle = |u\bar{d}\rangle \; . \quad (8.63)$$

For ϕ, this is easy. For ρ^0 and ω^0, you should write different linear combinations of $|u\bar{u}\rangle$ and

$|d\bar{d}\rangle$. To obtain the correct combinations, you will need to use the fact that ω^0 is an isospin 0 state, while ρ^0 is part of an isospin 1 multiplet.

(d) The matrix element for a vector meson to decay to e^+e^- is proportional to

$$\langle 0|\, j_{EM}^\mu(0)\,|V^0\rangle \;, \tag{8.64}$$

where $V = \rho, \omega, \phi$ and

$$j_{EM}^\mu = \sum_{f=u,d,s} Q_f \bar{\psi}_f \gamma^\mu \psi_f \;. \tag{8.65}$$

Work out the relative size of the matrix elements (8.64) for ρ, ω, ϕ, using the approximation that the three quarks u, d, s have the same masses and strong-interaction dynamics, so that they differ only in their electric charges. Notice that both the u and d quark terms in j_{EM}^μ contribute in the ρ and ω cases, with a different sign for the interference in the two cases.

(e) In this same approximation, find the ratios of the decay rates. Compare to the ratios of the partial widths found in (a).

Deep Inelastic Electron Scattering

9

In the previous chapter, I showed that the main features of the reaction of e^+e^- annihilation to hadrons could be described to quite a good approximation by a naive model in which we ignore the fact that quarks have strong interactions. The discovery that quarks can be described by spin-$\frac{1}{2}$ particles with simple electromagnetic interactions was actually made, not with this process, but in an earlier experiment studying a reaction in which this conclusion was even more surprising.

When electrons are scattered from protons, the simplest reaction that can take place is elastic scattering, $ep \to ep$. As electron scattering is observed with larger transfers of momentum to the proton, elastic collisions become infrequent. Most scattering events break the proton open and produce a large number of hadrons. When the total mass of the hadrons is much larger than the original proton mass, the reaction is refered to as *deep inelastic* electron-proton scattering.

We will see in this chapter that the deep inelastic regime of electron-proton scattering is well described using a picture in which electrons scatter from free quarks inside the proton. If it is surprising that strongly-interacting quarks behave as free particles when they are created out of nothing in e^+e^- annhilation, it is more surprising that it is possible to ignore the strong interaction, to a first approximation, in the scattering of electrons from quarks inside protons.

Deep inelastic electron scattering was first studied in the 1960's, at the SLAC linear electron accelerator. In this chapter, I will describe the results of this experiment, carried out by a SLAC-MIT collaboration (Bloom *et al.* 1969), and its interpretation.

9.1 The SLAC-MIT experiment

The original motivation of the SLAC linear accelerator was to provide very high energy electrons to study the structure of the proton through elastic scattering. In the 1950's, Robert Hofstadter at Stanford studied the elastic scattering process $e^-p \to e^-p$ and similar elastic scattering reactions for nuclei (Hofstadter 1957). He mapped out the size of the

Concepts of Elementary Particle Physics. Michael E. Peskin.
© Michael E. Peskin 2019. Published in 2019 by Oxford University Press.
DOI: 10.1093/oso/9780198812180.001.0001

Fig. 9.1 Layout of the SLAC-MIT deep inelastic scattering experiment (figure courtesy of SLAC). Electrons strike a hydrogen target just under the cylinder on the left of the figure. Scattered electrons, moving left to right, pass through a string of magnets that measure their momenta, and then into the large electromagnetic calorimeter on the right.

proton and the shapes of nuclei. SLAC was built to continue these studies to higher energy and perhaps identify structure within the proton.

Figure 9.1 shows a photograph of the deep inelastic scattering experiment. In this description, "deep" means very large momentum transfer. The cross section for elastic electron scattering from a proton falls off rapidly above 1 GeV momentum transfer, indicating that the smallest structures visible in this reaction are of size $\hbar/1$ GeV or larger. To see down to smaller distances, we must analyze scattering with a momentum transfer above 1 GeV, which would necessarily be inelastic scattering. The idea of the SLAC-MIT experiment was very simple: Bring in an electron beam with as high an energy as possible. Let electrons disrupt protons in a hydrogen target, giving up energy and momentum in the process. Then measure the energy and momentum of the outgoing electron to find the energy-momentum transfer in the reaction.

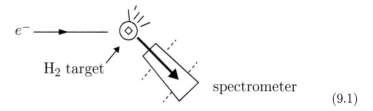

$$(9.1)$$

The odd genius of this experiment was that it ignored the hadronic final state and instead concentrated on measuring the 4-momentum of the outgoing electron with high precision. In Fig. 9.1, the electrons enter from the left. The figure shows the line of magnets used to bend and momentum-analyze the electron. The large orange box on the right is an electromagnetic calorimeter used to discriminate electrons from pions produced in the scattering reactions. The detector was mounted on railroad tracks that allowed it to be swung around to any angle with respect to the beam.

A Feynman diagram for the process of electron scattering has the form

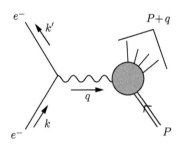

Kinematics of deep inelastic scattering.

$$(9.2)$$

The electron interacts through a simple current matrix element

$$\langle e^-(k')|\, j^\mu\, |e^-(k)\rangle \ . \tag{9.3}$$

The current couples to a virtual photon, which then couples to another current acting on the proton. The current matrix element between the proton and the particular hadronic final states are probably not simple.

Denote the initial electron momentum by k and the final electron momentum by k'. We prepare k and measure k', so we know that the momentum of the virtual photon is

$$q = (k - k') \ . \tag{9.4}$$

The mass W of the final hadronic system is given by

$$W^2 = (P + q)^2 = m_p^2 + 2P \cdot q + q^2 \ . \tag{9.5}$$

In my discussion here, I will use the simplifying approximation that the energy transfer in the scattering process is much larger than the mass of the proton, so that we can ignore both the electron and proton masses. For a scattering process, q is spacelike, that is, there is a frame where the energy transfer is zero and only momentum is transfered. It is convenient to write

$$q^2 = -Q^2 \ . \tag{9.6}$$

Definition of Q^2 for deep inelastic scattering.

Large Q^2 indicates large momentum transfer to the proton.

The cross sections as a function of W for increasing values of Q^2 are shown in Fig. 9.2. As W increases from left to right in each plot, we see the Δ, N^*, *etc.*, baryon resonances. However, at large Q^2, the

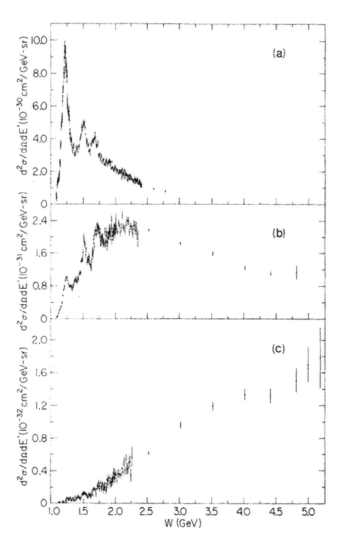

Fig. 9.2 Cross section for deep inelastic *ep* scattering as a function of the final hadronic mass W, measured by the SLAC-MIT experiment, at low, medium, and high values of Q^2, from (Bloom *et al.* 1969).

resonances become less visible over a smooth continuum rising with W.

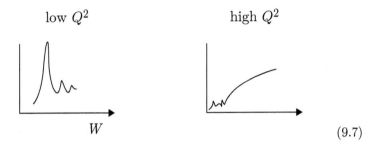

low Q^2 high Q^2

W (9.7)

The whole collection of data is quite complex, so it was a challenge to understand how to interpret it.

9.2   The parton model

The crucial clue for understanding the deep inelastic scattering data came from an important observation by James Bjorken that I will describe in Section 9.5. Feynman was then able to describe deep inelastic scattering using a simple picture based on free quarks and antiquarks that he called the *parton model*. In this section, I will describe the model; in the remainder of this chapter, we will work out its predictions for deep inelastic scattering and compare those to data.

At very high energy, we may analyze the $e^- p$ scattering reaction from the CM frame.

$$e^- \longrightarrow \qquad \longleftarrow P$$
(9.8)

Feynman modeled the proton as a collection of constituents, called *partons*. Some of these partons might be the quarks, which we already expect are constituents of the proton. At high energy, all partons are moving approximately in the direction of the proton. That is, all partons have a large component of momentum along the direction of the proton, while their momenta transverse to the proton direction remain of the order of the momenta within the proton bound state. In the simple parton model, we ignore these transverse momentum components, and the masses of the partons. We might expect that these approximations would be good for very high energy scattering processes. Then the momentum vector of a parton can be written

Statement of the parton model description of the proton wavefunction.

$$p^\mu = \xi \, P^\mu \, ,$$
(9.9)

where P is the total energy-momentum of the proton and ξ is the fraction of this energy-momentum carried by that parton. The parameter ξ runs over the values

$$0 < \xi < 1 \, .$$
(9.10)

Let $f_i(\xi)d\xi$ be the probability of finding a parton of type i carrying the momentum fraction ξ. In the following, I will assume that the partons that scatter electromagnetically are quarks and antiquarks. I will denote these by a label f (the flavor) for quarks and $\bar{f}$ for antiquarks. There might be additional partons that do not have electric charge. The whole set of partons carry the total energy-momentum of the proton. This implies the sum rule

$$\int_0^1 d\xi \; \sum_i f_i(\xi) \cdot \xi = 1 \; . \tag{9.11}$$

In the parton model, deep inelastic scattering is described by the Feynman diagram

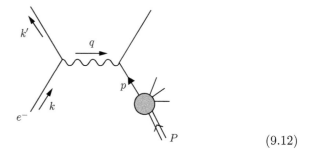

$$\tag{9.12}$$

We take each quark or antiquark in the proton and consider it to scatter from the electron as a pointlike spin-$\frac{1}{2}$ particle. As in e^+e^- annihilation, the outgoing quark cannot be seen in isolation. Rather, it must turn into a jet of hadrons through processes that involve the strong interactions in a nontrivial way. Here again, we will ignore the effects of the strong interactions when we compute the cross section. We will interpret the parton model cross section as giving the sum of the cross sections for all possible hadronic final states. The parton model cross section is written

The cross section for deep inelastic electron scattering according to the parton model.

$$\sigma(e^-p \to e^-X) = \int d\xi \sum_f [f_f(\xi) + f_{\bar{f}}(\xi)] \; \sigma(e^-q(\xi P) \to e^-q) \; . \tag{9.13}$$

The symbol X stands for any collection of hadrons in the final state.

9.3   Crossing symmetry

To compute the cross section required for (9.13), we need to evaluate the matrix elements for electron-quark scattering, a process described by the Feynman diagram

$$\tag{9.14}$$

The form of this diagram is similar to that of (8.1), and so we can immediately write down an expression for the corresponding matrix element,

$$\mathcal{M}(e^- q_f \to e^- q_f) = (-e) \langle e^- | j^\mu | e^- \rangle \frac{1}{q^2} (Q_f e) \langle q_f | j_\mu | q_f \rangle . \quad (9.15)$$

It is straightforward to evaluate this matrix element explicitly using the methods described in Section 8.3. I describe this method in Exercise 9.1. However, there is a much easier way to determine the value of the matrix element. This method requires explanation of a new concept, called *crossing symmetry*. This concept ties to important general properties of scattering matrix elements, so it will be worth a detour to explain it.

To begin, compare the diagram in (9.14) with the diagram for $e^+ e^- \to q\bar{q}$ computed in the previous chapter

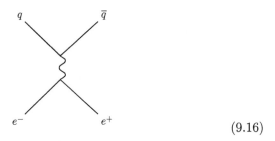

$$(9.16)$$

The two Feynman diagrams actually show the same process, laid out in different ways in space-time. In eq scattering, there is an electron in the final state and a quark in the initial state. In e^+e^- annihilation, the final electron is exchanged for the antiparticle of the electron, a positron, in the initial state, and the initial quark is exchanged for a final-state antiquark. The situations with a final electron and an initial positron, and that with a final quark and an initial antiquark, are strongly related, because the same quantum field that creates the electron destroys the positron, and similarly for a quark and antiquark. This translates into the simplest possible relation of the corresponding matrix elements: The matrix elements have the same functional form with appropriate identification of the external momenta. This relation of processes is called *crossing symmetry*. It is another theorem of quantum field theory that processes related by crossing symmetry are described by the same function of the external momenta. Thus, given our results for the $e^+e^- \to q\bar{q}$ matrix elements from the previous chapter, we are able to write down the matrix elements for $e^- q \to e^- q$ without further calculation.

To use crossing symmetry most easily, it is useful to introduce a standard notation for the kinematic invariants of 2-body scattering processes. Consider a general 2-particle scattering process $1 + 2 \to 3 + 4$. To write maximally symmetric expressions, I will write all momenta as directed

Crossing symmetry relates matrix elements for the reaction with outgoing particles to those with incoming antiparticles. This relation is true quite generally for matrix elements in relativistic quantum field theory.

outward,

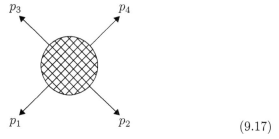

$$(9.17)$$

For the reaction $1 + 2 \to 3 + 4$, we will have $p_3^0, p_4^0 > 0$ and $p_1^0, p_2^0 < 0$. Negative energy here means that the particles are annihilated rather than created. The same amplitude, evaluated for $p_1^0 > 0, p_3^0 < 0$ will describe the reaction with the antiparticle of 3 in the initial state and the antiparticle of 1 in the final state. Energy-momentum conservation in the reaction implies

$$p_1 + p_2 + p_3 + p_4 = 0 \ . \tag{9.18}$$

The matrix element $\mathcal{M}$ can depend only on Lorentz-invariant combinations of the momenta. I will now count and classify these.

First of all, the squares of the 4-vectors are Lorentz-invariant. But these are fixed to the masses of the particles

$$p_1^2 = m_1^2 \ , \quad p_2^2 = m_2^2 \ , \quad p_3^2 = m_3^2 \ , \quad p_4^2 = m_4^2 \ . \tag{9.19}$$

The remaining Lorentz invariants have the form $p_i \cdot p_j$. To express these, we define the *Mandelstam invariants*,

Definition of the *Mandelstam invariants s, t, u* for a 2-body scattering reaction.

$$\begin{aligned} s &= (p_1 + p_2)^2 = (p_3 + p_4)^2 \\ t &= (p_1 + p_3)^2 = (p_2 + p_4)^2 \\ u &= (p_1 + p_4)^2 = (p_2 + p_3)^2 \ . \end{aligned} \tag{9.20}$$

Each variable has two definitions, related by (9.18). This implies that the six products $p_i \cdot p_j$ actually reduce to three; for example,

$$2p_1 \cdot p_2 + m_1^2 + m_2^2 = 2p_3 \cdot p_4 + m_3^2 + m_4^2 \ . \tag{9.21}$$

There is one further relation. When we add up the three invariants,

$$\begin{aligned} s + t + u = \frac{1}{2} \Big[&p_1^2 + 2p_1 p_2 + p_2^2 + p_3^2 + 2p_3 p_4 + p_4^2 \\ &p_1^2 + 2p_1 p_3 + p_3^2 + p_2^2 + 2p_2 p_4 + p_4^2 \\ &p_1^2 + 2p_1 p_4 + p_4^2 + p_2^2 + 2p_2 p_3 + p_3^2 \Big] \ , \end{aligned} \tag{9.22}$$

and gather up the terms in the square of $p_1 + p_2 + p_3 + p_4$,

$$s + t + u = \frac{1}{2} \Big[(p_1 + p_2 + p_3 + p_4)^2 + 2p_1^2 + 2p_2^2 + 2p_3^2 + 2p_4^2 \Big] \ . \tag{9.23}$$

we find, using (9.19),

$$s + t + u = m_1^2 + m_2^2 + m_3^2 + m_4^2 \ . \tag{9.24}$$

An important identity linking s, t, and u.

So, finally, there are only two independent Lorentz invariants, specified by any two of s, t, u. This is a general result for any 2-particle scattering process.

To understand s, t, and u better, we can evaluate them for the scattering of massless particles in the CM frame. The four momenta are

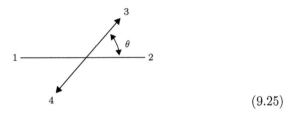

$$\tag{9.25}$$

$$
\begin{aligned}
p_1 &= (-E, 0, 0, -E) & p_3 &= (E, E\sin\theta, 0, E\cos\theta) \\
p_2 &= (-E, 0, 0, E) & p_4 &= (E, -E\sin\theta, 0, -E\cos\theta) \quad (9.26)
\end{aligned}
$$

Note that I am still writing the negative of the momentum for the initial state particles. We see that

$$s = (2E)^2 = E_{CM}^2 \ . \tag{9.27}$$

In the rest of this book, I will often write $\sqrt{s}$ for the center of mass energy E_{CM}.

In fact, even for general masses, $s = (p_1+p_2)^2 = E_{CM}^2$. It is conventional in particle physics to write the center of mass energy of any reaction as $\sqrt{s}$. For massless particles, t and u also have simple expressions.

$$
\begin{aligned}
t = (p_1 + p_3)^2 &= (0, E\sin\theta, 0, E(\cos\theta - 1))^2 \\
&= -E^2(\sin^2\theta + 1 - 2\cos\theta + \cos^2\theta) \ . \quad (9.28)
\end{aligned}
$$

Then

$$t = -2E^2(1 - \cos\theta) \ , \tag{9.29}$$

and, similarly

$$u = -2E^2(1 + \cos\theta) \ . \tag{9.30}$$

Note that the relation $s + t + u = \sum_i m_i^2 = 0$ is satisfied. The two independent variables represented by s, t, u correspond to the CM energy and the CM scattering angle.

An easy way to implement crossing symmetry is to permute the three invariants s, t, and u as the legs of the diagram are switched between the initial and the final state.

In Chapter 7, and again in Chapter 8, I argued that we could represent an intermediate state in a Feynman diagram with a Breit-Wigner denominator

The quantities s, t, u provide a Lorentz-invariant way to parametrize the two key variables of a scattering process—the center of mass energy and the scattering angle.

$$(p_1+p_2)\ \frac{1}{(p_1 + p_2)^2 - m_R^2 + im_R\Gamma_R} \ . \tag{9.31}$$

Feynman diagram for an *s*-channel process:

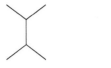

Feynman diagram for an *t*-channel process:

Feynman diagram for an *u*-channel process:

When the intermediate state separates the initial and the final state, the denominator depends on $(p_1 + p_2)^2 = s$:

$$\frac{1}{s - m_R^2 + im_R\Gamma_R} \, . \tag{9.32}$$

We call this type of reaction an *s-channel process*. Crossing symmetry relates this amplitude to other processes in which the virtual particle exchange appears in other configurations. If the resonance amplitude depends on t,

$$\frac{1}{t - m_R^2 + im_R\Gamma_R} \, , \tag{9.33}$$

we have a *t-channel process*. Similarly, when the resonance is a function of u,

$$\frac{1}{u - m_R^2 + im_R\Gamma_R} \, . \tag{9.34}$$

we have a *u-channel process*.

For each type of process, the channel determines the qualitative form of the expression for the scattering cross section. As a simple example, consider the implications for massless particles interacting through a virtual photon exchange ($m_R = 0$). The *s*-channel diagram leads to the term

$$\left| \begin{array}{c} q \updownarrow \end{array} \right|^2 \sim \left| \frac{1}{q^2} \right|^2 \sim \frac{1}{s^2} = \frac{1}{E_{CM}^4} \, . \tag{9.35}$$

The dependence of the cross section on $\cos\theta$ comes only from the numerator terms, as in e^+e^- annihilation. The *t*-channel diagram leads to

$$\left| \begin{array}{c} q \end{array} \right|^2 \sim \frac{1}{t^2} = \frac{4}{E_{CM}^4(1 - \cos\theta)^2} = \frac{1}{E_{CM}^4 \sin^4\theta/2} \, . \tag{9.36}$$

This expression is strongly peaked in the forward direction. You might recognize this factor as the forward peak in the differential cross section for Coulomb scattering. The *u*-channel diagram leads to

$$\left| \begin{array}{c} q \end{array} \right|^2 \sim \frac{1}{u^2} = \frac{4}{E_{CM}^4(1 + \cos\theta)^2} \, , \tag{9.37}$$

which has a strong peak for backward scattering. It is illuminating, and very pleasing, that we can infer the qualitative angular distribution of the elementary particle reaction simply by looking at the form of the corresponding Feynman diagram.

9.4 Cross section for electron-quark scattering

Crossing symmetry allows us to convert the calculations we did in the previous chapter for e^+e^- annihilation into calculations of the invariant amplitudes for electron-quark scattering. In Chapter 8, using in particular (8.44), we derived the results

$$|\mathcal{M}(e_R^- e_L^+ \to q_R \bar{q}_L)|^2 = |\mathcal{M}(e_L^- e_R^+ \to q_L \bar{q}_R)|^2 = Q_f^2 e^4 (1 + \cos\theta)^2 \, ,$$
$$|\mathcal{M}(e_R^- e_L^+ \to q_L \bar{q}_R)|^2 = |\mathcal{M}(e_L^- e_R^+ \to q_R \bar{q}_L)|^2 = Q_f^2 e^4 (1 - \cos\theta)^2 \, ,$$
$$(9.38)$$

where Q_f is the electric charge of the quark in question. Using (9.27), (9.29), and (9.30), we can write these expressions in a Lorentz invariant form as

$$|\mathcal{M}(e_R^- e_L^+ \to q_R \bar{q}_L)|^2 = |\mathcal{M}(e_L^- e_R^+ \to q_L \bar{q}_R)|^2 = 4Q_f^2 e^4 \frac{u^2}{s^2} \, ,$$

$$|\mathcal{M}(e_R^- e_L^+ \to q_L \bar{q}_R)|^2 = |\mathcal{M}(e_L^- e_R^+ \to q_R \bar{q}_L)|^2 = 4Q_f^2 e^4 \frac{t^2}{s^2} \, . \quad (9.39)$$

These expressions are correct in any frame. And, in addition, they yield the expressions for the crossed amplitudes after an appropriate permutation of variables. For example, consider the crossing

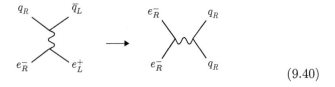

$$(9.40)$$

The eq scattering diagram on the right is obtained by moving the final antiquark $\bar{q}_L$ to the initial state, where it becomes the quark q_R, and moving the initial positron e_L^+ to the final state, where it becomes the electron e_R^-. Note that the final process continues to respect helicity conservation.

The interchange of momenta is

$$p_1 \to p_1 \quad p_2 \to p_3 \quad p_3 \to p_4 \quad p_4 \to p_2 \, . \quad (9.41)$$

This interchanges

$$s \to t \quad t \to u \quad u \to s \, . \quad (9.42)$$

The matrix element for $e_R^- q_R \to e_R^- q_R$ is then given by

$$|\mathcal{M}(e_R^- q_R \to e_R^- q_R)|^2 = 4Q_f^2 e^4 \frac{s^2}{t^2} \, . \quad (9.43)$$

This calculation demonstrates the use of crossing symmetry to calculate one matrix element from another known matrix element.

Similarly, the crossing

$$\tag{9.44}$$

produces

$$|\mathcal{M}(e_R^-\bar{q}_L \to e_R^-\bar{q}_L)|^2 = 4Q_f^2 e^4 \frac{u^2}{t^2} \ . \tag{9.45}$$

The scattering amplitude for the polarized reaction $e_R^-q_L \to e_R^-q_L$ has a zero required by angular momentum conservation.

Notice that this matrix element is proportional to

$$u^2 \sim (1 + \cos\theta)^2 \tag{9.46}$$

and vanishes for backward scattering, $\cos\theta = -1$. If we look at the flow of spin angular momentum,

$$\tag{9.47}$$

we see that, in this case, backward scattering is forbidden by angular momentum conservation. The matrix elements for the other helicity combinations allowed by helicity conservation can be obtained in the same way,

$$|\mathcal{M}(e_L^-q_L \to e_L^-q_L)|^2 = 4Q_f^2 e^4 \frac{s^2}{t^2} \ ,$$

$$|\mathcal{M}(e_L^-q_R \to e_L^-q_R)|^2 = 4Q_f^2 e^4 \frac{u^2}{t^2} \ . \tag{9.48}$$

We can now assemble the cross section for eq scattering. Averaging over the spins in the initial state and summing over the spins in the final state, the cross section is given by

$$\sigma(eq \to eq) = \frac{1}{2E\,2E\,2}\frac{1}{8\pi}\int \frac{d\cos\theta}{2}\frac{1}{4}\sum_{\text{spins}}|\mathcal{M}(e^-q \to e^-q)|^2. \tag{9.49}$$

Note that there is no color factor of 3 in this equation. Whatever color the quark has in the initial state, that color is passed to the quark in the final state.

Summing over the matrix elements for the allowed processes, we find

$$\frac{d\sigma}{d\cos\theta} = \frac{1}{2s}\pi\alpha^2\frac{2}{4}\left(4Q_f^2\frac{s^2+u^2}{t^2}\right) \ , \tag{9.50}$$

or

$$\frac{d\sigma}{d\cos\theta} = \frac{\pi Q_f^2\alpha^2}{s}\frac{s^2+u^2}{t^2} \ . \tag{9.51}$$

We can write this result completely invariantly by using (9.29) to replace the integral over $\cos\theta$,

$$dt = \frac{1}{2}s \, d\cos\theta \ . \tag{9.52}$$

Then

$$\frac{d\sigma}{dt}(eq \to eq) = \frac{2\pi Q_f^2 \alpha^2}{s^2} \frac{s^2 + u^2}{t^2} \ . \tag{9.53}$$

The final expression for the differential cross section for electron-quark scattering.

9.5   The cross section for deep inelastic scattering

Using the formula (9.13) together with (9.53), we obtain the parton model prediction for the deep inelastic scattering cross section

$$\sigma(e^- p \to e^- X) = \int d\xi \int d\hat{t} \sum_f [f_f(\xi) + f_{\bar{f}}(\xi)] \frac{2\pi Q_f^2 \alpha^2}{\hat{s}^2} \left(\frac{\hat{s}^2 + \hat{u}^2}{\hat{t}^2} \right) \ . \tag{9.54}$$

In this formula, I use $\hat{s}$, $\hat{t}$, $\hat{u}$ to denote the invariants for the electron-parton scattering process, reserving the symbols without hats for the full electron-proton scattering reaction.

In a hadron reaction described by the parton model, I will denote the parton-level kinematic invariants by $\hat{s}$, $\hat{t}$, $\hat{u}$.

It is not so obvious how to interpret this formula, since it is not clear how to measure the parton-level invariants. However, it is a beautiful feature of deep inelastic scattering that each of the parton-level invariants has a precise physical interpretation. We will now work these out.

First of all, $\hat{t} = q^2 = -Q^2$. I have already pointed out that this quantity is directly measured in the deep inelastic scattering experiment.

Next, compare s for the full $e^- p$ reaction

$$s = (k + P)^2 = 2k \cdot P \tag{9.55}$$

and for the parton reaction

$$\hat{s} = (k + p)^2 = 2k \cdot p = 2k \cdot \xi P \ . \tag{9.56}$$

We see that

$$\hat{s} = \xi s \ . \tag{9.57}$$

It is useful to define

$$y = \frac{2P \cdot q}{2P \cdot k} \ . \tag{9.58}$$

Definition of the deep inelastic scattering variable y.

In the proton rest frame, this is

$$y = \frac{q^0}{k^0} \ . \tag{9.59}$$

That is, y is the fraction of the initial electron energy that is transfered to the proton. This implies that

$$0 < y < 1 \ . \tag{9.60}$$

We can equally well evaluate

$$y = \frac{2\xi P \cdot q}{2\xi P \cdot k} = \frac{2p \cdot (k - k')}{2p \cdot k} = \frac{\hat{s} + \hat{u}}{\hat{s}} \ . \tag{9.61}$$

Then

$$\frac{\hat{u}}{\hat{s}} = -(1 - y) \ , \tag{9.62}$$

or

$$\hat{s}^2 + \hat{u}^2 = \hat{s}^2(1 + (1 - y)^2) \ . \tag{9.63}$$

At this point, we have expressed

$$\sigma(e^- p \to e^- X) = \int d\xi \int dQ^2 \sum_f [f_f(\xi) + f_{\bar{f}}(\xi)] \, 2\pi Q_f^2 \alpha^2 \left(\frac{1 + (1 - y)^2}{Q^4} \right) \ . \tag{9.64}$$

There is one more important kinematic relation. In the parton model, we assumed that the quark is a free pointlike Dirac particle and that the electron-quark scattering is elastic

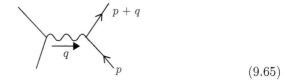

$$ \tag{9.65}$$

If the final quark is treated as massless, then

$$0 = (p + q)^2 = 2p \cdot q + q^2 = 2\xi P \cdot q - Q^2 \ . \tag{9.66}$$

Definition of the deep inelastic scattering variable x.

Thus, the parameter ξ becomes identified with an observable combination of momenta

$$x = \frac{Q^2}{2P \cdot q} \ . \tag{9.67}$$

This is quite amazing. In the parton model, a deep inelastic scatter at a fixed value of x is due to an initial parton carrying the fraction x of the initial proton momentum. By measuring x, we sample the momentum distribution of quarks in the proton wavefunction.

Finally, using (9.55) and (9.58), we see that

$$Q^2 = xys \ . \tag{9.68}$$

Then, with x fixed,

$$d\hat{t} = dQ^2 = xs \ dy \ . \tag{9.69}$$

This gives as our final formula for the deep inelastic scattering cross section

The final parton model formula for the cross section for deep inelastic electron scattering.

$$\frac{d^2\sigma}{dx dy}(e^- p \to e^- X) = \sum_f xQ_f^2[f_f(x) + f_{\bar{f}}(x)] \cdot \frac{2\pi\alpha^2 s}{Q^4}(1 + (1 - y)^2) \ . \tag{9.70}$$

Notice that both of the kinematic variables used here range over the interval $0 < x, y < 1$.

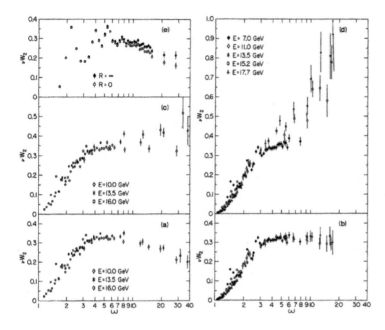

Fig. 9.3 Measurements of the quantity F_2, defined by (9.71), by the SLAC-MIT experiment, at different energy and angle settings, plotted as a function of $\omega = 1/x$, from (Breidenbach *et al.* 1969).

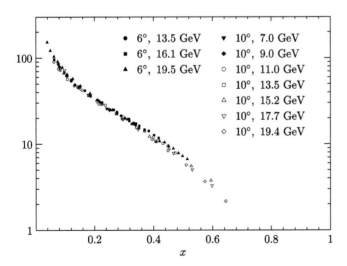

Fig. 9.4 Measurements of the quantity F_2 by the SLAC-MIT experiment, at energy and angle settings giving $Q^2 > 1$ GeV^2, plotted as a function of x, from (Peskin and Schroeder 1995).

9.6   Bjorken scaling

It is conventional to write the expression for the deep inelastic cross section as the product of an elementary QED cross section and an unknown *form factor* F_2 that contains the information about the proton structure,

$$\frac{d^2\sigma}{dxdy}(e^-p \rightarrow e^-X) = F_2 \cdot \frac{2\pi\alpha^2 s}{Q^4}(1 + (1-y)^2) . \tag{9.71}$$

In principle, the factor F_2 could depend on the general kinematics of the problem; that is, it could be a general function of x and Q^2. However, comparing (9.70) and (9.71), we see that the parton model prediction for F_2 is

$$F_2(x) = \sum_f Q_f^2 x[f_f(x) + f_{\bar{f}}(x)] . \tag{9.72}$$

Definition of *Bjorken scaling*.

It is striking that the predicted form depends only on x and is independent of Q^2. This behavior is called *Bjorken scaling*. Bjorken predicted this simple dependence based on more advanced hypotheses about the behavior of current matrix elements at high energy (Bjorken 1966). Bjorken encouraged the experimenters to plot the data shown in Fig. 9.2 as a function of x, or, rather, $\omega = 1/x$. The result is shown in Fig. 9.3. The deep inelastic cross sections from many settings of the beam energy and scattering angle come together into a single function of ω. Figure 9.4 shows the plot of F_2 versus x for the events with $Q^2 > 1$ GeV^2. All of the data falls on a single curve as a function of x!

Over the past decades, F_2 has been measured repeatedly at higher energies, using muons and neutrinos produced by proton beams of hundreds of GeV. Most recently, F_2 has been measured at the HERA colliding beam facility at the German high-energy physics laboratory DESY, which collided 820 GeV protons with 30 GeV electrons. The full world data set, collected by the Particle Data Group (Patrignani *et al.* 2016), is shown in Fig. 9.5.   Each row of points shows the value of F_2 at a

In fact, F_2 is only approximately independent of Q^2, varying slowly on a logarithmic scale.

different value of Q^2. There is a dependence on Q^2, but it is very slow, evolving on a logarithmic scale, so Bjorken scaling is still correct as a first approximation. F_2 decreases for large values of x and increases for small values of x. This slow evolution of F_2 with Q^2 requires an explanation that goes beyond the simple parton model. I will discuss the physical origin of this behavior in Chapter 12.

Figure 9.6 shows an event display from a typical deep inelastic scattering event at $Q^2 = (100 \text{ GeV})^2$, from the ZEUS experiment at the high energy electron-proton collider HERA at DESY. The electrons enter from the left and the protons from the right. We see the final electron scattered toward the upper left, shown as one track plus energy in the electromagnetic calorimeter. Going downward, there is a jet with four high energy hadronic tracks plus energy in the electromagnetic and hadron calorimeters. The calorimeter hits on the left show the energetic hadrons from the remnants of the proton left after one quark is ejected.

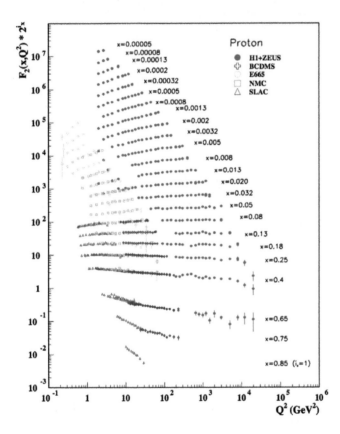

Fig. 9.5 Measurements of the quantity F_2 at increasing values of x as a function of Q^2, compiled by the Particle Data Group (Patrignani *et al.* 2016).

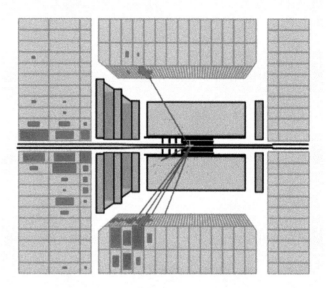

Fig. 9.6 Event display for an *ep* scattering event at $Q^2 = 5800$ GeV^2, from the ZEUS experiment at the *ep* collider HERA (figure courtesy of DESY and the ZEUS collaboration). Electrons entering from the left collide with a protons entering from the right. The single track going upward in the figure, associated with energy in the electromagnetic calorimeter, is the scattered electron. The tracks moving downward from the collision point form a jet of hadrons.

This precisely visualizes the parton model Feynman diagram that we drew at the beginning of the chapter.

Exercises

(9.1) In Section 9.3, we derived expressions for the electron-quark scattering amplitudes using results from $e^+e^- \to q\bar{q}$ plus crossing symmetry. We found

$$|\mathcal{M}(e_R^- q_R \to e_R^- q_R)|^2 = 4Q_f^2 e^4 s^2/t^2 ,$$
$$|\mathcal{M}(e_R^- q_L \to e_R^- q_L)|^2 = 4Q_f^2 e^4 u^2/t^2 . \quad (9.73)$$

Check these results by deriving them directly. Treat both the electron and the quark as massless fermions.

(a) Draw the Feynman diagram for electron-quark scattering and argue that

$$\mathcal{M} = Q_f e^2 \left\langle e^-(p') \middle| \overline{\psi}_e \gamma^\mu \psi_e \middle| e^-(p) \right\rangle$$
$$\cdot \frac{1}{t} \left\langle q(k') \middle| \overline{\psi}_q \gamma_\mu \psi_q \middle| q(k) \right\rangle , \quad (9.74)$$

where p, p' are the initial and final electron momenta and k, k' are the initial and final quark momenta.

(b) Draw a diagram showing the kinematics of the process. Work in the center of mass frame, with the electron and the quark having initial energy E. Take the initial electron and quark directions to be along the $\hat{3}$ axis and the final electron and quark directions to be along the vector $\hat{n} = cos\theta \, \hat{3} + \sin\theta \, \hat{1}$. Write out the four momentum 4-vectors. Write the values of s, t, and u.

(c) Show that the spinors with spin up and down

along the direction $\hat{n}$ are

$$\xi_+ = \begin{pmatrix} \cos\theta/2 \\ \sin\theta/2 \end{pmatrix}, \qquad \xi_- = \begin{pmatrix} -\sin\theta/2 \\ \cos\theta/2 \end{pmatrix}.$$
(9.75)

(d) Construct the initial and final spinors $u(p)$, $u(k)$, $u(p')$, $u(k')$ for the electron-quark scattering reaction, both for $e_R^- q_R \to e_R^- q_R$ and for $e_R^- q_L \to e_R^- q_L$.

(e) Compute the matrix elements

$$\langle e^-(p') | \bar{\psi}_e \gamma^\mu \psi_e | e^-(p) \rangle$$
(9.76)

for the cases of right- and left-handed electrons. There are 4 cases, but 2 are zero.

(f) Compute the matrix elements

$$\langle q(k') | \bar{\psi}_q \gamma^\mu \psi_q | q(k) \rangle$$
(9.77)

for the two nonzero cases.

(g) Compute the matrix elements in (9.73) and verify the results given.

(9.2) Consider the deep inelastic scattering of polarized electrons on polarized protons. There are four independent different possible initial states: $e_L^- p_L$, $e_L^- p_R$, $e_R^- p_L$, and $e_R^- p_R$. Analyze these cases in the parton model. Ignore antiquarks and the heavier quarks s, c, ... in the proton wavefunction. Then the proton state p_R is described by four parton distribution functions:

$$f_{uR}(x), \quad f_{uL}(x), \quad f_{dR}(x), \quad f_{dL}(x),$$
(9.78)

corresponding to quark partons with flavor u, d and with spin parallel or antiparallel to the spin of the proton.

(a) Derive expressions, within the parton model, for the cross sections

$$\frac{d^2\sigma}{dxdy}(e_R^- p_R) \quad \text{and} \quad \frac{d^2\sigma}{dxdy}(e_L^- p_R)$$
(9.79)

(b) The cross sections for e scattering from p_L are related to these by parity. Write the pdfs for p_L in terms of the spin-dependent pdfs for p_R defined in (9.78). Compute the deep inelastic scattering cross sections for e_L^- and e_R^- scattering on a p_L, the analogs of (9.79). Check that the average over all initial state spins gives the expression (9.70) for the unpolarized deep inelastic scattering cross section.

(c) Show that, because u and d quarks in the proton are antisymmetrized in color, the only spin 0 state that can be build from these quarks is a ud state with total $I = 0$ and $S = 0$. Then, when one quark carries most of the momentum of the proton, the proton wavefunction is likely to be described by this ud state at low momentum, plus an energetic u quark with its spin parallel to the spin of the proton. What predictions does this model make for the limiting forms of the cross sections computed in (a) as $x \to 1$?

The Gluon

10

At the end of the previous chapter, we saw that e^-p deep inelastic scattering allows us to meaure a quantity $F_2(x)$, interpreted as a sum over parton distributions for quarks and antiquarks in the proton. In this interpretation, x is the fraction of the momentum of a proton carried by a quark and $f_f(x)$, $f_{\bar{f}}(x)$ are the parton distribution functions, the probability distributions of quarks and antiquarks of flavor f in the proton as a function of x. Then our simple model for the deep inelastic scattering cross section gave

$$F_2(x) = \sum_f Q_f^2 x [f_f(x) + f_{\bar{f}}(x)] \tag{10.1}$$

For the rest of this book, I will represent the proton as bag of quarks, antiquarks, and gluons, each governed by its parton distribution function (pdf).

In this chapter, I will describe additional data on F_2 that makes this picture more concrete, and other measurements that reveal an additional parton component of the proton.

From here on, I will refer to parton distribution functions, for brevity, as *pdfs*. Any hadron will have a set of pdfs describing its wavefunction in terms of quarks and antiquarks. However, when I write pdfs without any further labels, I will be referring specifically to those of the proton.

10.1 Measurement of parton distribution functions

In the quark model, we would expect the major contributions to $F_2(x)$ to be those from the two u quarks and one d quark in the proton wavefunction. At this level, the formula (10.1) would read

$$F_2(x) = \frac{4}{9} x f_u(x) + \frac{1}{9} x f_d(x) \ . \tag{10.2}$$

These three quarks account for the proton electric charge and isospin quantum numbers. Any additional quarks in the proton must appear as quark-antiquark pairs. In quantum field theory, there are processes that create a quark-antiquark pair of any flavor, so we expect nonzero values for all of the possible pdfs

$$f_u(x), f_d(x), f_s(x), \cdots \qquad f_{\bar{u}}(x), f_{\bar{d}}(x), f_{\bar{s}}(x), \cdots \tag{10.3}$$

Concepts of Elementary Particle Physics. Michael E. Peskin.
© Michael E. Peskin 2019. Published in 2019 by Oxford University Press.
DOI: 10.1093/oso/9780198812180.001.0001

To give the correct quantum numbers $Q_p = +1$, $I_p^3 = +\frac{1}{2}$, $S = 0$, etc., the pdfs must satisfy the sum rules

$$\int_0^1 dx[f_u(x) - f_{\bar{u}}(x)] = 2$$

$$\int_0^1 dx[f_d(x) - f_{\bar{d}}(x)] = 1$$

$$\int_0^1 dx[f_s(x) - f_{\bar{s}}(x)] = 0 , \qquad (10.4)$$

and similarly for the distributions of c, b, t. Notice that parton distributions are not generally normalized to 1. They represent the probability of finding a quark of a given species in a small momentum interval dx. If there are several quarks of the same species in the proton, the parton distributions integrated over x will reflect that.

The pdfs $f_f(x)$ must be determined from experiment. The deep inelastic scattering process $e^- p \to e^- X$ gives us one combination of these distributions. But there are other reactions that give us access to other, orthogonal, combinations. From deep inelastic scattering on a deuterium target, we can extract the cross section for deep inelastic electron scattering from a neutron. In the parton model, this process is described by the same formulae (9.71), (9.72), but with the pdfs of the proton replaced by those of the neutron. These two sets of quantities are related by an isospin rotation

$$f_u^{(n)}(x) = f_d(x) , \qquad f_d^{(n)}(x) = f_u(x) ,$$
$$f_{\bar{u}}^{(n)}(x) = f_{\bar{d}}(x) , \qquad f_{\bar{d}}^{(n)}(x) = f_{\bar{u}}(x) , \qquad (10.5)$$

where the unlabeled pdfs are those of the proton. The pdfs for heavier quarks should be identical between the proton and the neutron. In the same approximation as in (10.2)

$$F_2^{(n)}(x) = \frac{4}{9}x f_d(x) + \frac{1}{9}x f_u(x) . \qquad (10.6)$$

so these two sets of measurements already give us a first determination of the separate pdfs for u and d.

Another important source of information is deep inelastic scattering by neutrinos. Neutrinos interact with protons through the weak interaction, and so we will need to understand the structure of that interaction to interpret this data in detail. I will discuss neutrino interactions in Chapter 15. It will be useful to give here a few details that will be explained there. We will see in Chapter 15 that neutrinos also interact through a form of the current-current interaction, and that, at the level of the parton model, neutrino and antineutrino deep inelastic scattering is also described by a formula similar to (9.71). In the dominant processes in neutrino scattering experiments, the neutrino converts to a muon. The four most important parton-level processes are

$$\nu + d \to u + \mu^- , \qquad \bar{\nu} + u \to d + \mu^+ ,$$
$$\nu + \bar{u} \to \bar{d} + \mu^- , \qquad \bar{\nu} + \bar{d} \to \bar{u} + \mu^+ . \qquad (10.7)$$

As we will see in Chapter 15, the distributions in y are different for scattering from quarks and antiquarks. So, by measuring the sign of the final muon each event and the distribution of events in y, we can separately measure u and d quark and antiquark distributions. By looking for strange or charmed particles in the final states of deep inelastic electron and neutrino scattering, we can also estimate the heavy quark distributions

$$f_s(x), f_{\bar{s}}(x), \cdots \tag{10.8}$$

The sum rules (10.4) imply that the total numbers of heavy quarks and antiquarks in the proton are equal, but they do not imply that $f_f(x) = f_{\bar{f}}(x)$. In fact, some processes that add quark-antiquark pairs lead to different distributions. For example, the quantum fluctuation

Physics considerations that explain the differences in the shapes of quark and antiquark pdfs for different flavors.

$$p \leftrightarrow \Lambda^0 + K^+ \tag{10.9}$$

adds a strange quark in a distribution similar to the u quark pdf of the proton, and an $\bar{s}$ at smaller x. The fluctuation

$$p \leftrightarrow n + \pi^+ \tag{10.10}$$

adds a $d + \bar{d}$, but no $\bar{u}$, so we might expect more $\bar{d}$ than $\bar{u}$ antiquarks in the proton. A proton at high momentum has a component in which one u quark carries the proton spin and most of the momentum, while the remaining ud pair form a low-energy $I = 0$, $S = 0$ state, This leads to the expectation that, on average, the u quarks have larger momentum fractions x than the d quark.

Using data from all of these reactions, it is possible to assemble a quantitative model of the full set of pdfs. In setting up such a model, we typically divide the u and d pdfs into *valence* and *sea* components. The valence component contains exactly two u quarks and one d quark, at values of x of order 1. These distributions will have the general form

The pdfs of the proton may be viewed as *valence* pdfs containing 2 u quarks and 1 d quark, plus a *sea* with equal numbers of quarks and antiquarks of each flavor.

$$\tag{10.11}$$

These valence quarks are accompanied by a sea of quarks and antiquarks. The sea distributions are largest at much smaller values of x. They are found to be divergent as $x \to 0$, so that the proton contains a very large number of quark-antiquark pairs carrying very small fractions of

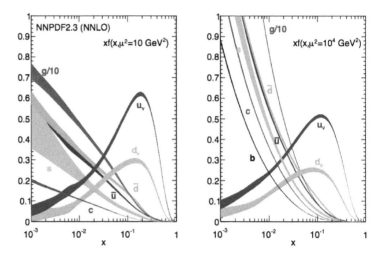

Fig. 10.1 Parton distribution functions $xf_i(x)$ at $Q = 3.1$ GeV and at $Q = 100$ GeV, according to the fit of the NNPDF collaboration, from (Foster *et al.* 2016).

the total proton momentum.

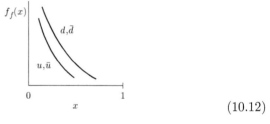

$$\tag{10.12}$$

The divergences of the quark and antiquark pdfs must match so that the integrals in (10.4) can be finite. Feynman called the partons at very small x the *wee partons*. He argued that a $1/x$ behavior of the sea pdfs would lead to the approximately constant value of the proton-proton total cross section at high energies (Feynman 1972). This model of the total cross section (now in a more modern form) is still debated.

These ideas can be incorporated into a model for the pdfs whose parameters are then fit to the relevant data. In performing this fit, we must take into account the physics of the Q-dependence of the pdfs that I will discuss in Chapter 12. However, when this ingredient is included, the entire data set is seen to be well described by the parton model. The fit gives explicit forms for the valence and sea pdf functions. Figure 10.1 shows the functions extracted by the NNPDF collaboration (Ball *et al.* 2015). Two other collaborations, MSTW and CTEQ, also have produced pdf fits to the global dataset, and all three collaborations have quite consistent results. The current status of pdf fits is summarized in (Rojo *et al.* 2015, Buckley *et al.* 2015).

As we have already seen in Fig. 9.5, the pdfs extracted from these fits have a slow dependence on Q, visible when the data are considered on a logarithmic scale. The left and right-hand figures show the pdfs at

Results of an explicit fit to data from deep inelastic scattering and other sources for valence and sea pdfs.

$Q = 3.1$ GeV and at $Q = 100$ GeV. These figures illustrate the valence plus sea form of the pdfs, and indicate clearly the growth of the sea component as Q becomes very large.

The pdfs obey one more sum rule, already stated in (9.11). The parton momenta must sum up to the total energy and momentum of the proton. Since each parton carries a fraction x of the proton's energy-momentum,

$$\int_0^1 dx \; x \sum_i f_i(x) = 1 \; . \qquad (10.13)$$

The fraction of the total energy-momentum of the proton carried by quarks is given by the integral

Momentum sum rule for pdfs.

$$P_{q+\bar{q}}/P = \int_0^1 dx \; x \sum_f [f_f(x) + f_{\bar{f}}(x)] \qquad (10.14)$$

With the extra factor of x relative to (10.4), this integral easily converges as $x \to 0$. The parton distributions determined as I have described give

The quarks and antiquarks alone do not account for the total energy-momentum of the proton.

$$P_{q+\bar{q}}/P \approx 0.5 \; . \qquad (10.15)$$

So, something is still missing. We need additional partons of another type, one that does not participate in deep inelastic scattering. Presumably, the proton must also contain the particle responsible for the binding of quarks into hadron bound states. I will call this particle the *gluon*. If gluons lead to the strong interaction, then, also, there should be a field equation for the gluon field, and there should be physical gluon particles. These particles should appear in the proton wavefunction and should carry some fraction of its momentum.

Introduction of the gluon as a quantum of the strong interaction.

If there is a gluon that interacts with quarks, it should be produced in the reaction $e^+e^- \to$ hadrons. Even if the photon does not couple directly to gluons, the gluon should be radiated from the outgoing quarks and antiquarks. We have seen that quarks and antiquarks appear in experiments as *jets* of hadrons, and that typical events in $e^+e^- \to$ hadrons at high energy are 2-jet events. If a gluon also appears as a jet, we should also see 3-jet events, in which one jet is the product of a gluon,

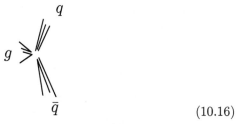

$$(10.16)$$

In fact, when the e^+e^- collider PETRA at the laboratory DESY began to operate at $E_{CM} = 30$ GeV, events of this type appeared. Figure 10.2 shows a 3-jet event recorded by the TASSO experiment (Brandelik *et al.* 1979). Figures 10.3 and 10.4 show events recorded by the SLD experiment at 91 GeV, a 3-jet event and also a 4-jet event.

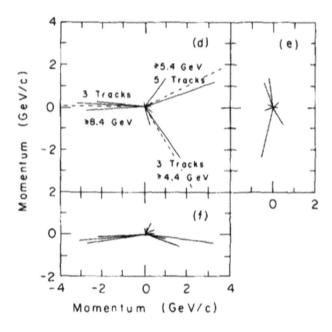

Fig. 10.2 Event display, in three views, of the tracks from a 3-jet event observed by the TASSO experiment at the e^+e^- collider PETRA, from (Brandelik *et al.* 1979).

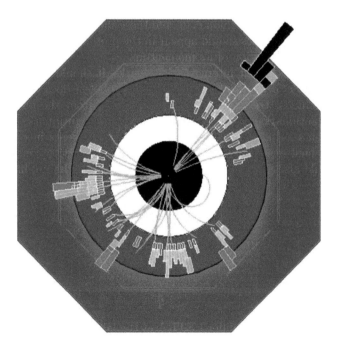

Fig. 10.3 Event display of a 3-jet event observed by the SLD experiment at the e^+e^- collider SLC (figure courtesy of SLAC and the SLD collaboration).

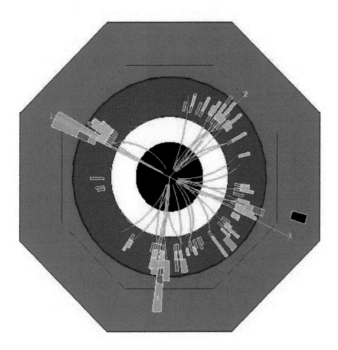

Fig. 10.4 Event display of a 4-jet event observed by the SLD experiment at the e^+e^- collider SLC (figure courtesy of SLAC and the SLD collaboration).

10.2   Photon emission in $e^+e^- \to q\bar{q}$

To understand three-jet events in e^+e^- annihilation quantitatively, it would be good to have a reference theory of gluon emission by quarks, which we could then compare to the data. The simplest hypothesis is that gluons are spin 1 particles like photons, and that they couple to the conserved quark current in the same manner as the photon. The theory of photon emission from relativistic charged particles is rather straightforward. We dipped into the theory of this emission—bremsstrahlung— in our discussion of detectors. Now we have the tools to work out the predictions of this theory more precisely. I will now compute the rate of photon emission from the final-state quarks in $e^+e^- \to q\bar{q}$. In this discussion, I will continue to assume that quarks are structureless spin-$\frac{1}{2}$ fermions, and that I can ignore their masses in high energy processes.

In the discussion of bremsstrahlung in Section 6.2, I explained that it is easy for relativistic particles to radiate additional relativistic particles with order-1 energy sharing, as long as the radiated particles are approximately collinear with the original particles. The final state of two collinear particles has a momentum very close to that of the original particle, so only a small momentum transfer is required. This process is called *collinear splitting*. In particle detectors, splitting is induced by the interaction of the electron or photon with an atomic nucleus.

The theory of photon emission from final state fermions in e^+e^- annihilation is a useful model for the description of gluon emission in e^+e^- annihilation. We can work out this theory through calculations similar to those done in Chapter 8.

A relativistic quark emerging from a particle reaction can readily convert into a pair of collinearly moving relativistic particles. Photon emission provides our first example of such a process.

However, when a relativistic particle is produced in a hard-scattering reaction, that reaction can give the small amount of extra momentum needed to allow splitting. In this section, I will explain how this works for a splitting that converts a quark to a collinear quark and photon.

Consider, then, Feynman diagram with $e^+e^- \to q\bar{q}$ followed by photon emisson,

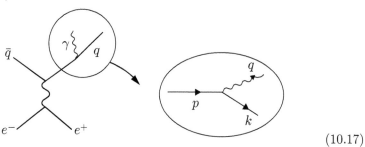

$$(10.17)$$

In the full process $e^+e^- \to q\bar{q}+\gamma$, photons can also be emitted from the initial-state electron and positron, and all of these emissions must be accounted to compare with data. However, it turns out that the dominant contribution to the cross section consists of separate contributions from each of the initial and final legs, so it makes sense to study these separately.

To begin, I will analyze the kinematics of the splitting process. Notice that the initial quark corresponds to an internal line of the Feynman diagram, so it is described as a resonance and it can be slightly off the mass shell. I will use coordinates in which the quark emerges from the e^+e^- reaction moving in the $\hat{3}$ direction. Then

Kinematic analysis of a model collinear splitting process. We see that the initial quark must be slightly off the mass shell to allow the splitting.

$$p \approx (E, 0, 0, E) . \tag{10.18}$$

We can divide this momentum between the final photon and quark, each moving at a small angle with respect the initial quark direction. If the photon carries off a fraction z of the momentum of the original quark, the two momentum vectors can be written

$$q = (zE, q_\perp, 0, zE - \frac{q_\perp^2}{2zE})$$

$$k = ((1-z)E, -q_\perp, 0, (1-z)E - \frac{q_\perp^2}{2(1-z)E}) \tag{10.19}$$

I have modified the $\hat{3}$ components of these 4-vectors to put the final photon and quark on mass shell, $q^2 = k^2 = 0$, up to corrections of relative order $(q_\perp/E)^4$. Energy-momentum conservation implies that the original quark cannot be on its mass shell. Rather, the $\hat{3}$ component of momentum must be

$$E - \frac{q_\perp^2}{2zE} - \frac{q_\perp^2}{2(1-z)E} . \tag{10.20}$$

Then (10.18) can be written more precisely as

$$p = (E, 0, 0, E - \frac{q_\perp^2}{2z(1-z)E}) . \tag{10.21}$$

Squaring this 4-vector, we find

$$p^2 = \frac{q_\perp^2}{z(1-z)} \, . \qquad (10.22)$$

The quark of momentum p is an intermediate state in the process illustrated by the Feynman diagram. It makes sense to treat this particle as a resonance, assigning it the Breit-Wigner factor

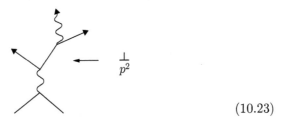

$$(10.23)$$

Notice that, if $q_\perp$ is small, the denominator is small, and thus the quantum amplitude is very large. There is then a high probability that a photon will be emitted in this process, with higher probability for photons more collinear with the original quark.

The fact that the initial quark is close to its mass shell produces a small denominator in the evaluation of the Feynman diagram, leading to a large rate for photon emission.

The Feynman diagram suggests that the full amplitude can be computed as a product of terms

$$\mathcal{M}(e^+e^- \to q\bar{q}\gamma) = \mathcal{M}(e^+e^- \to q\bar{q}) \cdot \frac{1}{p^2} \cdot \mathcal{M}(q(p) \to \gamma(q)q(k)) + \cdots$$
$$(10.24)$$

where the omitted terms contain the amplitude for emissions from the antiquark and the initial e^+ and e^-. I will now analyze the emission from the quark in some detail.

For definiteness, I assume that the initial quark is right-handed. (The final answer for the rate will be the same for a left-handed quark.) Then,

$$\mathcal{M}(q_R \to \gamma q_R) = Q_f e \, \langle q_R(k) | \, j^\mu \, | q_R(p) \rangle \, \epsilon_\mu^*(q) \, , \qquad (10.25)$$

where $\epsilon_\mu(q)$ is the polarization vector of the photon. Using the right-hand part of the current $j^\mu = \psi_R^\dagger \sigma^\mu \psi_R$, we find

$$\mathcal{M}(q_R \to \gamma q_R) = Q_f e u_R^\dagger(k) \, \sigma^\mu \, u_R(p) \, \epsilon_\mu^*(q) \, . \qquad (10.26)$$

Matrix element computation for the splitting process $q \to q + \gamma$. This computation is a model for other computations of splitting amplitudes that we will meet later in this book. So, please follow the steps carefully.

The needed spinors are

$$u_R(p) = \sqrt{2E} \begin{pmatrix} 1 \\ 0 \end{pmatrix}, \qquad u_R(k) = \sqrt{2(1-z)E} \begin{pmatrix} 1 \\ -q_\perp/2(1-z)E \end{pmatrix} \, .$$
$$(10.27)$$

Notice that I have rotated the spinor $u_R(k)$ so that it is the spin-up spinor in the direction of the momentum k in (10.19). It will suffice to work to $\mathcal{O}(q_\perp)$ in the calculation of $\mathcal{M}$. The possible photon polarizations are

$$\epsilon_R = \frac{1}{\sqrt{2}}(0, 1, i, -\frac{q_\perp}{zE}) \qquad \epsilon_L = \frac{1}{\sqrt{2}}(0, 1, -i, -\frac{q_\perp}{zE}) \qquad (10.28)$$

I have rotated these vectors to be orthogonal to q in (10.19) Assembling the pieces, we can compute the matrix elements. First, for emission of a left-handed photon,

$$
\begin{aligned}
u_R^\dagger \sigma \cdot \epsilon_L^* u_R &= -2E\sqrt{1-z}\left(1 \quad -\frac{q^\perp}{2(1-z)E}\right)\frac{1}{\sqrt{2}}[\sigma^1 + i\sigma^2 - \frac{q_\perp}{zE}\sigma^3]\begin{pmatrix}1\\0\end{pmatrix}\\
&= -\frac{2E\sqrt{1-z}}{\sqrt{2}}\left(-\frac{q_\perp}{2(1-z)E} + \frac{q_\perp}{2(1-z)E} - \frac{q_\perp}{zE}\right)\\
&= \sqrt{2}q_\perp\frac{\sqrt{1-z}}{z}\ .
\end{aligned}
\tag{10.29}
$$

For emission of a right-handed photon,

$$
\begin{aligned}
u_R^\dagger \sigma \cdot \epsilon_R^* u_R &= -2E\sqrt{1-z}\left(1 \quad -\frac{q^\perp}{2(1-z)E}\right)\frac{1}{\sqrt{2}}[\sigma^1 - i\sigma^2 - \frac{q_\perp}{zE}\sigma^3]\begin{pmatrix}1\\0\end{pmatrix}\\
&= -\frac{2E\sqrt{1-z}}{\sqrt{2}}\left(-\frac{q_\perp}{2(1-z)E} - \frac{q_\perp}{2(1-z)E} - \frac{q_\perp}{zE}\right)\\
&= \sqrt{2}q_\perp\frac{\sqrt{1-z}}{z(1-z)}\ .
\end{aligned}
\tag{10.30}
$$

Summing the squared amplitudes over photon polarizations,

The final result for the splitting amplitude.

$$
\sum_\epsilon |\mathcal{M}(q \to \gamma q)|^2 = 2Q_f^2 e^2 q_\perp^2 (1-z)\frac{1}{z^2(1-z)^2}(1 + (1-z)^2)\ .
\tag{10.31}
$$

Now we need to combine the result (10.31) with the amplitude for the production of the $q\bar{q}$ system and integrate the complete amplitude over phase space. The complete formula for the cross section is

$$
\sigma(e^+e^- \to q\bar{q}\gamma) = \frac{1}{2E_A \cdot 2E_B \cdot 2}\int d\Pi_3\ |\mathcal{M}(e^+e^- \to q\bar{q}\gamma)|^2\ .
\tag{10.32}
$$

If $\bar{p}$ is the momentum of the antiquark and Q is the total center of mass momentum, the phase space integral is

$$
\int d\Pi_3 = \int \frac{d^3\bar{p}\,d^3k\,d^3q}{(2\pi)^9 2\bar{p}2k2q}(2\pi)^4\delta^{(4)}(Q - \bar{p} - q - k)
\tag{10.33}
$$

Since $k = p - q$, $d^3k = d^3p$. Also, to first approximation, $k = (1-z)p$, $q = zp$. Then we can rearrange the phase space integral as

$$
\int \frac{d^3\bar{p}\,d^3p}{(2\pi)^6 2\bar{p}2p}(2\pi)^4\delta^{(4)}(Q - \bar{p} - p) \cdot \frac{d^3q}{(2\pi)^3 z(1-z)\,2p}\ .
\tag{10.34}
$$

The d^3q integral can be divided into collinear and perpendicular terms,

$$
d^3q = dq^3 d^2q_\perp = p\,dz\ \pi dq_\perp^2\ .
\tag{10.35}
$$

We can assemble the expression (10.32) by using the approximation (10.24) to evaluate the amplitude,

$$
\begin{aligned}
\sigma(e^+e^- \to q\bar{q}\gamma) &= \frac{1}{2E_A \cdot 2E_B \cdot 2}\int d\Pi_2 |\mathcal{M}(e^+e^- \to q\bar{q})|^2\\
&\quad \cdot \int \frac{dz\ \pi dq_\perp^2}{(2\pi)^3 2z(1-z)}|\frac{1}{p^2}|^2|\mathcal{M}(q \to \gamma q)|^2\ .
\end{aligned}
\tag{10.36}
$$

We recognize the first half of (10.36) as the cross section for $e^+e^- \to q\bar{q}$. The second half of (10.36) is

$$\int \frac{dz dq_\perp^2}{16\pi^2 z(1-z)} \left(\frac{z(1-z)}{q_\perp^2} \right)^2 2Q_f^2 e^2 q_\perp^2 \frac{1}{z^2(1-z)} (1 + (1-z)^2) \quad (10.37)$$

Then, finally, we find

$$\sigma(e^+e^- \to q\bar{q}\gamma) = \sigma(e^+e^- \to q\bar{q}) \cdot \frac{Q_f^2 \alpha}{\pi} \int dz \int \frac{dq_\perp}{q_\perp} \frac{1 + (1-z)^2}{z} \,. \quad (10.38)$$

The computation of the cross section for $e^+e^- \to q\bar{q}\gamma$ approximately factorizes into a piece associated with the production process $e^+e^- \to q\bar{q}$ and a piece associated with the photon emission.

This equation gives the cross section for emission of a photon approximately collinear with the final quark. For the full cross section for photon emission in e^+e^- annihilation to hadrons, we must add similar expressions for photon emission from the final antiquark and from the initial electron and positron.

It is hard not to notice that the $q_\perp$ and z integrals are divergent as $q_\perp, z \to 0$. So the photon emission is strongly peaked for photons that are soft and also collinear with respect to the original quark direction. I will discuss the treatment of these singularities in the next chapter. We will see that the reaction rates are not actually infinite; instead, the divergent integrals reflect the fact that a very large number of photons are emitted into these regions of small and collinear momentum.

I have derived this formula for photon emission from an outgoing quark, but actually, the formula is correct for radiation from any charged spin $\frac{1}{2}$ fermions, either in the initial or the final state, as long as the energies involved are high enough that we can ignore the fermion mass. In general, then,

$$\sigma(A \to B + f + \gamma) \approx \sigma(A \to B + f) \cdot \int dz \int \frac{dq_\perp}{q_\perp} \frac{Q_f^2 \alpha}{\pi} \frac{1 + (1-z)^2}{z} \,,$$

$$\sigma(A + f \to B + \gamma) \approx \sigma(A + f \to B) \cdot \int dz \int \frac{dq_\perp}{q_\perp} \frac{Q_f^2 \alpha}{\pi} \frac{1 + (1-z)^2}{z} \,, \quad (10.39)$$

The final formulae for photon emission from very relativistic initial- and final-state fermions.

where the approximation is correct for photons emitted approximately collinearly with the fermion f. This formula is called the *Weizsacker-Williams distribution*. For an electron, we can estimate the integrals as running over the ranges m_e to E_{CM} or m_e/E_{CM} to 1. Then

Photon emission from a very relativistic particle is strongly peaked in the direction collinear with that particle. To a first approximation, we can treat the radiation from each external relativistic particle separately.

$$\sigma(A \to B + e^- + \gamma) \approx \sigma(A \to B + e^-) \cdot \frac{2\alpha}{\pi} \log^2 \frac{E_{CM}}{m_e} \,. \quad (10.40)$$

The radiation pattern is peaked in the directions collinear with the initial and final particles. Then we can associate photon emission with each relativistic particle in the intial and final state. We refer to this collinear radiation as *initial-state radiation* and *final-state radiation*. In each case, the radiation follows the path of the relativistic particle,

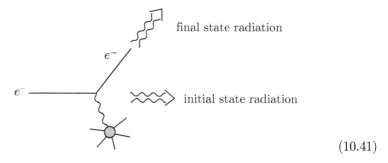

final state radiation

initial state radiation

$$(10.41)$$

10.3   Three-jet events in e^+e^- annihilation

Prediction of the theory in the previous section for the rate of gluon emission in e^+e^- annihilation, assuming that gluons are massless spin 1 particles.

If gluons are massless spin 1 particles coupling to the vector current, and we can treat quarks as massless at high energy, the same formula applies to the emission of gluons from quarks. Let g_s be the strong interaction coupling constant, and let $\alpha_s = g_s^2/4\pi$. In the theory of strong interactions that I will discuss in the next chapter, there is an additional numerical factor $\frac{4}{3}$ in the emission formula, associated with the way that g_s is defined in that theory. The rate of gluon emission from a quark emitted into the final state of a strong interaction reaction would then be

$$\sigma(A \to B + q + g) \approx \sigma(A \to B + q) \cdot \int dz \int \frac{dq_\perp}{q_\perp} \frac{4}{3} \frac{\alpha_s}{\pi} \frac{1 + (1 - z)^2}{z}.$$
$$(10.42)$$

This formula applies only when the gluon is emitted into the collinear region. With more work, one can assemble the complete formula for gluon emission in e^+e^- annihilation to leading order in α_s without making the approximation of collinear emission. To do this, we must consider the processes of gluon emission from the final quark and antiquark,

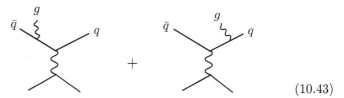

$$+$$

$$(10.43)$$

and add these contributions to $\mathcal{M}$ coherently. The processes interfere constructively when the gluon is radiated into the region between the quark and antiquark.

To actually carry out this computation would take us beyond the scope of this book. However, the result is fairly simple to write in the CM frame for the reaction $e^+e^- \to q\bar{q}g$. The final state has three particles, and so we can use the kinematic relations for three-body phase space that lead to the formula (7.36). Let the CM energies of q, $\bar{q}$, and g be

$$E_q, \quad E_{\bar{q}}, \quad E_g.$$
$$(10.44)$$

For $E_{CM} \equiv Q = E_q + E_{\bar{q}} + E_g$, let

$$x_q = \frac{2E_q}{Q} , \quad x_{\bar{q}} = \frac{2E_{\bar{q}}}{Q} , \quad x_g = \frac{2E_g}{Q} , \tag{10.45}$$

so that $x_q + x_{\bar{q}} + x_g = 2$. The variables x_i have maximum value 1. For example, $x_{\bar{q}} = 1$ corresponds to a configuration in which the antiquark recoils against the quark and gluon, which share the recoil momentum,

$$\tag{10.46}$$

The complete 3-body phase space is a triangle, with these collinear configurations at the edges

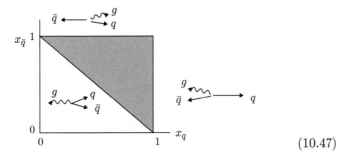

$$\tag{10.47}$$

In terms of these variables, the distribution of events with gluon emission predicted by the sum of diagrams in (10.43) can be shown to be

$$\sigma(e^+e^- \to q\bar{q}g) = \sigma(e^+e^- \to q\bar{q})$$
$$\cdot \int dx_q dx_{\bar{q}} \frac{2\alpha_s}{3\pi} \frac{x_q^2 + x_{\bar{q}}^2}{(1-x_q)(1-x_{\bar{q}})} . \tag{10.48}$$

We can readily check that this agrees with the previous computation in the limit of collinear splitting. Take the limit $x_{\bar{q}} \to 1$, and label the g, q, and $\bar{q}$ momenta as q, k, and $\bar{p}$, as above. We have

$$x_g \approx z , \quad x_q \approx (1 - z) \tag{10.49}$$

Also

$$p^2 = (q+k)^2 = (Q - \bar{p})^2 = Q^2 - 2\bar{p} \cdot Q = Q^2(1 - x_{\bar{q}}) . \tag{10.50}$$

Then we can replace

$$\frac{dx_{\bar{q}}}{(1 - x_{\bar{q}})} = \frac{dq_\perp^2}{q_\perp^2} = 2\frac{dq_\perp}{q_\perp} . \tag{10.51}$$

In this limit, the general expression becomes

$$\sigma(e^+e^- \to q\bar{q}g) = \sigma(e^+e^- \to q\bar{q}) \cdot \int dz \int \frac{dq_\perp}{q_\perp} \frac{4\alpha_s}{3\pi} \frac{1 + (1-z)^2}{z} , \tag{10.52}$$

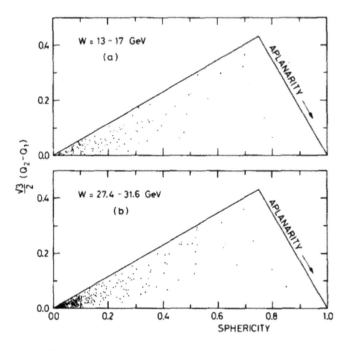

Fig. 10.5 Distribution of e^+e^- annhilation events observed by the TASSO experiment in a two variables related to event shapes, from (Brandelik *et al.* 1979). The lower left-hand corner of this plot corresponds to the upper right-hand corner of the triangle in (10.53).

just as we found in the approximate computation for the collinear region.

The distribution of events predicted by (10.48) over the phase space triangle has the form

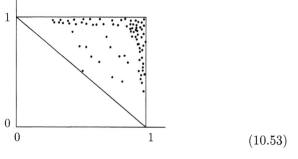

$$\tag{10.53}$$

In observed $e^+e^- \to 3$ jet events, a pair of jets tend to be collinear, as we would expect if the third jet results from a collinear splitting process.

Figure 10.5 shows the distribution of events in a related phase space description used by the TASSO experiment (Brandelik *et al.* 1979). The bottom left-hand corner of the plot contains 2-jet-like events; the region just below the diagonal in the plot contains planar events.

A more detailed comparison of this theory with data is shown in Figs. 10.6 and 10.7 (Abe *et al.* 1997). For a sample of $e^+e^- \to$ 3-jet events analyzed using x_i variables defined to be ordered in energy,

$$x_1 > x_2 > x_3 , \tag{10.54}$$

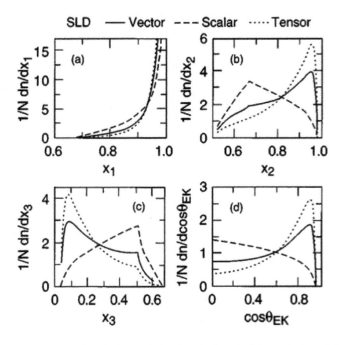

Fig. 10.6 Expectation for the form of the plots in Fig. 10.7 for emission of gluons of spin 0 (scalar), 1 (vector), 2 (tensor), from (Abe *et al.* 1997).

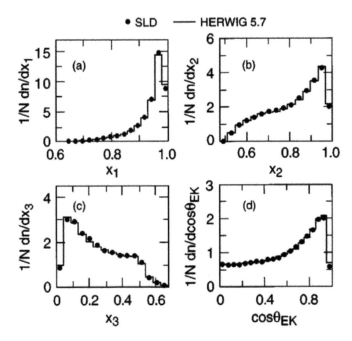

Fig. 10.7 Measurement of the rates of e^+e^- annihilation events as a function of the variables x_1, x_2, x_3 defined in (10.45), (10.54) by the SLD experiment, from (Abe *et al.* 1997).. The data are compared to the simulation program HERWIG, which gives the prediction of the spin 1 gluon model.

The energy distributions for collinear splitting differ depending on the spin of the emitted particle. For 3-jet events in e^+e^- annihilation to hadrons, experiment favors the case of spin 1.

Figure 10.6 shows the predictions for jet production rates as a function of x_1, x_2, x_3. The predictions of the spin 1 gluon model are shown as the solid curves, and they are compared to the predictions from alternative models with spin 0 and spin 2 gluons. Figure 10.7 shows the data from the SLD experiment, which is in excellent agreement with the spin 1 case. In the spin 1 model, the jet with the lowest energy is typically the gluon. You can see that the measured x_3 distribution has the expected shape

$$\frac{1 + (1 - x_3)^2}{x_3} \tag{10.55}$$

up to the point at large x_3 where the gluon is no longer the least energetic particle. The predicted distributions for spin 0 and spin 2 gluons are significantly different, and are not in good agreement with the data.

10.4   Effects of gluon emission on pdfs

Just as we can radiate gluons from final-state quarks, we can radiate gluons from initial-state quarks. In deep inelastic scattering, this process is represented by the Feynman diagram

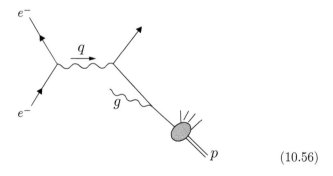

$$\tag{10.56}$$

A parton radiates a gluon, and then scatters from the electron at a lower momentum than it had previously. The effect is proportional to

$$\frac{4}{3}\frac{\alpha_s}{\pi} \int \frac{dq_\perp}{q_\perp} \sim \frac{4}{3}\frac{\alpha_s}{\pi} \log \frac{Q}{m_p}, \tag{10.57}$$

where $q_\perp$ runs over the range $m_p < q_\perp < Q$. Thus, the modification of the parton distribution is proportional to $\log Q$. The effect of gluon emission is to shift the quark parton momenta to lower values of x, since the quarks lose energy and momentum to the emitted gluons. Then we

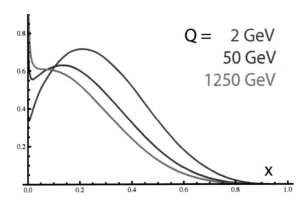

Fig. 10.8 Evolution of the u quark pdf $xf_u(x)$ from $Q = 2$ GeV to $Q = 1250$ GeV, showing the flow of valence quark energy-momentum into gluons. The distributions are computed using the global fit to pdfs by the NNPFD collaboration (Ball *et al.* 2015).

expect the evolution

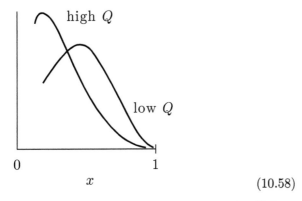

Quark pdfs should evolve with Q due to collinear gluon emission, which increases as Q increases. This physics predicts that quark pdfs should decrease at large x as a function of Q.

$$(10.58)$$

The effect can be seen in Fig. 10.8, which gives the u quark pdf from the fit by the NNPDF collaboration (Ball *et al.* 2015) at $Q = 2$, 50, 1250 GeV.

There are other effects in pdf evolution that are still missing from this description. Gluon emission alone does not produce the strong peaking of pdfs as $x \to 0$, and it does not directly generate the antiquark distributions. For this, we need the feedback of gluons into the quark and antiquark distributions provided by the conversion of gluons to $q\bar{q}$ pairs, just as photons in a detector can convert to electron-positron pairs.

$$(10.59)$$

Including this process of gluon splitting not only produces the antiquark distributions for light quarks; it also correctly predicts the quark and

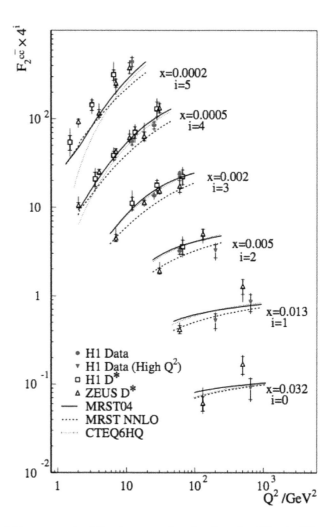

Fig. 10.9 Measurements of the charm quark and antiquark pdf by the H1 and ZEUS experiments at the HERA collider, at increasing values of x as a function of Q, from (Aktas *et al.* 2006). The measurements are compared to expectations from the formulae that will be presented in Chapter 12.

antiquark distributions for c and b quarks. The comparison of measurements of c and b production from the HERA collider to this parton evolution model is shown in Fig. 10.9. In Chapter 12, I will add this effect to our formalism in the context of a complete theory of the strong interaction.

Exercises

(10.1) A very rough approximation to the u and d valence parton distributions shown in Fig. 10.1 is given by

$$f_{val}(x) = A(1 + ax)(1 - x)^{\beta} , \qquad (10.60)$$

where

	u	d
$Q = 3.1$ GeV	$a = 0.5$, $\beta = 4$	$a = -0.6$, $\beta = 4$
$Q = 100$ GeV	$a = 0.3$, $\beta = 5$	$a = -0.8$, $\beta = 5$

$$(10.61)$$

(a) For u and d at each value of Q, determine A such the pdf satisfies its sum rule.

(b) For u and d and each value of Q, compute the average value of the momentum fraction x.

(c) For each value of Q, determine the fraction of the proton's momentum carried by valence quarks.

(10.2) If the gluon were a spin-0 particle G, it would couple to quarks according to the interaction

$$\Delta \mathcal{L} = g_s G \bar{\psi}\psi = g_s G \left(\psi_L^\dagger \psi_R + \psi_R^\dagger \psi_L \right) . \quad (10.62)$$

This gives a different pattern of gluon radiation that is straightforward to work out.

(a) Show that the interaction (10.62) leads to the emission matrix element

$$\mathcal{M}(q_R \to G q_L) = g_s u_L^\dagger(k) u_R(p) \quad (10.63)$$

(b) Show that, in the coordinates used in Section 10.2, the spinors that should be used in this calculation are

$$u_R(p) = \sqrt{2E} \begin{pmatrix} 1 \\ 0 \end{pmatrix} ,$$

$$u_L(k) = \sqrt{2(1-z)E} \begin{pmatrix} q_\perp/2(1-z)E \\ 1 \end{pmatrix} .$$

$$(10.64)$$

(c) Following the derivation of (10.31), show that

$$|\mathcal{M}(q \to Gq)|^2 = g_s^2 q_\perp^2 (1 - z) \frac{1}{(1-z)^2} .$$

$$(10.65)$$

Remember that a spin 0 particle has only one polarization state.

(d) Work out the analogue of (10.38) for emission of G from a quark. Show that the function $(1 + (1 - z)^2)/z$ in the integrand of (10.38) and (10.42) is replaced by the function z. This difference is apparent in the x_3 distributions plotted in Fig. 10.6. The data clearly favors the choice of spin 1 for the gluon.

Quantum Chromodynamics

11

We have now accumulated enough clues to guess at the underlying theory of the strong interaction. This theory should be a theory of massless spin 1 bosons — the gluons. The basic equations of the theory should be some generalization of Maxwell's equations. It would be good if this theory accounted for two of the odd properties of the hadrons that we have already noticed. First, the quarks carry the 3-valued quantum number color, which still needs a physical interpretation. Second, there is a mystery that, although the strong interactions are strong enough to bind quarks permanently into hadrons, we can ignore the strong interactions to first order in analyzing the dynamics of quarks in e^+e^- annihilation and deep inelastic scattering.

It turns out that these clues suggest a unique proposal for the fundamental theory that describes the strong interaction. This theory is called *Quantum Chromodynamics* (QCD). In this chapter, I will describe some new theoretical ideas that we will need to understand this theory. Then I will write down the Lagrangian for QCD and discuss some of its properties.

11.1 Lagrangian dynamics and gauge invariance

To introduce QCD, I must first take a step away from the data and continue the discussion of the Lagrangian dynamics of relativistic field that we began in Chapter 3. We realized in Chapter 3 that Maxwell's equations provide a consistent quantum theory of spin 1 bosons. It is logical to ask what other theories share the same advantages. In this section, we will study the properties of Quantum Electrodynamics that will allow us to construct natural generalizations of that theory.

In (3.75), I wrote the Lagrangian for QED as

$$\mathcal{L} = -\frac{1}{4}F^{\mu\nu}F_{\mu\nu} + \overline{\Psi}(i\gamma^\mu D_\mu - m)\Psi \ , \tag{11.1}$$

where

$$D_\mu = \partial_\mu + ieA_\mu \ . \tag{11.2}$$

Concepts of Elementary Particle Physics. Michael E. Peskin.
© Michael E. Peskin 2019. Published in 2019 by Oxford University Press.
DOI: 10.1093/oso/9780198812180.001.0001

The tensor $F_{\mu\nu}$ contains the electromagnetic field strengths. This Lagrange density is manifestly Lorentz invariant. It is also invariant under the symmetries P, C, and T. We checked in Chapter 3 that this Lagrangian leads to Maxwell's equations coupled to the electron current and to the Dirac equation coupled to the Maxwell A_μ field.

We should now look more closely at (11.1), and, in particular, at the symmetries of this Lagrangian. In addition to the space-time symmetries just listed, the Lagrangian is invariant with respect to a phase rotation of the Dirac field

$$\Psi(x) \to e^{i\alpha}\Psi(x) \qquad \overline{\Psi}(x) \to e^{-i\alpha}\overline{\Psi}(x) \ , \tag{11.3}$$

This symmetry is known as *global gauge invariance*. You know from your study of classical mechanics that every symmetry of the Lagrangian yields a conservation law (Noether's theorem). In this case, we obtain a conservation law of QED associated with (11.3) by varying (11.1) with respect to the transformation

$$\delta\Psi(x) = i\delta\alpha(x)\Psi(x) \qquad \delta\overline{\Psi}(x) = -i\delta\alpha(x)\overline{\Psi}(x) \ . \tag{11.4}$$

This is the infinitesimal form of (11.3), but now with $\alpha(x)$ depending on x. The action is not invariant under this transformation. The derivative in the Dirac Lagrangian leads to a leftover term in which the derivative acts on $\alpha(x)$,

$$\delta\mathcal{L} = \overline{\Psi}(i\gamma^\mu\Psi)(i\partial_\mu\alpha) \ , \tag{11.5}$$

Putting this under the action integral and integrating by parts, this becomes

$$\delta S = \int d^4x(\delta\alpha(x))\partial_\mu(\overline{\Psi}\gamma^\mu\Psi) \ . \tag{11.6}$$

Then $\delta S = 0$ implies the field equation

$$\partial_\mu j^\mu = 0 \ , \tag{11.7}$$

the conservation of the vector current.

In fact, the Lagrangian (11.1) contains a larger symmetry. We can combine the transformation under a local phase transformation with a transformation of the A_μ field

The local gauge transformation of QED.

$$\delta\Psi(x) = i\delta\alpha(x)\Psi(x) \qquad \delta\overline{\Psi}(x) = -i\delta\alpha(x)\overline{\Psi}(x)$$
$$\delta A_\mu(x) = -\frac{1}{e}\partial_\mu\delta\alpha(x) \tag{11.8}$$

The change in the action from this variation is

$$\delta\mathcal{L} = \overline{\Psi}(i\gamma^\mu\Psi)(i\partial_\mu\delta\alpha) + \overline{\Psi}(i\gamma^\mu)(+ie)(-\frac{1}{e}\partial_\mu\delta\alpha)\Psi = 0 \ . \tag{11.9}$$

Notice that $F_{\mu\nu}$ is invariant to the transformation (11.8),

$$\delta F_{\mu\nu} = -\frac{1}{e}[\partial_\mu\partial_\nu\delta\alpha - \partial_\nu\partial_\mu\delta\alpha] = 0 \ . \tag{11.10}$$

So the entire Lagrangian (11.1) is invariant under (11.8). The transformation (11.8) is called a *local gauge transformation*. We say that the QED Lagrangian has *local gauge invariance*.

Local gauge invariance is a powerful, even magical, constraint on the properties of the quantum theory of electromagnetism. Even at the classical level, it requires the field equations to take the form of Maxwell's equations. It is also the principle that allows the 4-vector A_μ to contain only two polarization states, a principle that we saw in Section 3.3 was necessary for the consistency of the quantum theory. A related problem is the question of why the photon does not gain mass in the quantum theory from its interaction with quantum fluctuations. This feature of QED, which is absolutely necessary for its consistency, is actually quite subtle to understand. The explanation, which can be found in textbooks of quantum field theory, makes essential use of local gauge invariance.

Local gauge invariance is a powerful principle responsible for many of the important and nontrivial features of Quantum Electrodynamics.

11.2 More about Lie groups

In searching for a theory of the gluon, a massless spin 1 particle with only the two transverse polarizations, it is natural to build on the idea of local gauge invariance. But, the strong interaction is not simply a slightly modified version of QED. QED, even with a stronger coupling constant, does not have 3-fermion bound states. Also, if the QED coupling were strong enough to bind quarks, it would not be possible to ignore the effects of the QED interactions as we did in our discussions of e^+e^- annihilation and deep inelastic scattering. We need a different generalization that can change these properties.

In QED, the local symmetry is based on the group $U(1)$ of phase rotations, as in (11.3). In principle, we can find larger theories that generalize QED by enlarging the local symmetry to a larger Lie group. It turns out that the change from an Abelian to a non-Abelian local symmetry group changes the theory profoundly. It will be interesting, then, to develop the theory of spin 1 particles with non-Abelian local symmetry.

In Section 2.4, we discussed some simple aspects of non-Abelian continuous groups. I explained that the action of of a non-Abelian group G is generated by the action of Hermitian operators T^a. The number of such operators is d_G, and these operators obey the *Lie algebra* of G, a set of commutation relations that can be written

Please look back at Section 2.4 and review the concepts and notation presented there. We will need these concepts to describe local gauge invariance under non-Abelian groups.

$$[T^a, T^b] = i f^{abc} T^c , \qquad a,b,c = 1, \ldots, d_G , \qquad (11.11)$$

The *structure constants* f^{abc} are totally antisymmetric in their indices. A d_R-dimensional unitary representation of G is generated by a set of d_G Hermitian matrices of size $d_R \times d_R$ that obey this algebra,

$$[t_R^a, t_R^b] = i f^{abc} t_R^c . \qquad (11.12)$$

The Lie algebra satisfied by the representation matrices t_R^a for the generators of a Lie group.

These matrices act on d_R-dimensional complex vectors. The infinitesimal group transformation of such a vector takes the form

$$\Phi \rightarrow (1 + i\alpha^a t_R^a)\, \Phi \qquad (11.13)$$

As I described in Section 2.4, the Hermitian matrices t_R^a generate a set of unitary transformations

$$U(\alpha) = \exp[i\alpha^a t_R^a] \, . \qquad (11.14)$$

These matrices satisfy the group multiplication law of G. They also tranform the vector Φ by the actions

$$\Phi \rightarrow U(\alpha)\, \Phi \, , \qquad (11.15)$$

and these transformations also form a representation of G.

We have already met the Lie groups $SU(N)$, at least for the cases $N = 2, 3$. For any N such that $N \geq 2$, $SU(N)$ is the group of $N \times N$ unitary matrices with determinant 1. $SU(2)$ has the same structure as the rotation group in 3 dimensions $SO(3)$, and it has the same finite-dimensional representations. The smallest nontrivial representation is the 2-dimension spinor representation, with generators given by (2.57). For $SU(3)$, the smallest nontrivial representations are 3-dimensional. There are two inequivalent representations **3** and **$\overline{3}$**, which are complex conjugates. In general, the smallest representations of $SU(N)$ are N-dimensional. The corresponding generator matrices t_N^a are the $N^2 - 1$ traceless $N \times N$ Hermitian matrices. I will make the convention that these matrices are normalized to

The simplest representations N and $\overline{N}$ of $SU(N)$ groups.

$$\text{tr}[t_N^a t_N^b] = \frac{1}{2}\delta^{ab} \, . \qquad (11.16)$$

This convention fixes the normalization of f^{abc} in (11.12) and, through this, fixes the normalization of the representation matrices for all other irreducible representations. It is useful to define a scalar quantity $C(R)$ associated with each representation by

The definition of $C(R)$, a normalization associated to each irreducible representation R.

$$\text{tr}[t_R^a t_R^b] = C(R)\delta^{ab} \, . \qquad (11.17)$$

Some properties of $C(R)$ are worked out in Exercise 11.4.

It is an interesting problem in algebra to find the complete set of finite-dimensional irreducible representations of a Lie algebra (11.11). In the easiest case of $SU(2)$, these irreducible representations are the representations of spin j, with j integer or half-integer. For more general Lie groups, the solution to this problem is discussed in (Georgi 1999).

For our discussion of non-Abelian gauge theories, we will need to know about one other representation of $SU(N)$. This the *adjoint representation*, the representation under which the generators of the Lie algebra transform. In Exercise 11.3, it is shown that the representation matrices in the adjoint representation of any Lie group G can be written as

The *adjoint representation* of a Lie group.

$$(t_G^b)_{ac} = if^{abc} \, . \qquad (11.18)$$

Note that f^{abc}, with b fixed, is a Hermitian $d_G \times d_G$ matrix, as required. These matrices are normalized to

$$\text{tr}[t_G^a t_G^b] = f^{acd} f^{bcd} = C(G)\delta^{ab} . \tag{11.19}$$

For $SU(N)$, $C(G)$ has the value $C(G) = N$. For example, for $SU(2)$, $f^{abc} = \epsilon^{abc}$, and

$$\epsilon^{acd}\epsilon^{bcd} = 2\delta^{ab} . \tag{11.20}$$

For $SU(3)$, the adjoint representation is 8-dimensional (the *octet* representation, corresponding to the set of 8 3×3 traceless Hermitian matrices, and $C(G) = 3$. The formula is proved for general N in Exercise 11.4.

11.3   Non-Abelian gauge symmetry

With this preparation in the formalism of group theory, we can work out the Lagrangian for fermions whose local gauge symmetry is a non-Abelian Lie group G. Consider a Dirac field that also transforms under the symmetry G according to the representation R. An example might be the nucleon field Ψ_i, $i = p, n$, which is rotated by isospin transformations. This was the example considered by Yang and Mills in their original construction of a non-Abelian generalization of QED (Yang and Mills 1954). Honoring this contribution, a non-Abelian gauge theory is also called a *Yang-Mills theory*.

Consider, then, the Lagrangian

$$\mathcal{L}_0 = \overline{\Psi}_j i\gamma^\mu \partial_\mu \Psi_j . \tag{11.21}$$

Let Ψ transform according to a representation R of the gauge group G. Generalizing (11.13), an infinitesimal local gauge transformation of Ψ would take the form

$$\Psi_j(x) \to \Psi'_j(x) = (1 + i\alpha^a(x)t_R^a)_{jk}\Psi_k . \tag{11.22}$$

Then, as before

$$\delta\mathcal{L}_0 = \overline{\Psi}_j i\gamma^\mu (i\partial_\mu \alpha^a(x)t_{R\,jk}^a)\Psi_k . \tag{11.23}$$

We can compensate this transformation by replacing the derivative ∂_μ by the covariant derivative

$$D_\mu = \partial_\mu - igA_\mu^a t_R^a . \tag{11.24}$$

To build a non-Abelian gauge theory, we introduce one spin 1 field for each generator of the gauge group.

Note that we must introduce one vector field for each generator of the group G. We will see in a moment that the variation of the Lagrangian is compensated if we assign the field A_μ^a the transformation law

$$A_\mu^a(x) \to A_\mu^a(x) + \frac{1}{g}\partial_\mu \alpha^a(x) + A_\mu^b f^{abc}\alpha^c(x) . \tag{11.25}$$

This is very similar to the transformation of the A_μ field in (11.8), except that it includes one additional nonlinear term. The parameter

$\alpha^a(x)$ transforms according to the adjoint representation of G, so we might expect that derivatives acting on $\alpha^a(x)$ should also be promoted to covariant derivatives. Using (11.18) for t^a in (11.24), the covariant derivative on $\alpha^a(x)$ takes the form

$$D_\mu \alpha^a(x) = \partial_\mu \alpha^a(x) + g A_\mu^b f^{abc} \alpha^c(x) . \qquad (11.26)$$

So (11.25) can be written more clearly as

$$A_\mu^a(x) \to A_\mu^a(x) + \frac{1}{g} D_\mu \alpha^a(x) . \qquad (11.27)$$

The extra nonlinear term in (11.25) is needed to make the Lagrangian invariant. The variation of the Lagrangian

$$\mathcal{L} = \overline{\Psi} i \gamma^\mu D_\mu \Psi \qquad (11.28)$$

is

$$\delta\mathcal{L} = \overline{\Psi}(i\gamma^\mu D_\mu)(i\alpha^a t_R^a)\Psi + \overline{\Psi}(-i\alpha^a t_R^a)(i\gamma^\mu D_\mu)\Psi$$
$$+ \overline{\Psi}(i\gamma^\mu)(-ig)\Big[\frac{1}{g}\partial_\mu \alpha^a t_R^a + f^{abc} t_R^a A_\mu^b \alpha^c\Big]\Psi . \qquad (11.29)$$

The terms involving $\partial_\mu \alpha^a$ cancel as before. However, there are now three terms involving α with no derivatives,

$$\delta\mathcal{L} = \overline{\Psi}(i\gamma^\mu)(-ig A_\mu^b t_R^b)(i\alpha^a t_R^a)\Psi + \overline{\Psi}(-i\alpha^a t_R^a)(i\gamma^\mu)(-ig A_\mu^b t_R^b)\Psi$$
$$+ \overline{\Psi}(i\gamma^\mu)(-ig)(-f^{abc}\alpha^a A_\mu^b t_R^c)\Psi . \qquad (11.30)$$

In the last term, I have used the antisymmetry of the f^{abc} symbol. Since t^a and t^b do not commute, the first two terms cancel only up to a commutator. Using (11.12), we see that the third term cancels this last piece.

The algebra of the previous paragraph is not the simplest, so let me give it in another version. The finite transformation of Ψ is

$$\Psi \to e^{+i\alpha^a t_R^a}\Psi. \qquad (11.31)$$

The transformation (11.27) of A_μ^a can be written as the finite local transformation of A_μ^a

$$D_\mu[A] \to e^{+i\alpha^a t_R^a} D_\mu[A] e^{-i\alpha^a t_R^a} . \qquad (11.32)$$

It is not so difficult to expand this equation and see that the terms of order α^a reproduce (11.27). Combining (11.32) and (11.31), we have

$$(D_\mu \Psi) \to e^{+i\alpha^a t_R^a} (D_\mu \Psi) . \qquad (11.33)$$

That is, the transformation (11.27) gives the covariant derivative $D_\mu \Psi$ a simple transformation law. It is one that is easily compensated by the transformation of $\overline{\Psi}$ so that the Lagrangian (11.28) is invariant.

We can also use the formula (11.32) to discover the gauge-invariant kinetic term for A_μ^a. According to (11.32), the commutator of covariant derivatives $[D_\mu, D_\nu]$ also transforms as

$$[D_\mu, D_\nu] \to e^{i\alpha^a t_R^a} [D_\mu, D_\nu] e^{-i\alpha^a t_R^a} . \qquad (11.34)$$

It is interesting to compute the commutator of covariant derivatives (11.24) more explicitly, taking account that derivatives ∂_μ act on all fields to the right of them. We find

$$[D_\mu, D_\nu] = \left[(\partial_\mu - igA_\mu^a t_R^a), (\partial_\nu - igA_\nu^a t_R^a) \right]$$
$$= -ig\partial_\mu A_\nu^a t_R^a + ig\partial_\nu A_\mu^a t_R^a + (-ig)^2 [A_\mu^a t_R^a, A_\nu^b t_R^b] . \qquad (11.35)$$

Note that the resulting expression has no derivatives acting to the right; it is a pure field. In its form, it bears a strong similarity to (3.31). In fact, it suggests that we should define the *Yang-Mills field strength* as

$$[D_\mu, D_\nu] = -igF_{\mu\nu}^a t_R^a , \qquad (11.36)$$

so that

$$F_{\mu\nu}^a = \partial_\mu A_\nu^a - \partial_\nu A_\mu^a + gf^{abc} A_\mu^b A_\nu^c . \qquad (11.37)$$

Note that $F_{\mu\nu}^a$ does not depend on the representation R used in the construction. The transformation of this field strength tensor is

The field strength for a non-Abelian gauge field contains nonlinear terms. Then also, the non-Abelian generalization of Maxwell's equations will contain nonlinear interactions.

$$F_{\mu\nu}^a t_R^a \to e^{+i\alpha^b t_R^b} F_{\mu\nu}^a t_R^a e^{-i\alpha^b t_R^b} . \qquad (11.38)$$

This transformation law implies that the quantity

$$\text{tr}[(F_{\mu\nu}^a t_R^a)(F^{\mu\nu a} t_R^a)] \qquad (11.39)$$

is invariant to local gauge transformations. On the other hand, using (11.17), we see that (11.39) is proportional to $(F_{\mu\nu}^a)^2$. Then

$$\mathcal{L} = -\frac{1}{4} F^{\mu\nu a} F_{\mu\nu}^a \qquad (11.40)$$

is a gauge- and Lorentz-invariant Lagrangian for the Yang-Mills field.

A complete locally gauge-invariant Lagrangian with both vector bosons and fermions is

The locally gauge-invariant Lagrangian with non-Abelian gauge symmetry for Dirac fermions and gauge bosons.

$$\mathcal{L} = -\frac{1}{4} F^{\mu\nu a} F_{\mu\nu}^a + \overline{\Psi}(i\gamma^\mu D_\mu - m)\Psi . \qquad (11.41)$$

The Dirac fields Ψ must be assigned to transform in some finite-dimensional representation of G. This Lagrangian leads to the Dirac equation as the field equation of Ψ_i. For the field equation of A_μ^a, it gives a set of equations very similar to Maxwell's equations. However, because of the extra, nonlinear term in (11.37), these equations are nonlinear. This makes the dynamics of non-Abelian gauge theories more complex, and more interesting, than that of ordinary electrodynamics.

11.4   Formulation of QCD

I am now in a position to make a proposal for the underlying theory of the strong interaction. I propose that this should be a non-Abelian gauge theory, with quarks as the fermions and gluons as the spin 1 bosons. For the gauge group G, I will choose the $SU(3)$ symmetry acting on the *color* quantum number that we found in hadron spectroscopy. The quark field of flavor f is $\Psi_{f\alpha i}$, where α is a Dirac index, $\alpha = 1, \ldots, 4$, and i runs over colors $1, 2, 3$. From here on, I will write the representation matrices t_3^a simply as t^a.

The covariant derivative acting on quark fields is

$$D_\mu = \partial_\mu - ig_s A_\mu^a t^a \tag{11.42}$$

where t^a is a 3×3 traceless Hermitian matrix. The parameter g_s is the (dimensionless) strong interaction coupling constant. I will write

$$\alpha_s = \frac{g_s^2}{4\pi} . \tag{11.43}$$

The index a runs over the 8 generators of the $SU(3)$ gauge group.

Finally, the Lagrangian of the theory is

$$\mathcal{L} = -\frac{1}{4} F^{\mu\nu a} F_{\mu\nu}^a + \overline{\Psi}_f (i\gamma^\mu D_\mu - m)\Psi_f . \tag{11.44}$$

The Lagrangian of Quantum Chromo-dynamics—QCD.

The index f runs over the quark flavors; This theory is called *Quantum Chromodynamics*, or QCD.

11.5   Gluon emission in QCD

To understand the theory (11.44) more concretely, we can make contact with the formula (10.42) for the rate for gluon emission in high energy processes. The amplitude is described by the Feynman diagram

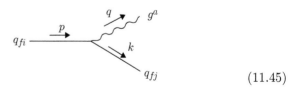

$$\tag{11.45}$$

where I have included all color indices. The corresponding matrix element is

$$\mathcal{M}(q_{f\,L,R\,i}(p) \to g_{L,R}^a(q) + q_{f\,L,R\,j}(k)) = g_s \overline{u}(k)\gamma^\mu t_{ji}^a u(p)\, \epsilon_\mu(q) , \tag{11.46}$$

where $u(p)$, $u(k)$, $\epsilon(q)$ depend on the helicities of the quarks and gluon in the way that I described in Section 10.2. When we compute a rate, we square the amplitude, sum over final colors, and average over initial colors. This gives the extra factor

$$\frac{1}{3} \sum_{ija} g_s^2 |t_{ij}^a|^2 = \frac{g_s^2}{3} \mathrm{tr}[t^a t^a] = \frac{g_s^2}{3} \frac{1}{2} \cdot \delta^{aa} \tag{11.47}$$

or, summing over 8 values of a,

$$\frac{4}{3} g_s^2 \ . \tag{11.48}$$

The color factor for the emission of a gluon from a quark.

This is the origin of the extra factor $\frac{4}{3}$ that I introduced in the discussion just above (10.42). For particles in the **8** or adjoint representation of $SU(3)$, this factor would be

$$\frac{1}{8} \sum_{abc} g_s^2 \, |(t_G^b)_{ac}|^2 = \frac{g_s^2}{8} \mathrm{tr}[t_G^a t_G^a] = \frac{g_s^2}{8} 3\delta^{aa} \ , \tag{11.49}$$

or

$$3 \, g_s^2 \ . \tag{11.50}$$

The color factor for the emission of a gluon from a gluon.

Up to this factor for the color indices, all of the results of Section 10.2 apply to QCD, provided that it is a good approximation to work only to first order in α_s. The formula that we derived gives a reasonable description of the distribution of 3-jet events. But, it emphasizes the question: If the strong interactions are strong, how could it possibly be valid to treat α_s as a small parameter?

11.6   Vacuum polarization

The answer to this question comes from theory, and it is a very surprising one. It turns out that there is a special property of non-Abelian gauge theories that makes these theories unique among all quantum field theories. I will discuss that now.

To understand the uniqueness of non-Abelian gauge theories, we first need to discuss a property of the quantum corrections to QED. The leading contribution to electron-electron scattering is associated with the Feynman diagram

$$\tag{11.51}$$

Quantum corrections to this process include the diagram

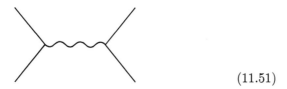

$$\tag{11.52}$$

in which the virtual photon converts to an electron-positron pair, which then reforms into the photon. This effect is called *vacuum polarization*. It is not so easy to compute the matrix element corresponding to this diagram, but it is not difficult to describe its qualitative effect. In the same way that a photon can convert to an electron-positron pair, any electromagnetic disturbance can create a *virtual* electron-positron pair, that is, a quantum state with an e^+e^- pair that contributes to the complete wavefunction of the state. This effect causes the vacuum state of QED to become a mixture of quantum states, most of which contain one or more e^+e^- pairs. Through the influence of these states, the vacuum in QED has properties of a dielectric medium. The virtual e^+e^- pairs can screen electric charge, so that apparent strength of electric charge is smaller than the original strength of the charge found in the Lagrangian.

The effect of vacuum polarization causes the apparent electric charge of a charged particle to become smaller at large distances or larger at larger momentum transfer.

The largest separation of a virtual electron-positron pair is the electron Compton wavelength $\hbar/m_ec$ or $1/m_e$. Pairs can be produced at all size scales smaller than this. At distances short compared to $1/m_e$, the screening influence of virtual electron-positron pairs is scale-invariant; charges are screened by the same factor at each length scale. Then, the apparent charge of the electron increases when the electron is probed at shorter distances or scattered with larger momentum transfer. This effect is described by the equation

For reasons that are not very obvious, (11.53) is called the *renormalization group equation*.

$$\frac{d}{d\log Q}e(Q) = \beta(e(Q)) \tag{11.53}$$

where Q is the momentum transfer in the process under study and $\beta(e)$ is a positive function that depends on e but not directly on Q. An explicit computation in quantum field theory, assuming that $Q \gg m_e$, gives (Peskin and Schroeder 1995)

$$\beta(e) = +\frac{e^3}{12\pi^2} \ . \tag{11.54}$$

To solve this equation, multiply by e and integrate with the initial condition $e(Q_0) = e_0$ to find

$$e^2(Q) = \frac{e_0^2}{1 - (e_0^2/6\pi^2)\log(Q/Q_0)} \ . \tag{11.55}$$

This can also be written as

$$\alpha(Q) = \frac{\alpha_0}{1 - (2\alpha_0/3\pi)\log(Q/Q_0)} \ . \tag{11.56}$$

The value of $\alpha(Q)$ changes on a logarithmic scale when $Q > m_e$. At distances larger than $1/m_e$, $\alpha = 1/137$, but at shorter distances, $\alpha(Q)$ is stronger,

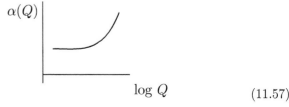

$$\tag{11.57}$$

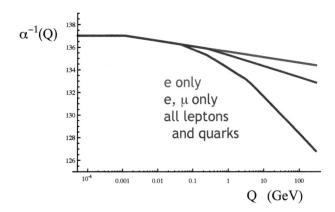

Fig. 11.1 Dependence of $\alpha^{-1}(Q)$ on the momentum transfer Q predicted by the vacuum polarization effect. The three curves show the vacuum polarization effect from electrons only, from electrons and muons, and from all leptons and quarks. The effect of each particle f turns on for $Q > 2m_f$.

Figure 11.1 gives is a more detailed look at the evolution of α. According to (11.56), α^{-1} should be a linear function of $\log Q$. However, at $Q \sim m_\mu$, states with virtual $\mu^+\mu^-$ pairs also come into play, doubling the slope of the linear function. As Q goes above the values of quark masses, the quarks provide additional contributions to vacuum polarization. In all, we have the picture shown in the figure. At low Q, the value of α is $1/137$, but at $Q \sim 30$ GeV, $\alpha = 1/130$ and at $Q \sim 91$ GeV, $\alpha = 1/129$.

This effect is observed experimentally. Figure 11.2 shows the cross section for Bhabha scattering, $e^+e^- \to e^+e^-$, at $E_{CM} = 29$ GeV, measured by the HRS experiment at the e^+e^- collider PEP at SLAC. The specific effect of vacuum polarization raises the predicted cross section by about 10%, giving good agreement with the data.

The increase in the value of α at larger values of Q predicted by QED is confirmed by experiment.

The idea that couplings are modified by QFT corrections on a logarithmic scale in momentum transfer or distance should not seem unfamiliar. We have already seen that pdfs evolve on a log scale in Q as the result of initial state gluon emission. We will see more examples of strong interaction quantities evolving with $\log Q$ in the next two chapters.

11.7 Asymptotic freedom

Non-Abelian gauge theories also have a vacuum polarization effect, corresponding to the Feynman diagram

$$ (11.58) $$

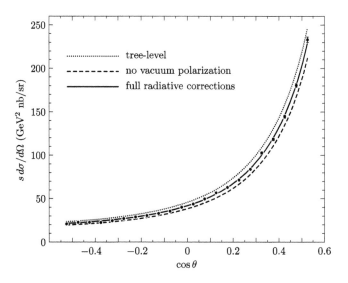

Fig. 11.2 Differential cross section for $e^+e^- \to e^+e^-$ measured by the HRS experiment (Derrick *et al.* 1986), showing the effect of vacuum polarization, from (Peskin and Schroeder 1995). The dotted curve gives the leading order prediction. The dashed curve shows the effect of omitting the vacuum polarization correction while keeping other relevant higher-order corrections.

However, this diagram actually contains two separable and distinct physical effects. The combination of effects is easiest to see if one considers the scattering of heavy particles, for which the exchanged gluon creates a Coulomb potential (Appelquist, Dine, and Muzinich 1977)

$$\tag{11.59}$$

The first effect is the creation of a virtual gluon pair by the Coulomb potential, using the nonlinear interaction of the non-Abelian theory.

$$\tag{11.60}$$

This effect contributes

$$\frac{dg_s}{d\log Q} = +\frac{1}{3}\frac{g_s^3}{16\pi^2}C(G) \,, \tag{11.61}$$

A new effect, present only in non-Abelian gauge theories, causes the apparent gauge charge to become *smaller* at small distances or at larger momentum transfer.

where $C(G)$ is the coefficient defined in (11.17), evaluated for the adjoint representation. In (11.19), we saw that $C(G) = N$ for $SU(N)$. The other contribution is of the form

$$\tag{11.62}$$

The Coulomb potential creates a virtual gluon, which then changes the color transferred by the Coulomb exchange. By explicit computation, the effect of this diagram is to confuse what colors the potential is carrying. At short distances, the color carried by the potential becomes indefinite, and, as a result, the apparent charge becomes smaller. The precise size of the effect is

$$\frac{dg_s}{d\log Q} = -\frac{12}{3}\frac{g_s^3}{16\pi^2}C(G) , \qquad (11.63)$$

In the non-Abelian case, this effect completely dominates the effect of vacuum polarization.

In all, the coupling constant of a non-Abelian gauge theory satisfies the equation

$$\frac{dg_s}{d\log Q} = \beta(g_s) , \qquad (11.64)$$

where

$$\beta(g_s) = -(\frac{11}{3}C(G) - \frac{4}{3}n_f C(R))\frac{g_s^3}{16\pi^2} . \qquad (11.65)$$

In (11.65), I have added the effect of n_f flavors of fermions in the fundamental representation of $SU(N)$.

For QCD, the equation (11.64) can be written

$$\frac{dg_s}{d\log Q} = -b_0 \frac{g_s^3}{16\pi^2} , \qquad (11.66)$$

with

$$b_0 = 11 - \frac{2}{3}n_f . \qquad (11.67)$$

The solution for the scale-dependent coupling is

$$\alpha_s(Q) = \frac{\alpha_s(Q_0)}{1 + (b_0\alpha_s(Q_0)/2\pi)\log(Q/Q_0)} . \qquad (11.68)$$

This can be written as

$$\alpha_s(Q) = \frac{2\pi/b_0}{\log(Q/\Lambda)} , \qquad (11.69)$$

defining $\Lambda = Q_0 \exp[-2\pi/b_0\alpha_s(Q_0)]$. Λ has the units of GeV. It is the mass scale at which the QCD coupling, with the value $\alpha_s(Q_0)$ at the scale Q_0, becomes strong.

The new dynamics of the non-Abelian gauge theory causes $\alpha_s(Q)$ to decrease and actually tend to zero as Q increases. On the other hand, for small Q or large distances, the coupling α_s increases, apparently without bound.

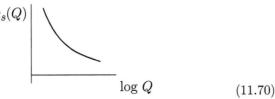

$$(11.70)$$

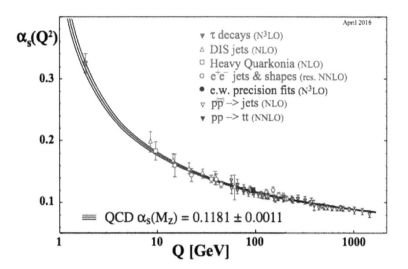

Fig. 11.3 Measured values of α_s from a variety of experiments, compiled in (Bethke *et al.* 2016). In the figure, each values of α_s is plotted at the value of the momentum transfer Q associated with that measurement.

This remarkable effect—discovered by Gerard 't Hooft, David Politzer, David Gross, and Frank Wilczek—is called *asymptotic freedom* ('t Hooft 1972, Gross and Wilczek 1973, Politzer 1973). Exhaustive analysis of other quantum field theories reveals that, in 4 dimensions, only non-Abelian gauge theories have the property that $\beta(g) < 0$ for small g, so that the coupling flows to zero as Q becomes large. The discovery of asymptotic freedom explained at a stroke how we are able to have quark dynamics that needs a large value of α_s for strongly coupled bound states but a small value of α_s to model hard-scattering processes.

I have already explained that we can measure $\alpha_s(Q)$ in a number of different ways. The rate of the emission of gluon jets or the appearance of 3-jet final states in e^+e^- annihilation is proportional to $\alpha_s(q_\perp)$. The rate of evolution of quark pdfs is proportional to $\alpha_s(Q)$. At short distances, gluon exchange produces a Coulomb potential between heavy quarks, of the form

$$V(r) = -\frac{4}{3}\frac{\alpha_s(r)}{r} \ . \tag{11.71}$$

The ψ and Υ bound states are sufficiently small that we can measure the coefficient of this term in the potential. The total cross section for $e^+e^- \to hadrons$, computed to the next order in α_s, is

$$\frac{\sigma(e^+e^- \to \text{hadrons})}{\sigma(e^+e^- \to \mu^+\mu^-)} = 3\sum_f Q_f^2 \left(1 + \frac{\alpha_s(s)}{\pi} + \cdots\right) \ . \tag{11.72}$$

The correction proportional to α_s explains the small difference that we saw in Fig. 8.1 between the measured cross section and the lowest order prediction.

Figure 11.3 shows a compilation of these measured values as presented in (Bethke, Dissertori, and Salam 2016). In the figure, each measurement of α_s is plotted at its appropriate value of Q. The values do become smaller as Q increases, exactly following the QCD prediction shown by the solid band.

From this set of measurements, the value of α_s can be quoted as

$$\alpha_s(91. \text{ GeV}) = 0.1181 \pm 0.0011 \approx 1/8.5 \ . \tag{11.73}$$

Thus, it seems, the strong interactions are actually weak when viewed at short distances, in a way that we can express quantitatively.

The fact that α_s becomes strong, and even formally goes to infinity, at large distances, tempts us to say that asymptotic freedom explains the permanent confinement of quarks into hadrons. This is handwaving. But in fact the permanent confinement of quarks is now understood through a more precise analysis that is, unfortunately, beyond the scope of this book. It is possible to compute the spectrum of QCD in an expansion for *large* values of the coupling constant g_s (Wilson 1974). In this expansion, the gauge fields emerging from each colored particle form a tube of fixed cross section. An isolated particle with color would then carry an infinite flux tube and would have infinite energy.

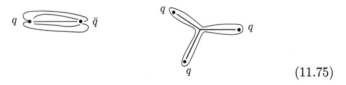

$$\tag{11.74}$$

The only finite-energy states are those with zero total color, in other words, states that are singlets of color $SU(3)$.

$$\tag{11.75}$$

We have seen in Chapter 5 that this principle gives the mesons and baryons as the bound states of quarks and antiquarks.

There is as yet no rigorous proof that the increasing $\alpha_s(Q)$ that we see from the weak-coupling side takes us into the regime where this strong-coupling analysis applies. But extensive numerical calculations have shown that these two regions are indeed smoothly connected. These numerical calculations also show that the low-lying energy eigenvalues of the QCD Hamiltonian are in excellent agreement with the masses of the light hadrons (Kronfeld 2012). Thus, the QCD Lagrangian gives a unified description of the properties of the strong interaction at both large and small distances.

The decrease in the value of α_s at larger values of Q predicted by QCD is confirmed by experiment.

From the compilation of $\alpha_s(Q)$ measurements, we have a precise value for the strength of the QCD coupling constant.

QCD accounts for the confinement of quarks into hadrons and the absence of strongly interacting particles with fractional charge.

Exercises

(11.1) Using (11.17), show that

$$C(R)f^{abc} = -i\,\mathrm{tr}\left[[t_R^a, t_R^b]\,t_R^c\right].\qquad (11.76)$$

Using the cylic property of the trace, show that it follows from this relation that f^{abc} is completely antisymmetric.

(11.2) This problem discusses the complex conjugate of a representation of a Lie group.

(a) The infinitesimal form of the group action of a Lie group is given by (11.13). Take the complex conjugate of this equation. Show that it is of the same form as (11.13), with

$$t_{\bar{R}}^a = -t_R^{a*}.\qquad (11.77)$$

Show that the matrices $(-t_R^{a*})$ satisfy the Lie algebra (11.12). . They form the *complex conjugate representation* of R, called $\bar{R}$. This is representation has dimension d_R. Note that, since t_R^a is Hermitian, also $t_{\bar{R}}^a = -t_R^{aT}$.

(b) Show that $C(R) = C(\bar{R})$.

(c) It is possible that the representation $\bar{R}$ is equivalent to the representation R. For the spinor representation of $SU(2)$, given by (2.57), show that

$$\sigma^2(-t^{a*})\sigma^2 = t^a.\qquad (11.78)$$

Show also that σ^2 is a unitary matrix. Then, in $SU(2)$, the spinor representation $\mathbf{2}$ is equivalent to $\bar{\mathbf{2}}$ by a unitary transformation. In $SU(N)$, $N > 2$, the representations $\mathbf{N}$ and $\bar{\mathbf{N}}$ are inequivalent.

(11.3) This problem justifies (11.18) as the representation matrices of the adjoint representation.

(a) Prove the Jacobi identity: If A, B, C are any matrices,

$$[[A, B], C] + [[C, A], B] + [[B, C], A] = 0.\qquad (11.79)$$

The method of proof is to write (11.79) at the top of a large piece of paper, expand the commutators, and notice many cancellations.

(b) Write out the Jacobi identity for $A = t^a$, $B = t^b$, $C = t^c$, where the t^a are representation matrices of some any representation of a Lie group. Write out the various commutators using (11.12). Using (11.17), show that

$$f^{abd}f^{dce} - f^{acd}f^{dbe} = f^{bcd}f^{ade}.\qquad (11.80)$$

(c) Rearrange (11.80) to show that

$$[t_G^b, t_G^c] = if^{bcd}t_G^d.\qquad (11.81)$$

Then (11.18) generates a d_G-dimensional representation of the Lie group.

(11.4) This problem derives some properties of the quantity $C(R)$ in (11.17).

(a) Show that, if $\mathcal{C}_2 = t_R^a t_R^a$, $[\mathcal{C}_2, t_R^b] = 0$ for all generators t_R^b. This implies that $\mathcal{C}_2$ acts as a constant on an irreducible representation. We write

$$t_R^a t_R^a = C_2(R)\mathbf{1},\qquad (11.82)$$

where $\mathbf{1}$ is the $d_R \times d_R$ unit matrix. $C_2(R)$ is called the *quadratic Casimir operator*.

(b) By taking the trace of (11.82), find a relation between $C_2(R)$ and $C(R)$.

(c) Consider the product of representations $R \otimes R'$. This is a $d_R d_{R'}$-dimensional representation, whose representation matrices are

$$\mathbf{t}_{RR'}^a = t_R^a \otimes \mathbf{1}_{R'} + \mathbf{1}_R \otimes t_{R'}^a.\qquad (11.83)$$

Show that $\mathbf{t}_{RR'}^a$ satisfies (11.12).

(d) The representation (11.83) might be reducible into irreducible representations $\{R_i\}$. Argue that $d_R d_{R'} = \sum_i d_{R_i}$.

(e) By studying $\mathrm{tr}[(\mathbf{t}_{RR'}^a)^2]$, show that

$$d_R d_{R'}(C_2(R) + C_2(R')) = \sum_i d_{R_i} C_2(R_i).\qquad (11.84)$$

(f) In $SU(N)$, the adjoint representation G is $(N^2 - 1)$-dimensional. Then $\mathbf{N} \otimes \bar{\mathbf{N}} = \mathbf{1} + G$, where $\mathbf{1}$ is the trivial representation with $t_1^a = 0$. Use this information and (11.84) to show that $C_2(G) = C(G) = N$.

(g) In $SU(3)$, $\mathbf{3} \otimes \mathbf{3} = \bar{\mathbf{3}} + \mathbf{6}$. Compute $C_2(\mathbf{6})$.

(11.5) This problem studies the QCD analogue of the Coulomb potential. For the QCD interaction of states in a representation R with states in a representation R', the QCD potential is given by the operator

$$V(r) = \frac{g_s^2}{4\pi r} \, t_R^a \otimes t_{R'}^a \, . \qquad (11.85)$$

The energies of states are found by diagonalizing this operator.

(a) Using results of Exercise 11.3, show that Coulomb energy depends on the breakdown of the states $R \times R'$ into irreducible representations. For a state in the irreducible representation R_i, show that

$$V(r) = \frac{g_s^2}{4\pi r} \cdot \frac{1}{2} \cdot \left(C_2(R_i) - C_2(R) - C_2(R') \right) . \qquad (11.86)$$

(b) Show that, in a color-singlet quark-antiquark state,

$$V(r) = -\frac{4}{3} \frac{g_s^2}{4\pi r} \, . \qquad (11.87)$$

(c) Show that, in a color **8** quark-antiquark state,

$$V(r) = +\frac{1}{6} \frac{g_s^2}{4\pi r} \, . \qquad (11.88)$$

Notice that the center of gravity of (11.87) and (11.88) is zero. Why?

(d) Show that a quark-quark (diquark) state in the $\bar{3}$ representation has the Coulomb energy

$$V(r) = -\frac{2}{3} \frac{g_s^2}{4\pi r} \, . \qquad (11.89)$$

That is, the diquark is bound. Notice that this diquark can have a bound state with a third quark; this model is sometimes used to describe a baryon.

(e) Show that a quark and a gluon in the $\bar{3}$ combination is bound.

Partons and Jets

12

In the previous chapter, I introduced QCD as a proposal for the theory of the strong interaction. We saw that QCD explains the main puzzling feature of the strong interaction, the fact that the strong interactions are strong, to bind hadrons, but can be neglected to first approximation in hard scattering processes.

This understanding motivates us to look more closely at high energy scattering to provide more evidence for the validity of QCD. Though the QCD interactions are weak at high energy, they are not ignorable. They produce an enhancement of the cross section for $e^+e^- \to hadrons$, required by the data. They give rise to 3-jet events. In our earlier discussion, I explained in intuitive terms how quark-gluon interactions give a theory of the evolution of pdfs with Q^2. I will now return to that theory and complete it, with insight from our new understanding of QCD.

12.1 Altarelli-Parisi evolution of parton distribution functions

In (10.42), we derived the expression for gluon emission from a quark in the approximation of collinear emission,

$$\text{Prob}(q \to gq) = \int dz \int \frac{dq_\perp}{q_\perp} \frac{4}{3} \frac{\alpha_s(q_\perp)}{\pi} \frac{1 + (1-z)^2}{z} . \tag{12.1}$$

The $q_\perp$ integral runs up to values where the momentum transfer is comparable to that in the hard-scattering process. In deep inelastic scattering, $q_\perp$ can take values up to Q within the approximation that we are using. If we start from a distribution of quarks in the proton parametrized by a pdf $f_f(\xi)$, the probability of finding a gluon emitted by one of these quarks in the proton is

$$\int dx f_g(x) = \int d\xi \int dz \int^Q \frac{dq_\perp}{q_\perp} \frac{4}{3} \frac{\alpha_s(q_\perp)}{\pi} \frac{1 + (1-z)^2}{z} f_f(\xi) . \tag{12.2}$$

The gluon will have a fraction of the proton's momentum $x = z\xi$. We can change variables from ξ to x using

$$d\xi = \frac{dx}{z} . \tag{12.3}$$

Concepts of Elementary Particle Physics. Michael E. Peskin.
© Michael E. Peskin 2019. Published in 2019 by Oxford University Press.
DOI: 10.1093/oso/9780198812180.001.0001

Then (12.2) above becomes

$$f_g(x) = \int \frac{dz}{z} \int^Q \frac{dq_\perp}{q_\perp} \frac{4}{3} \frac{\alpha_s(q_\perp)}{\pi} \frac{1+(1-z)^2}{z} f_f(\frac{x}{z}) \,. \tag{12.4}$$

The rate of gluon emission by an incoming or outgoing quark can be computed as the solution of a differential equation in the momentum transfer Q.

We recognize that $f_g(x)$ satisfies the differential equation

$$\frac{d}{\log Q} f_g(x) = \frac{4}{3} \frac{\alpha_s(Q)}{\pi} \int_x^1 \frac{dz}{z} \frac{1+(1-z)^2}{z} f_f(\frac{x}{z}) \,. \tag{12.5}$$

Note the limits of integration for the dz integral. The parent quark must come from a higher momentum fraction ξ satisfying

$$x < \frac{x}{z} < 1 \,. \tag{12.6}$$

QCD also predicts other collinear splitting processes for partons: $g \to q\bar{q}$, $g \to gg$.

Gluon partons of the proton can split into collinear $q\bar{q}$ pairs. Also, using the nonlinear 3-gluon coupling of QCD, a gluon parton can split into two collinear gluons. Then the evolution of pdfs must also contain the processes

$$\tag{12.7}$$

These processes have collinear enhancements very similar to the one that we found in $q \to g + q$. One can compute the rates for these splittings in the same way that we did for $q \to g + q$ and derive additional terms in the differential equations. Putting all of the pieces together, we find that pdfs obey a system of differential equations called the *Altarelli-Parisi equations* or the *DGLAP (Dokshitzer-Gribov-Lipatov-Altarelli-Parisi) equations* (Altarelli and Parisi 1977, Dokshitzer 1977)

The *Altarelli-Parisi equations* that describe the evolution of parton distributions by the emission of quarks, antiquarks, and gluons.

$$\frac{d}{\log Q} f_g(x) = \frac{\alpha_s(Q)}{\pi} \int_x^1 \frac{dz}{z} \left\{ P_{g\leftarrow g}(z) f_g(\frac{x}{z}) \right.$$
$$\left. + \sum_f P_{g\leftarrow q}(z)[f_f(\frac{x}{z}) + f_{\bar{f}}(\frac{x}{z})] \right\} \,,$$

$$\frac{d}{\log Q} f_f(x) = \frac{\alpha_s(Q)}{\pi} \int_x^1 \frac{dz}{z} \left\{ P_{q\leftarrow q}(z) f_f(\frac{x}{z}) + P_{q\leftarrow g}(z) f_g(\frac{x}{z}) \right\} \,,$$

$$\frac{d}{\log Q} f_{\bar{f}}(x) = \frac{\alpha_s(Q)}{\pi} \int_x^1 \frac{dz}{z} \left\{ P_{q\leftarrow q}(z) f_{\bar{f}}(\frac{x}{z}) + P_{q\leftarrow g}(z) f_g(\frac{x}{z}) \right\} \,. \tag{12.8}$$

The *Altarelli-Parisi splitting functions.*

The functions $P_{g\leftarrow g}(z)$, etc., are called the *Altarelli-Parisi splitting functions.* We have computed

$$P_{g\leftarrow q}(z) = \frac{4}{3} \frac{1+(1-z)^2}{z} \,. \tag{12.9}$$

The $q \to gq$ splitting also gives the $q \leftarrow q$ splitting function by exchanging the final quark and gluon, that is, exchanging $z \leftrightarrow (1 - z)$. Then

$$P_{q \leftarrow q}(z) = \frac{4}{3} \left[\frac{1 + z^2}{(1 - z)} + A\delta(z - 1) \right] . \qquad (12.10)$$

The second term here is new and needs some explanation. In the process $q \to gq$, a new gluon is created, but no new quark is created; rather, a quark is moved from higher x to lower x. This implies that

$$\int_0^1 dz \; P_{q \leftarrow q}(z) = 0 . \qquad (12.11)$$

So whatever the number of quarks that appear at $z < 1$, that number must be subtracted at $z = 1$. In fact, the integral over the first term of $P_{q \leftarrow q}(z)$ diverges as $z \to 1$. This actually makes no difference to the evolution of the quark pdf, since it corresponds to the quark emitting very soft gluons and changing its x value only infinitesimally. To control this in a quantitative calculation, we might cut off the integral at $z = (1 - \epsilon)$ and assign

$$A = -[2 \log \frac{1}{\epsilon} - \frac{3}{2}] . \qquad (12.12)$$

to satisfy the sum rule.

The Altarelli-Parisi functions for the gluon splitting processes are

$$P_{q \leftarrow g}(z) = \frac{1}{2}[z^2 + (1 - z)^2] ,$$

$$P_{g \leftarrow g}(z) = 3 \, [\frac{1 + z^4 + (1 - z)^4}{z(1 - z)} + B\delta(z - 1)] , \qquad (12.13)$$

where B includes a term that compensates the singularity of the first term in $P_{g \leftarrow g}(z)$ at $z = 1$. The derivation of the splitting function for $g \to gg$ is given in Exercise 12.1. The Altarelli-Parisi splitting functions are summarized, along with other important QCD formulae, in Appendix E.

We have discussed the splitting of a gluon to a quark-antiquark pair in Section 10.4. We saw there that this effect leads to the build-up of the quark and antiquark sea distributions at small x, and to the generation of heavy quarks and antiquarks in the proton wavefunction. The splitting of a gluon to two gluons is a new effect that comes from the nonlinear interactions of QCD. Because of the large coefficient in this splitting function and its singular nature at small z, this effect is typically the most important one in the generation of new partons by final-state radiation.

The Altarelli-Parisi equations give a precise model for relating pdfs measured at different values of Q^2. To compare this model to the data, we need knowledge of the evolution of α_s and a model for the gluon pdf $f_g(x)$. Typically, the gluon pdf is described by parameters that are then varied in the fit. For an accurate theoretical prediction, the corrections to the above formulae of order α_s should also be included. An example of a comparison of QCD theory and experiment for the

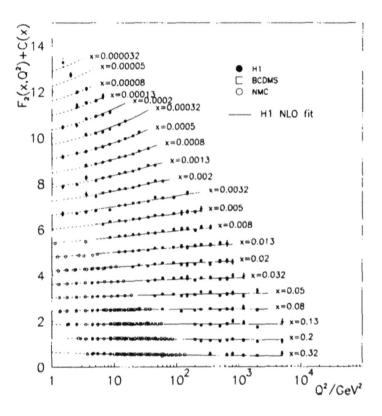

Fig. 12.1 Comparison of the values of F_2 measured in high-energy deep inelastic scattering experiments with the evolution in Q^2 predicted by QCD, from (Aid *et al.* 1996).

evolution of $F_2(x, Q)$ is shown in Fig. 12.1. The black circles are data from measurements by the H1 experiment at the HERA electron-proton collider; the open circles are data from muon deep inelastic scattering experiments at CERN.

The most singular terms in the splitting functions are

$$P_{g \leftarrow q}(z) \sim \frac{4}{3} \cdot \frac{2}{z} \,, \qquad P_{g \leftarrow g}(z) \sim 3 \cdot \frac{2}{z} \,, \qquad (12.14)$$

QCD predicts that an emitted quark or gluon will emit a cloud of soft gluons that will surround it. The number of soft gluons is larger for a primary gluon than for a primary quark by a factor of 9/4.

as $z \to 0$. The ratio of these terms is $9/4$, which is the ratio of the SU(3) group theory factors for quarks and gluons, (11.48) vs. (11.50), that appear with the the squared charge g_s^2. These terms imply, first, that quarks and gluons both accumulate a cloud of soft gluons as their structure evolves with Q, and, second, that the number of these soft gluons is larger for a primary gluon by more than a factor of 2. We will see direct consequences of this in the next section and in Section 13.4.

12.2   The structure of jets

The physics of quark and gluon splitting gives us a picture of the evolution from quarks and antiquarks produced as primary particles in $e^+e^- \to hadrons$ to the pions, kaons, etc. that form the hadronic final states. Begin from the initial $q\bar{q}$ pair. The quark will radiate a gluon, with the highest probability of radiation in the collinear region, $q_{1\perp} \ll \sqrt{s}$. This gluon, and also the recoiling quark, emits additional gluons, with $q_{2\perp} \ll q_{1\perp}$. Occasionally, a gluon splits to a quark-antiquark pair. We obtain a shower of gluons, quarks, and antiquarks

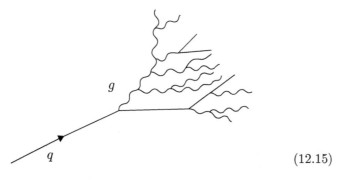

$$(12.15)$$

At each stage, the momentum transfer decreases. So, the quarks and gluons in the shower are all roughly collinear. Eventually, the $q_\perp$ in the splittings falls below 1 GeV, the value of $\alpha_s(q_\perp)$ becomes large, and the strong interaction effects of QCD take over, combining quarks and antiquarks into mesons and baryons. This gives a *jet* of hadrons, similar to those we have seen in e^+e^- event displays.

Collinear QCD splitting of quarks, antiquarks, and gluons naturally generates the stream of collinear hadrons that we observe in event displays as a *jet*.

To test this picture, we need to define a jet more precisely and quantify its structure. This is not so easy. The coupling constant $\alpha_s(q_\perp)$ changes only slowly with $q_\perp$. If α_s were constant, the quark-gluon splitting process would be scale-invariant. Each parton in a jet would split to produce a subjet with the same structure as the overall jet. Then we would see subjets inside jets, with smaller subjets inside the subjets. This is the structure of a fractal. The true behavior of QCD is not so far from this limit. To work with jets, we need to define observables that capture their behavior, recognizing that higher resolution will unveil more levels of structure inside each jet. An excellent introduction to the study of jets can be found in (Salam 2010).

There are two solutions to the problem of giving quantitative predictions for the structure of jets. First, we can define variables for a whole event that are sensitive to the jet structure inside the event. Second, we can attempt to find the jets in an event by clustering particles according to some algorithm. In the latter case, a higher-resolution algorithm will produce more jets, so to compare data to QCD theory we will need to take care that the same resolution is used on each side.

To compare either approach to QCD calculations, we must take account of the fact that, when we integrate over soft and collinear emissions, we encounter formal infinities in the limits where the emissions are extremely soft or exactly collinear. The full predictions of QCD are not actually infinite, but the infinities indicate regions where $\alpha_s(q_\perp)$ has become large and a description in terms of weakly coupled quarks and gluons breaks down. To be computable in a quark-gluon picture, an observable should be defined in such a way that its value is not affected by these limiting cases of soft and collinear emissions. Such an observable is called infrared and collinear safe or just *IR-safe*.

The earliest study of e^+e^- event shapes (Hanson *et al.* 1975) searched for jets in the particle distributions by computing the *sphericity tensor*, defined by

$$Q^{ab} = \left[\frac{\sum_i p_i^a p_i^b}{\sum_i |\vec{p}_i|^2} \right] . \tag{12.16}$$

This definition is very convenient, because the tensor can be directly computed from the particle momenta and then diagonalized. The principal axis with the largest eigenvector of Q is called the *sphericity axis*. Comparing with simulation, the sphericity axis was a good indicator of the initial quark direction in $e^+e^- \to$ hadrons.

Sphericity has the defect, though, that it is not IR-safe and so is difficult to use for quantitative comparison to QCD predictions. For example, a collinear splitting affects the diagonal elements by converting

$$|\vec{p}|^2 \to |\vec{q}|^2 + |\vec{k}|^2 = (z^2 + (1-z)^2) \cdot |\vec{p}|^2 , \tag{12.17}$$

giving a factor that is generally less than 1. The evolution of sphericity can still be modeled using simulation programs, but the comparison of the results of these programs to data depends on the model used for conversion of quarks and gluons to hadrons.

The variable *thrust*, an IR-safe observable that measures the 2-jet nature of final states produced in $e^+e^- \to$ hadrons.

A more useful observable for measuring the jettiness of an event is the *thrust*. For an $e^+e^- \to$ hadrons event, thrust is defined as follows: Go to the CM frame. Choose an axis, represented by a unit vector $\hat{n}$. The thrust is then given by

$$T = \max_{\hat{n}} \left[\frac{\sum_i |\hat{n} \cdot \vec{p}_i|}{\sum_i |\vec{p}_i|} \right], \qquad (12.18)$$

where i runs over all particles or observed energy depositions in the event. T is the maximum of the ratio in (12.18) over all possible choices of $\hat{n}$. The corresponding $\hat{n}$ is called the *thrust axis*. This axis can be measured in each event and used as a proxy for the initial quark and antiquark direction.

In a collinear splitting $p \to q + k$, it is approximately true that $q = zp$, $k = (1-z)p$. Then

$$|\vec{p}| \approx |\vec{q}| + |\vec{k}| \qquad |\hat{n} \cdot \vec{p}| \approx |\hat{n} \cdot \vec{q}| + |\hat{n} \cdot \vec{k}| . \qquad (12.19)$$

so the value of the thrust is not affected by the splitting.

At the lowest order of approximation, an $e^+ e^- \to hadrons$ event contains only a quark and antiquark

QCD physics that determines the shape of the thrust distribution.

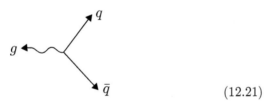

$$\bar{q} \longleftarrow \quad \longrightarrow q \qquad (12.20)$$

This state has $T = 1$. When one gluon is emitted, most events are still near $T = 1$, though planar events with maximum energy sharing

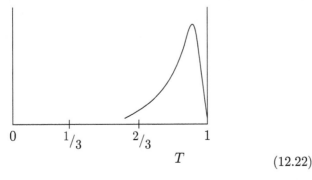

$$(12.21)$$

can have values of T as low as $\frac{2}{3}$. As more gluons are emitted, still lower values of T can be produced, but the probability to find such a low value is proportional to many powers of α_s. Also, any emission that is not precisely collinear will move the event away from $T = 1$; then the final distribution of T will have a zero at $T = 1$. This effect is called *Sudakov suppression*. The final QCD prediction for the distribution of T has the form

$$(12.22)$$

The peak near $T = 1$ should become steeper at higher E_{CM}, since the probability of emitting the first, hardest, gluon is proportional to

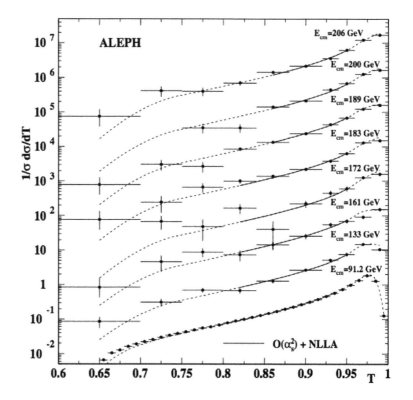

Fig. 12.2 Distribution of the thrust variable (12.18) in e^+e^- annihilation events at energies from 91 GeV to 206 GeV, as measured by the ALEPH experiment, from (Heister *et al.* 2004). The continuous lines show the distributions predicted by QCD. Note that the vertical axis is a log scale; these distributions do peak sharply near $T = 1$.

$\alpha_s(E_{CM})$. Figure 12.2 shows the thrust distribution measured by the ALEPH experiment at the e^+e^- collider LEP at energies from 91 GeV to 206 GeV, and a comparison to QCD theory (Heister *et al.* 2004).

The other approach to the quantitative analysis of jets is to identify the jets in each event by a clustering algorithm. If this algorithm is IR-safe, the same algorithm can be applied to particles observed in an e^+e^- event and to quarks and gluons in the QCD model. Many studies use the *JADE algorithm*, developed in the JADE experiment at the PETRA collider at DESY. For all pairs of particles i, j, compute

$$y_{ij} = \frac{(p_i + p_j)^2}{s} \ . \tag{12.23}$$

This is the ratio of the invariant mass of the pair to the invariant mass of all particles in the event. Choose a value y_{cut} that will set the resolution with which we observe the jets. To begin the clustering, choose the pair i, j with the smallest value of y_{ij}, and combine these into a single particle

$$p_i + p_j \to p_k \ . \tag{12.24}$$

Repeat until all values of y_{ij} are greater than y_{cut}. The jets in the event are defined to be the (composite) particles remaining at this stage. For jet analysis at 100 GeV, a typical value of y_{cut} used to count jets is $y_{cut} = 10^{-2}$. Then we resolve the 100 GeV event into jets that are clusters of particles with mass roughly 10 GeV.

The fractal nature of QCD is revealed when we change the value of y_{cut}. For large values of y_{cut}, essentially all events are clustered into 2 jets. As y_{cut} is lowered, the number of jets increases as jets at one level are resolved into pairs of jets at the next level. Figure 12.3 shows the fraction of 2-, 3-, 4-, and 5-jet events in e^+e^- annihilation at 206 GeV as a function of y_{cut}, as measured by the ALEPH experiment.

Given the probabilities for quark and gluon splitting in QCD, it is possible to write a computer program that models the physics of jet production by emitting quarks and gluons stochastically according to these laws. The transition from quarks and gluons to hadrons, at the momentum scale of 1 GeV, is treated by an *ad hoc* model with many adjustable parameters. Once these parameters are fit to low-energy e^+e^- data, these simulation programs give predictions for the shapes and numbers of jets in higher-energy reactions. The codes PYTHIA and HERWIG, built according to this strategy, have been in development since the 1980's (Bahr *et al.* 2008, Sjöstrand *et al.* 2015). These and a more recent competitor Sherpa (Gleisberg 2009) are used today to model events for all high energy collider experiments. They fit the data very effectively. Figure 12.3 shows the comparison of PYTHIA and HERWIG simulations with the ALEPH data on the number of jets as a function of resolution.

Jets are identified in a particle production event by the use of a clustering algorithm

The *JADE algorithm* for clustering observed particles in an $e^+e^- \to$ hadrons event into jets.

Applying the JADE algorithm to $e^+e^- \to$ hadrons events reveals the fractal substructure of jets, in agreement with the predictions of QCD.

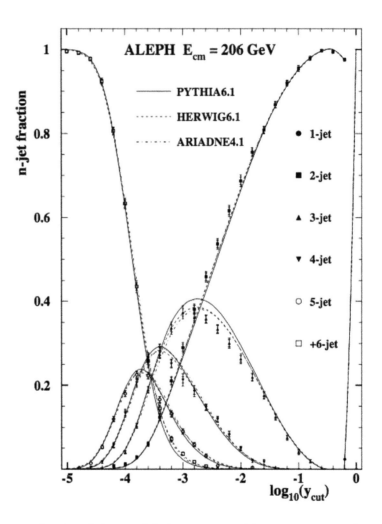

Fig. 12.3 Fraction of e^+e^- annihilation events containing 1, 2, 3, *etc.* jets shown as a function of the jet criterion y_{cut}, measured by the ALEPH experiment at 206 GeV, from (Heister *et al.* (2004).

Exercises

(12.1) The nonlinear terms in the QCD Lagrangian lead to the following expression for the matrix element for a gluon to radiate a gluon:

$$\mathcal{M}\left(g^a(p,\epsilon_p) \to g^b(q,\epsilon_q) + g^c(k,\epsilon_k)\right)$$
$$= g_s f^{abc}\left[(k+p)\cdot\epsilon_q^*\ \epsilon_k^*\cdot\epsilon_p - (p+q)\cdot\epsilon_k^*\ \epsilon_p\cdot\epsilon_q^*\right.$$
$$\left. +(q-k)\cdot\epsilon_p\ \epsilon_q^*\cdot\epsilon_k^*\right], \qquad (12.25)$$

where f^{abc} is the structure constant in the commutator of two group generators. For $SU(2)$, $f^{abc} = \epsilon^{abc}$.

(a) To make the expression (12.25) more symmetric among p, q, k, replace the incoming momentum p by an outgoing momentum $(-p)$ and ϵ_p by ϵ_p^*. Show that the resulting expression is completely symmetric under interchange of any two gluons, as required by Bose symmetry.

(b) Set up the kinematics of almost collinear gluon radiation for $g(p) \to g(q)g(k)$, following the calculation in Section 10.2 of the matrix elements for $q(p) \to \gamma(q)q(k)$. To order q_T, the three 4-vectors are

$$p \approx (E, 0, 0, E),$$
$$q \approx (zE, q_\perp, 0, zE),$$
$$k \approx ((1-z)E, -q_\perp, 0, (1-z)E),$$
$$(12.26)$$

Modify the 3 component of q and k, as we did in (10.19), so that these vectors satisfy $q^2 = k^2 = 0$ to order $q_\perp^2$, and then modify the 3 component of $p = q + k$.

(c) Write the polarization vectors (for L and R polarizations) for the three gluons. These polarization vectors should be transverse to the corresponding momenta (so that, for example, $\vec{\epsilon}(q) \cdot \vec{q} = 0$), correct to order $q_\perp^1$. Make tables of the values of $(\epsilon \cdot p)$ and $(\epsilon \cdot \epsilon)$ needed to compute the matrix elements in (b) for all possible polarizations.

(d) Work out the matrix element for $g_R \to g_L g_L$, to order $q_\perp^1$, and show that it is zero.

(e) Work out the matrix element for $g_R \to g_R g_L$. And, changing just what needs to be changed, work out the matrix element for $g_R \to g_L g_R$.

(f) Work out the matrix element for $g_R \to g_R g_R$.

(g) Show that the matrix elements for $g_L \to gg$ are given by the results of (d), (e), and (f) with all polarizations reversed.

(h) Square the matrix elements, sum over final spins and colors, and average over initial spins and colors. To compute the sum over colors, you will need the identity (11.19).

(i) Write the analogous sum over matrix elements for $q \to gq$, given by (10.31) with $Q_f^2 e^2 \to \frac{4}{3}g_s^2$, and compare this to the result found in (h).

(j) Finally, following the derivation for $q \to gq$ and just changing what needs to be changed, derive the expression for the emission of an almost-collinear gluon from a gluon,

$$\text{Prob}(g \to gg)$$
$$= \frac{3\alpha_s}{\pi}\int dz \int \frac{dq_\perp}{q_\perp} \frac{1 + z^4 + (1-z)^4}{z(1-z)}. \qquad (12.27)$$

(12.2) Derive the leading-order QCD prediction for the thrust distribution in $e^+e^- \to$ hadrons events. For $e^+e^- \to q\bar{q}$ events, $T = 1$. For $e^+e^- \to q\bar{q}g$ events, we can use the differential cross section (10.48) to determine the distribution of T for $T < 1$.

(a) To compute the thrust for a 3-parton configuration, we will need to identity the thrust as a function of the variables (10.45). Show that $T = \max\{x_q, x_{\bar{q}}, x_g\}$. Show that $T \geq \frac{2}{3}$ for any $q\bar{q}g$ configuration.

(b) Consider the region of phase space where x_q is the largest of the three variables. Show that the contribution to $d\sigma/dT$ from this region is given by a one-dimensional integral over $x_{\bar{q}}$ over the interval $2(1 - x_q) < x_{\bar{q}} < x_q$. Compute this integral.

(c) Consider the region of phase space where $x_{\bar{q}}$ is the largest of the three variables. Show that the contribution to $d\sigma/dT$ from this region is given by a one-dimensional integral over x_q

over the interval $2(1 - x_{\bar{q}}) < x_q < x_{\bar{q}}$. Compute this integral.

(d) Consider the region of phase space where x_g is the largest of the three variables. Substitute $x_{\bar{q}} = 2 - x_q - x_g$ into the integrand of (10.48). Show that the contribution to $d\sigma/dT$ from this region is then given by a one-dimensional integral over x_q over the interval

$2(1 - x_g) < x_q < x_g$. Compute this integral.

(e) Assemble the pieces and show that

$$\frac{d\sigma}{dT} = \sigma(e^+e^- \to q\bar{q})$$
$$\cdot \frac{2\alpha_s}{3\pi} \Big[\frac{2(3T^2 - 3T + 2)}{T(1-T)} \log \frac{2T-1}{1-T}$$
$$- \frac{3(3T-2)(2-T)}{(1-T)} \Big] . \quad (12.28)$$

QCD at Hadron Colliders

13

The understanding of pdfs and jets presented in the previous chapter, gives us the conceptual tools to understand the basic features of proton-proton and proton-antiproton collisions at very high energy. Protons are composite states that contain quarks, antiquarks, and gluons. In a high-energy pp or $p\overline{p}$ collision, these particles can interact softly or through individual hard collisions. The soft collisions should be described by low-momentum transfer QCD forces. The hard collisions should be described by the QCD interactions of quarks and gluons.

Because QCD interactions are strong at low momentum transfer, the dominant feature of high-energy proton-proton collisions should be soft scattering events. It is useful to picture the protons as bags containing quarks, antiquarks, and gluon. The soft collisions can rip these bags open, liberating many partons, which then reform into hadrons. Soft collisions then should produce large numbers of final-state particles, but all with small transverse momentum relative to the original collision axis. Hard scattering of quarks and gluons should occur much more rarely. However, these hard-scattering reactions should be quite distinctive, since they should generate jets with very large momentum components transverse to the beam direction.

To the extent that we can separate the hard collisions from the soft reactions, we can test QCD in high-energy hadron-hadron collisions. In this chapter, I will describe the various levels of a hadron-hadron collision and methods for finding jets in hadron-hadron collision events. From this, we will see that QCD has characteristic predictions for hard-scattering processes in hadron-hadron collisions that are confirmed by experiment.

13.1 Hadron scattering at low momentum transfer

Figure 13.1 shows the total cross sections for a variety of hadron-hadron scattering reactions as a function of $\sqrt{s}$. The top curves are the $p\overline{p}$ and pp total cross sections. Notice that these cross sections are dominated by s-channel resonances up to about 2 GeV. At higher energies, the behavior is smooth, almost constant. Also, particle and antiparticle

The behavior of total cross sections for hadron-hadron scattering.

Concepts of Elementary Particle Physics. Michael E. Peskin.
© Michael E. Peskin 2019. Published in 2019 by Oxford University Press.
DOI: 10.1093/oso/9780198812180.001.0001

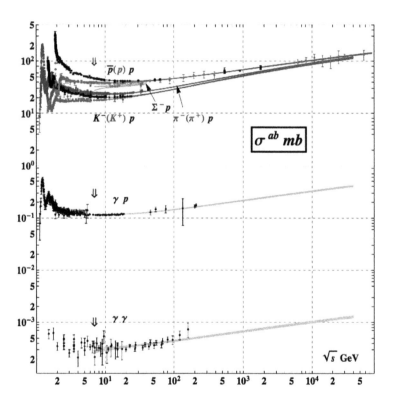

Fig. 13.1 Total cross sections for hadron-hadron collision processes, $\gamma p \to$ hadrons, and $\gamma\gamma \to$ hadrons, as a function of center of mass energy, from (Patrignani *et al.* 2016).

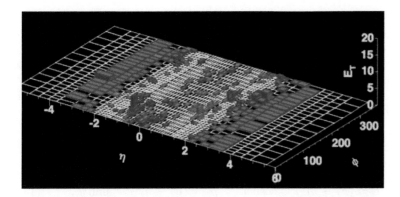

Fig. 13.2 Lego plot of a typical $p\bar{p}$ collision event recorded by the CDF experiment at the Tevatron collider at 2 TeV (figure courtesy of Fermilab and the CDF collaboration). The plot shows the distribution of E_T recorded by the calorimeters (pink for the electromagnetic calorimeter, blue for the hadronic calorimeter) as towers in the (η,ϕ) plane.

cross sections become identical:

$$\frac{\sigma(\bar{p}p)}{\sigma(pp)} \to 1 \qquad \frac{\sigma(\pi^- p)}{\sigma(\pi^+ p)} \to 1 \; . \qquad (13.1)$$

This behavior is called the *Pomeranchuk theorem* (Pomeranchuk 1958). It indicates that, at very high energies, the bulk of the total cross section is not generated by valence quarks but rather by soft collisions of the sea quarks and gluons. The $p\bar{p}$ and pp total cross sections at TeV energies are about 100 mb in size.

A typical $p\bar{p}$ event recorded by the CDF experiment at the Fermilab Tevatron collider at 1.96 TeV is shown in Fig. 13.2. The vertical scale shows the momentum transverse to the beam direction, called p_T or E_T, in GeV. Many particles are produced, but few have $E_T > 2$ GeV. The physics is that of many soft scatterings among the proton constituents.

I must pause for a moment to explain the coordinates used in this figure. To make this plot, we wrap a cylinder around the beam axis,

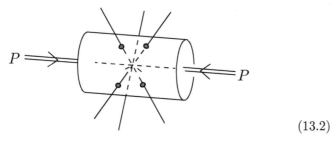

$$(13.2)$$

We then divide this cylinder into cells, measure the calorimetric energy deposition in each cell, and record the quantity

$$E_T = E \sin\theta \; , \qquad (13.3)$$

the deposited energy projected onto the direction transverse to the beam axis. The quantity E_T is called the *transverse energy*. Finally, we un-

E_T is defined as the energy deposition in a calorimeter, projected onto directions transverse to the beam direction:

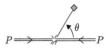

wrap the cylinder and plot the measured E_T in each cell as the height of a tower. This event display is called a *Lego plot*, for obvious reasons. One coordinate is the azimuthal angle ϕ around the cylinder. We might take the other coordinate to be the polar angle θ, but there is a better choice.

A particle momentum (for simplicity, in the $\hat{1}$, $\hat{3}$ plane, where $\hat{3}$ is the beam axis) has the form

$$(E, p_T, 0, p_\parallel) \qquad E^2 = p_T^2 + p_\parallel^2 + m^2 . \tag{13.4}$$

We can represent the components of this vector as

$$E = (p_T^2 + m^2)^{1/2} \cosh y \qquad p_\parallel = (p_T^2 + m^2)^{1/2} \sinh y . \tag{13.5}$$

The variable y is called the *rapidity*,

Definition of the *rapidity* y of a particle.

$$y = \tanh^{-1} \frac{p_\parallel}{E} . \tag{13.6}$$

A boost along the $\hat{3}$ axis gives

$$E' = \gamma(E + \beta p_\parallel) \qquad p'_\parallel = \gamma(p_\parallel + \beta E) \qquad p'_T = p_T . \tag{13.7}$$

For a boost by β, $\gamma^2(1 - \beta^2) = 1$, so we can represent the magnitude of the boost by writing

$$\gamma = \cosh \alpha , \qquad \gamma\beta = \sinh \alpha . \tag{13.8}$$

Then

$$E' = (p_T^2 + m^2)^{1/2}[\cosh y \cosh \alpha + \sinh y \sinh \alpha]$$
$$p'_\parallel = (p_T^2 + m^2)^{1/2}[\sinh y \cosh \alpha + \cosh y \sinh \alpha] \tag{13.9}$$

so that

$$E' = (p_T^2 + m^2)^{1/2} \cosh(y + \alpha) \qquad p'_\parallel = (p_T^2 + m^2)^{1/2} \sinh(y + \alpha) . \tag{13.10}$$

We see that a boost along the beam axis is a simple translation of y. In the parton model, the components of the proton have all values of momentum, up to the total momentum of the proton. For most parton-parton collisions, the CM system is boosted along $\hat{3}$ relative to the lab frame. So it is useful to use a variable that is transformed very simply by a boost.

At large values of the energy, we can often ignore the mass of the particle. Then $E = |\vec{p}|$ and the above relations become

$$p = p_T \cosh y \qquad p_\parallel = p_T \sinh y . \tag{13.11}$$

Since

$$p + p_\parallel = p_T e^y , \tag{13.12}$$

the quantity y can be computed as

$$y = \log \frac{p + p_\parallel}{p_T} = \log \frac{1 + \cos\theta}{\sin\theta} = \frac{1}{2} \log \frac{1 + \cos\theta}{1 - \cos\theta} . \tag{13.13}$$

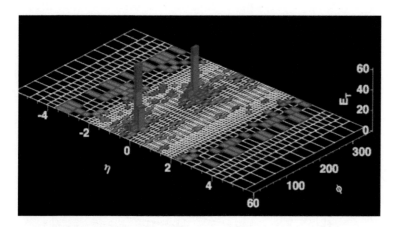

Fig. 13.3 Lego plot of a $p\bar{p}$ collision event recorded by the CDF experiment at the Tevatron collider at 2 TeV with two jets in the final state (figure courtesy of Fermilab and the CDF collaboration).

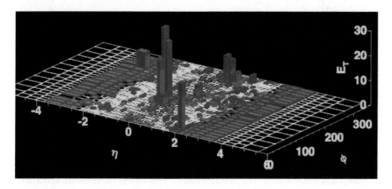

Fig. 13.4 Lego plot of a $p\bar{p}$ collision event recorded by the CDF experiment at the Tevatron collider at 2 TeV with four jets in the final state (figure courtesy of Fermilab and the CDF collaboration).

It then makes sense to define the *pseudo-rapidity* η of a particle or an energy deposition as

$$\eta = \frac{1}{2}\log\frac{1+\cos\theta}{1-\cos\theta} \ . \tag{13.14}$$

Definition of the *pseudo-rapidity* of a particle.

This quantity is directly computable from the particle's polar angle θ, η without identifying the particle or even separating particles from one another. For example, η can be computed from the location of an energy deposition in a calorimeter. For pions, or for other hadrons at high momentum, η is a good proxy for the rapidity. This is then the natural variable to use instead of the polar angle in analyzing hadron-hadron collisions. In typical pp collisions, the particle production is roughly uniform in η and ϕ, at least for $|\eta| < 3$.

More rarely, $p\bar{p}$ collision events have the form shown in Fig. 13.3. We see two jets with $E_T \sim 50$ GeV standing out above the soft debris from the $p\bar{p}$ collision. It is not hard to imagine that this event contains a hard quark-antiquark collision. Multijet final states are also seen, as in

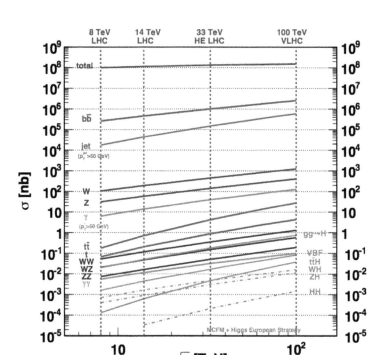

Fig. 13.5 QCD predictions as a function of center of mass energy for cross sections for a variety of processes observable in pp collisions, from (Campbell *et al.* 2013).

Fig. 13.4.

Figure 13.5 shows a QCD prediction of the rates of various components of the pp total cross sections as a function of CM energy for LHC energies and above. At the 13 TeV CM energy of the LHC, the total cross section is about 100 mb. The cross secton for production of a jet with $p_T > 50$ GeV is 20 μb, smaller by a factor of 10^{-4}. This rate is still enormous compared to the rates for more exotic processes such as the production of weak-interaction bosons and top quarks. To study the whole range of physics processes available at a hadron collider, it is necessary to accumulate huge quantities of data and to filter this data very effectively to find rare classes of events.

13.2   Hadron scattering at large momentum transfer

I have already remarked that it is difficult to build a quantitative theory of the pp total cross section. However, we can build models of jet production and other hard processes by combining the parton model of the proton with scattering cross sections computed from QCD. I will now sketch the theory for the rate of 2-jet events in pp collisions.

To construct this theory, imagine taking one parton—a quark, anti-

quark, or gluon—from each proton using its known pdf. These partons
can then scatter, as shown in the process

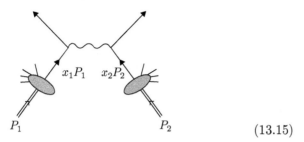

$$(13.15)$$

Let θ_* be the polar angle for scattering measured in the parton-parton
CM frame. Since the final jets are observed, we can boost to this frame
and measure θ_*. Each 2-jet reaction can then be viewed as a 2-parton
scattering process in QCD with known kinematics. The 2-jet production
cross section is then given by the formula

$$\sigma(pp \to 2 \text{ jets}) = \sum_{ijk\ell} \int dx_1 f_i(x_1) \int dx_2 f_j(x_2)$$

$$\int d\cos\theta_* \cdot \frac{d\sigma}{d\cos\theta_*}(ij \to k\ell) , \quad (13.16)$$

The parton model, combined with
quark-gluon scattering cross sections
from QCD, gives a definite prediction
for the rate of production of 2 jet events
in proton-proton scattering.

The cross sections needed on the right-hand side can be computed in
QCD as quark-gluon reactions. The indices i, j, k, ℓ run over all possible
quarks, antiquarks, and gluons.

The kinematics of the parton-parton scattering reaction is described
by the Lorentz invariants $\hat{s}, \hat{t}, \hat{u}$. The CM energy of the parton reaction
is related to the total CM energy by

$$\hat{s} = (p_i + p_j)^2 = 2p_i \cdot p_j = 2x_1 x_2 P_1 \cdot P_2 , \quad (13.17)$$

where P_1, P_2 are the initial proton momenta. We can thus identify

$$\hat{s} = x_1 x_2 s . \quad (13.18)$$

By measuring the momenta of the final jets, we can determine $\hat{s}$ and θ_*
and, from these, obtain $\hat{t}, \hat{u}$. To evaluate the formula, we need to supply
the values of the differential cross sections for the various scattering
processes of quarks and gluons. This computation is straightforward
but somewhat beyond the level of this book. In the rest of this section,
I will sketch some accessible properties of these cross sections. The full
expressions for these cross sections, at leading order in QCD, are given
in Appendix E.

In Section 9.4, we computed the cross section for $e^- q$ scattering (9.51),

$$\frac{d\sigma}{d\cos\theta}(eq \to eq) = \frac{\pi Q_f^2 \alpha^2}{s} \frac{s^2 + u^2}{t^2} . \quad (13.19)$$

The scattering of quarks of different flavor, for example $ud \to ud$, is
described by the same formula, replacing α by α_s and supplying an

appropriate factor for color. We must average over initial colors and sum over final colors. The process

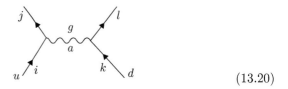

(13.20)

leads to the color factor

$$\frac{1}{3} \cdot \frac{1}{3} \sum_{ijk\ell a} |t_{ji}^a t_{\ell k}^a|^2 ,$$

(13.21)

Computation of the color factor for quark-quark scattering.

where here $ijk\ell$ are the color indices of the corresponding partons. Evaluating, (13.21) becomes

$$\frac{1}{9}(t^b ij t^b k\ell)(t_{ji}^a t_{\ell k}^a) = \frac{1}{9}\mathrm{tr}[t^b t^a]\mathrm{tr}[t^b t^a]$$
$$= \frac{1}{9}\frac{1}{2}\delta^{ab}\frac{1}{2}\delta^{ab} = \frac{1}{9} \cdot \frac{1}{2} \cdot \frac{1}{2} \cdot 8 ,$$

(13.22)

so, finally, the color factor (13.21) is

$$\frac{1}{3} \cdot \frac{1}{3} \sum_{\mathrm{colors}} |t_{ji}^a t_{\ell k}^a|^2 = \frac{2}{9} .$$

(13.23)

Then

$$\frac{d\sigma}{d\cos\theta_*}(ud \to ud) = \frac{2}{9}\frac{\pi\alpha_s^2}{\hat{s}}\frac{\hat{s}^2 + \hat{u}^2}{\hat{t}^2} .$$

(13.24)

For quarks of the same flavor, there is an additional Feynman diagram that must be added to the amplitude $\mathcal{M}$. In the process $uu \to uu$, because u quarks are identical particles, either final-state u quark can go into either observed jet. Then the complete scattering amplitude is a sum of the expressions for the two Feynman diagrams.

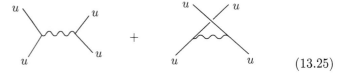

(13.25)

and the sum of diagrams is squared to give the total rate. This leads to the expression

$$\frac{d\sigma}{d\cos\theta_*}(uu \to uu) = \frac{2}{9}\frac{\pi\alpha_s^2}{\hat{s}}\left[\frac{\hat{s}^2 + \hat{u}^2}{\hat{t}^2} + \frac{\hat{s}^2 + \hat{t}^2}{\hat{u}^2} - \frac{2}{3}\frac{\hat{s}^2}{\hat{u}\hat{t}}\right].$$

(13.26)

In the forward direction, $\theta_* \to 0$, $\hat{t} \to 0$, and $\hat{u} \to -\hat{s}$. Then (13.24) and (13.26) both have the singular behavior

$$\frac{d\sigma}{d\cos\theta_*}(qq \to qq) \sim \frac{4}{9}\frac{\pi\alpha_s^2}{\hat{s}}\frac{\hat{s}^2}{\hat{t}^2} \sim \frac{4}{9}\frac{\pi\alpha^2}{\hat{s}}\frac{1}{\sin^4\theta/2} .$$

(13.27)

We recognize this as Coulomb scattering by the QCD potential. The formulae for scattering processes involving quarks and antiquarks are slightly more complicated, but in all cases the singular Coulomb term is the same. We must also include less singular processes such as $u\bar{u} \to d\bar{d}$,

$$(13.28)$$

The most singular terms for qg, $\bar{q}g$, and gg scattering follow in the same way. For qg or $\bar{q}g$ scattering, the cross section for Coulomb exchange is obtained by replacing one factor of the quark squared charge factor $\frac{4}{3}$ by the gluon squared charge factor 3. This gives

$$\frac{d\sigma}{d\cos\theta_*}(qq \to qq) \sim \frac{\pi\alpha_s^2}{\hat{s}}\frac{\hat{s}^2}{\hat{t}^2} .$$

$$(13.29)$$

The complete expression, obtained by summing the diagrams

$$(13.30)$$

is

$$\frac{d\sigma}{d\cos\theta_*}(qg \to qg) = \frac{\pi\alpha_s^2}{2\hat{s}}\left[\frac{\hat{s}^2+\hat{u}^2}{\hat{t}^2} - \frac{4}{9}\left(\frac{\hat{u}}{\hat{s}}+\frac{\hat{s}}{\hat{u}}\right)\right].$$

$$(13.31)$$

Less singular processes such as $gg \to q\bar{q}$ must also be included.

For gg scattering, we replace two factors of $\frac{4}{3}$ with two factors of 3 to find

$$\frac{d\sigma}{d\cos\theta_*}(gg \to gg) \sim \frac{9}{4}\frac{\pi\alpha_s^2}{\hat{s}}\frac{\hat{s}^2}{\hat{t}^2} .$$

$$(13.32)$$

The complete list of parton-parton QCD cross sections is given in Appendix E. By folding these expressions with the corresponding pdfs, we find the leading order QCD prediction of the jet production cross section in pp collisions. Figure 13.6 shows the leading-order QCD prediction for proton-proton collisions at the LHC at a center of mass energy of 13 TeV. The prediction has two important features. First, the differential cross section falls by 6 orders of magnitude as the jet p_T is increased from 100 GeV to 1000 GeV. Second, different parton reactions dominate in different regimes of p_T. At low p_T, gg scattering is most important, since this process has the largest intrinsic cross section. However, the valence quarks in the proton have higher energy than the gluons, so quark scattering processes are increasingly important at high p_T. As p_T increases across the figure, the dominant role is played, first by gg scattering, then by qg scattering, and, finally, at the highest values of p_T, by qq scattering.

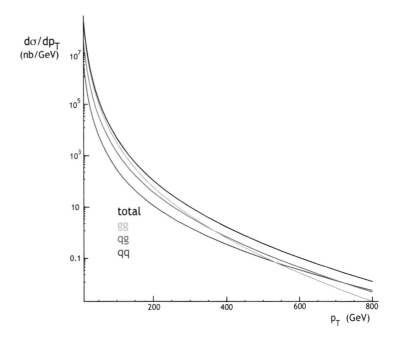

Fig. 13.6 Leading order QCD prediction for the rate of jet production in proton-proton collisions reactions as a function of the jet transverse momentum. The three lower curves show the contributions from gg, qg, and qq scattering processes.

13.3   Jet structure observables for hadron collisions

To compare the QCD prediction to data, we need precise definitions of the observables that we will relate between theory and experiment. If we wish to discuss jets in hadron-hadron collisions, we will need to define these jets in a robust way that we can apply to observed events. The jet algorithms used for hadron collisions are somewhat different from those we applied in the simpler environment of e^+e^- annihilation. I will then first describe methods for defining jets in hadronic collisions and then show how measurements on these jets compare to QCD predictions.

Hadron-hadron collisions contain many particles in the final state. Most of these particles are not associated with a hard scattering process but rather are liberated when the colliding protons are disrupted. These soft particles are produced roughly uniformly in pseudo-rapidity. Most of them, then, are emitted into angular regions near the beam direction. Collider detectors such as ATLAS and CMS are not sensitive to particles produced at very small angles with $|\eta| > 5$. So we cannot use jet definitions that require knowledge of all particles in the event. Instead, we need to define observables that are built from particles in the central rapidity region and that emphasize particle with large transverse momenta. Hadron collider experiments also typically measure calorimetric energy, which sums over particles, rather than individual particle

Difficulties for defining jets in hadron-hadron collisions.

momenta.

A useful approach is to look at the distribution of E_T over the (η,ϕ) plane, as we saw in the event displays. Instead of using y_{ij} as a criterion for clustering particles as we did in (12.23), we can use distance in the (η,ϕ) plane

$$\Delta R_{ij} = \left[(\Delta\eta_{ij})^2 + (\Delta\phi_{ij})^2\right]^{1/2}. \tag{13.33}$$

Again, we combine 4-vectors until all composite particles are separated by a distance larger than a predetermined quantity R, called the *cone size*. This clusters energy into jets that correspond roughly to circles in the (η,ϕ) plane or cones in 3 dimensions.

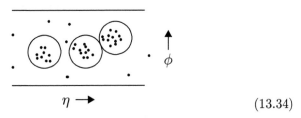

$$(13.34)$$

This *cone jet* algorithm is IR-safe. It remains IR-safe if the energy elements are weighted by their E_T. A convenient definition used by the LHC experiments is to use as the clustering criterion

IR-safe algorithms for defining jets that are well-adapted to the physics of hadron-hadron collisions.

$$\lambda_{ij} = \Delta R_{ij}^2 \cdot \min(E_{Ti}^{-2}, E_{Tj}^{-2}) \tag{13.35}$$

This defines the *anti-k_T* jet algorithm (Cacciari *et al.* 2008). It gathers elements with small E_T in a neat circle or cone around a nearby element with large E_T. Experiments in $p\bar{p}$ scattering at the Tevatron typically used cone jets with $R = 0.7$. Experiments in pp scattering at the LHC typically use anti-k_T jets with $R = 0.4$ or 0.5. It is important to remember that the cone size R is, in principle, arbitrary. Smaller R leads to more or more highly resolved jets.

Figure 13.7 shows a comparison of theory and experiment for the jet production rate from the ATLAS experiment at the LHC. Both for the theoretical calculation and for the analysis of the experimental data, the jets are defined to be $R = 0.4$ anti-k_T jets. The theory calculation is carried out to higher order in QCD, so that it includes final states with 2 and more partons. The cross section for producing a jet at fixed p_T, plus any other jet activity in the event, is compared, in intervals of η, with the QCD theory. It is evident that QCD correctly tracks the full dependence on p_T and η.

Using a fixed jet algorithm to define jets, we can compare the observed rate of jet production to that predicted by QCD.

13.4   The width of a jet in hadron-hadron collisions

The physics topics that we have discussed in this chapter can be combined to produce a very rich array of predictions for high energy QCD

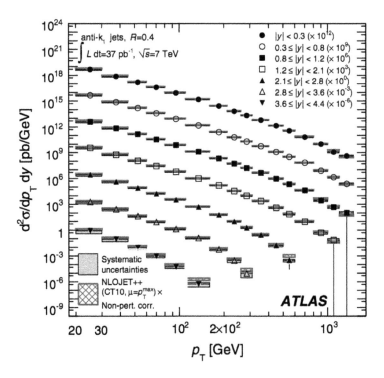

Fig. 13.7 Comparison of QCD theory and experiment for the differential cross section to produce a jet in pp collisions at 7 TeV over a large range of p_T and y ($= \eta$), measured by the ATLAS experiment at the LHC, from (Aad *et al.* 2012b).

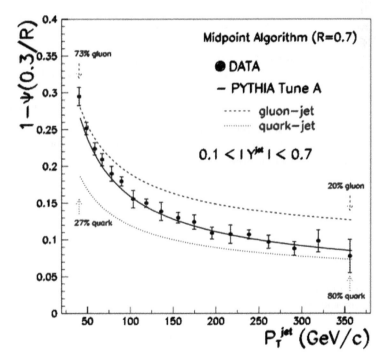

Fig. 13.8 Measurement of the jet size variable $\psi(\rho)$ as a function of p_T in $p\bar{p}$ collisions at 2 TeV by the CDF experiment at the Tevatron collider, and comparison to QCD predictions from the simulation program PYTHIA, from (Acosta *et al.* 2005).

scattering processes. To conclude our discussion of hadron collisions, I will present two of these of special interest.

The first of these involves the QCD prediction of the width of a jet observed in hadron-hadron collisions as a function of p_T. The prediction, and the comparison with experiment, is shown in Fig. 13.8.

This measurement, in $p\bar{p}$ collisions at 1.96 TeV by the CDF experiment, defines jets with a cone size $R = 0.7$. It is possible to define the width of the jet by looking at the flow of transverse energy inside the cone. To study this, we can construct the following quantity: Let ρ be a parameter with values between 0 and 1. Let E_{Ti} label the individual depositions of E_T used to construct the jet. Then let

A parameter that measures the width of a jet observed in hadron-hadron collisions.

$$\psi(\rho) = \sum_{\Delta R < \rho R} E_{Ti} \; / \sum_{\Delta R < R} E_{Ti} \; . \tag{13.36}$$

This quantity measures the fraction of the jet E_T that is contained within a narrower cone, of size $r = \rho R$ on the same axis as the original cone of size R. Clearly

$$0 < \psi(\rho) < 1 \; . \tag{13.37}$$

For fixed $\rho < 1$, smaller values of $(1 - \psi(\rho))$ indicate narrower jets.

In our lowest order description of QCD scattering, each jet contains one parton, the one giving rise to the jet.

$$(13.38)$$

At the next order in α_s, this parton can radiate, typically producing an almost collinear gluon but sometimes radiating a parton at larger angle. This gives the jet a width.

$$(13.39)$$

Emission of many quarks and gluons produces a range of widths. The variable $(1 - \psi(\rho))$ can then be computed by summing QCD processes or by a simulation program such as PYTHIA.

$$(13.40)$$

We might expect that, as p_T is increased, the width of the jet would decrease as $\alpha_s(p_T)$ decreases.

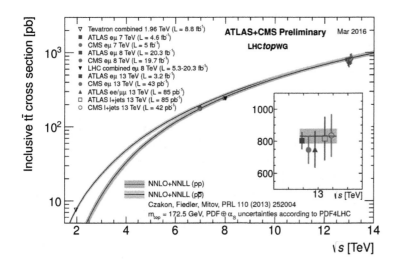

Fig. 13.9 Measurements of the cross section for production of a pair of top quarks in $p\bar{p}$ and pp collisions at the Tevatron collider and the LHC, as a function of energy, compared to predictions from QCD, from the LHC Top Quark Working Group (figure courtesy of CERN and the ATLAS collaboration).

However, there is another effect. As I explained in (12.14), gluon jets contain more radiation than quark jets. As a result, gluon jets are wider than quark jets. As we move from small p_T to large p_T, we move from the region of the 2-jet p_T distribution dominated by gg scattering to the region dominated by valence $q\bar{q}$ scattering. This also leads to narrower jets as a function of p_T.

> Two different effects predicted by QCD are needed to explain the dependence of jet width on p_T. The combination of these effects does successfully explain the data.

In Fig. 13.8, the average value of $(1 - \psi)$ for $\rho = 0.3$ and $R = 0.7$, measured as a function of the jet p_T by the CDF experiment at the Tevatron, is compared to the prediction from PYTHIA (Acosta *et al.* 2005). The top reference curve shows the variation in the width of gluon jets as a function of p_T. The bottom reference curve shows the variation in the width of quark jets as a function of p_T. The data interpolates between these limits, showing both the narrowing of jets with p_T and the change in the jet sample composition.

13.5 Production of the top quark

The second of these QCD predictions concerns the rate of production of the heavy quark t or top. The top quark has a mass of 173 GeV. In QCD, the top quark can be pair-produced from quark-antiquark annihilation

> Mechanisms for top quark pair production in hadron-hadron collisions.

(13.41)

or from gluon-gluon annihilation.

$$\text{(13.42)}$$

Figure 13.9 shows measurements of the top quark pair production cross section at the LHC at 7, 8, and 13 TeV and the average of measurements at the Tevatron at 1.96 TeV. The blue and green curves are the QCD theory predictions for $p\bar{p}$ and pp collisions as a function of energy.

The theory prediction has a quite unusual feature. As the energy increases by only a factor of 3.5 to 4 from the Tevatron to the LHC, the cross section increases by a factor of 20. The predicted LHC cross section is actually a factor of 100 higher than the QCD prediction for the cross section in pp collisions at 2 TeV. The measurements confirm this energy-dependence with high accuracy. But, what is the origin of this effect?

Two features of the QCD prediction come into play. First, top quark pair production requires a parton-parton CM energy of 350 GeV or more. So, at the Tevatron, it requires a collision of two partons, each of which is carrying more than 15% of the total energy of the proton or antiproton. This criterion is met only for the valence quarks and antiquarks. As the energy of the collider is increased, more of the partons can participate in $t\bar{t}$ production, increasing the predicted rate.

Two distinct QCD effects explain the sharp rise in the cross section for producing top quark pairs over the collider energy range 2 TeV to 13 TeV. The full QCD theory is in good agreement with the measurements.

In addition, the production of $t\bar{t}$ from gluons has an intrinsically larger cross section than the production from quarks of the same energy, by about a factor of 5. As the collider energy is increased, gluons from the parton sea have enough energy to produce top quarks. The large value of the cross section and the large value of the gluon pdf lead to a dramatically increased prediction. The dominance of production by gluons is seen in the theory prediction by the approximate equality of the predictions for pp at $p\bar{p}$ collisions at the highest energies.

These two examples give just a sampling of the wide variety of phenomena that are observed and explained by QCD in hadron collider physics.

Exercises

(13.1) Solve the relation (13.14) for $\cos\theta$ in terms of the pseudorapidity η. Find the corresponding values of θ for $\eta = 1,2,3,4,5,6$. What is the practical limit in η for measuring the tracks of charged particles produced in a hadron collider event?

(13.2) The W and Z bosons are massive spin 1 particles that are the basic quanta of the weak interactions. We will discuss their properties in detail in Chapter 17. Both the W and the Z were discovered by the UA1 experiment at the CERN proton-antiproton collider. The Z boson has a decay $Z^0 \to e^+e^-$, so it is possible to observe the Z as a resonance by identifying events with muone^+e^- pairs and plotting the distribution of the e^+e^- invariant mass. The corresponding decay of the W is $W^+ \to e^+\nu$. This decay cannot be observed as a resonance because neutrinos are not observed by collider detectors. This problem will explain how the mass of the W was measured and give tools for computing the mass of other particles that decay with unobservable decay products.

(a) Although neutrinos are not observed by hadron collider detectors, these detectors can observe other particles and measure the imbalance of observed momentum in the final state. Typically, the measurement of the missing momentum $\vec{p}_T$ in the two directions orthogonal to the beam direction is good, while the measurement of the missing p^3, parallel to the beam direction, is poor. Why is the imbalance of p^3 difficult to measure? In the following, we set $\vec{p}_T(\nu)$ equal to the missing transverse momentum.

(b) For a W^+ produced at a hadron collider and decaying to $e^+\nu$, we can treat the final e^+ and ν as massless. Write the 4-vectors for the momenta of these particles in terms of the η and $\vec{p}_T$ for each, where η is the pseudorapidity and $\vec{p}_T$ is a 2-component vector transverse to the beam asix.

(c) Using this representation of the 4-vectors, compute $m_W^2 = (p(e) + p(\nu))^2$.

(d) Show that
$$m_W^2 \geq m_{\text{tr}}^2 , \qquad (13.43)$$

where m_{tr}, the *transverse mass*, is given by
$$m_{\text{tr}}^2 = (|\vec{p}_T(e)| + |\vec{p}_T(\nu)|)^2 - (\vec{p}_T(e) + \vec{p}_T(\nu))^2 . \qquad (13.44)$$
In practice, the distribution of m_{tr} is strongly peaked toward this upper limit and thus allows an accurate estimate of m_W.

(13.3) The Drell-Yan process is the reaction in a pp or $p\bar{p}$ collision that produces a muon or electron pair. The underlying process is $q\bar{q} \to e^+e^-$ or $\mu^+\mu^-$. In this problem, we will work out the parton-model description of this process. Ignore all quark and lepton masses.

(a) Write the total cross section for $q_f\bar{q}_f \to \mu^+\mu^-$ as a function of the quark-antiquark center of mass energy, ignoring all fermion masses. The factor for color should be $\frac{1}{3}$ rather than the 3 in the formula for the $e^+e^- \to q_f\bar{q}_f$ cross section. Why? Aside from this factor, you can get the rest of the expression from our analysis of $e^+e^- \to q_f\bar{q}_f$.

(b) In the parton model, the cross section for $pp \to \mu^+\mu^- + X$ is given by
$$\sum_f \left[\int dx_1\, f_f(x_1) \int dx_2\, f_{\bar{f}}(x_2) \right.$$
$$\left. \cdot \sigma(q_f(x_1 P_1)\bar{q}_{\bar{f}}(x_2 P_2) \to \mu^+\mu^-) + (f \leftrightarrow \bar{f}) \right], \qquad (13.45)$$

where the sum runs over quark flavors. Write an expression for $\hat{s}$ for the parton reaction in terms of x_1, x_2, and s for the pp collision. Note that $\hat{s} = M^2$, the mass-squared of the observed $\mu^+\mu^-$ system.

(c) Working in the pp CM system, write the 4-vectors of the initial quark and antiquark. Let $(\mathcal{E}, 0, 0, \mathcal{P})$ be the sum of these momenta. This is also the total momentum of the $\mu^+\mu^-$ system. The *rapidity* y of the $\mu^+\mu^-$ system is defined by
$$\tanh y = \mathcal{P}/\mathcal{E} \qquad (13.46)$$
Verify that $\mathcal{E} = M \cosh y$ and $\mathcal{P} = M \sinh y$. Write an expression for y in terms of x_1 and x_2.

(d) Write the converse expressions for x_1 and x_2 in terms of y and M (with s fixed). Notice that in this process, as in deep inelastic scattering, we can determine the values of the parton momentum fractions by measuring only the lepton momenta.

(e) In the cross section formula, change variables from x_1, x_2 to M, y. Use the Jacobian determinant to convert $dx_1 dx_2$ to $dM dy$. Show that

$$\frac{d}{dM dy} \sigma(pp \to \mu^+\mu^- + X) =$$

$$\sum_f [x_1 f_f(x_1) x_2 f_{\bar{f}}(x_2) + x_1 f_{\bar{f}}(x_1) x_2 f_f(x_2)]$$

$$\cdot \frac{8}{9} \frac{\pi \alpha^2}{M^3}, \qquad (13.47)$$

where x_1 and x_2 are the values derived from the measurement of M and y.

(f) The equation (13.47) has two terms, one with a quark from proton 1 and an antiquark from proton 2, and the other in which the antiquark comes from proton 1. We might break this down further into contributions from valence quarks and sea quarks annihilating with antiquarks. For concreteness, think about the LHC, where the pp center of mass energy is 13000 GeV, a typical value of M is 90 GeV, valence quarks have $x > 0.05$, and sea quarks have $x < 0.01$ (see Fig. 10.1). Argue that, near $y = 0$, most annihilations are sea with sea. At what value of y should sea with valence annihilations be important?

(13.4) The Drell-Yan process discussed in Exercise 13.3 involves the annihilation of an initial-state q and $\bar{q}$. It is possible that a photon or a gluon could be radiated in this annihilation process. This problem will estimate the probability of this initial-state radiation. In Chapter 10, in (10.39) and (10.42), we wrote expressions for the probability that an initial-state highly relativistic quark emits an approximately collinear photon or gluon. The formula is a double integral, dominated by a term with two large logarithms. For the estimates in this problem, it is sufficient to keep only the term from the evaluation of the integrals with this double logarithm.

(a) Find an expression, using the approximation of keeping double logarithmic terms only, for the probability that a Drell-Yan event with a muon pair of mass M also contains a radiated photon with momentum transverse to the beam direction greater than p_T. The needed limits of integration can be obtained from the following considerations: For the q_T integral, the radiated photon is no longer approximately collinear if $q_T > M/2$. For the z integral, the photon is no longer approximately collinear if the longitudinal momentum of the photon, approximately $zM/2$ in the parton-parton center of mass frame, is less than q_T.

(b) Evaluate this expression for some typical parameters of the Drell-Yan cross section measurement at the LHC: $M = 300$ GeV, $p_T = 30$ GeV.

(c) In a similar way, estimate the probability for a Drell-Yan event to contain a radiated gluon with transverse momentum greater than p_T. In this case, the final state will contain a $\mu^+\mu^-$ pair and the gluon jet.

(d) Evaluate this expression for $M = 300$ GeV, $p_T = 30$ GeV, using $\alpha_s = 0.2$. What is the probability that one of these events will contain a gluon jet?

Chiral Symmetry

14

Before we finish with the strong interactions, there is one more aspect of QCD that we need to discuss. So far in this book, I have treated quark masses as parameters of the nonrelativistic quark model—or ignored them altogether. But, what are the values of the quark masses?

For heavy quarks, it is probably correct to use as a first approximation

$$m_c \approx \frac{1}{2}m(J/\psi) \approx 1.5 \text{ GeV} ,$$

$$m_b \approx \frac{1}{2}m(\Upsilon) \approx 4.5 \text{ GeV} . \tag{14.1}$$

These estimates of the masses can be refined using more accurate QCD descriptions of the heavy quark-antiquark bound states.

To quote the masses of light quarks, however, we will need to develop a better understanding of the properties of QCD at low energies. It turns out that there is a new principle at work here, called *chiral symmetry*, which gives additional insight into the nature of the lightest hadrons.

Chiral symmetry is a symmetry of QCD in the limit that the quark masses are set equal to zero. However, this symmetry is not manifest in the spectrum of hadron masses, even after we correct for the fact that the masses of quarks are not exactly zero. Instead, this symmetry is realized in a different way; it is said to be *spontaneously broken*. In the course of this chapter, I will introduce the notion of a spontaneously broken symmetry and discuss its consequences.

The concepts of chiral symmetry and spontaneous symmetry breaking both have an important role to play in the theory of the weak interaction that I will present in Part III. The study of this last aspect of QCD will give us a useful starting point for the more general understanding of these ideas.

14.1 Symmetries of QCD with zero quark masses

The Lagrangian of QCD was given in (11.44) as

$$\mathcal{L} = -\frac{1}{4}F^{\mu\nu a}F^a_{\mu\nu} + \overline{\Psi}_f(i\gamma^\mu D_\mu - m)\Psi_f . \tag{14.2}$$

Concepts of Elementary Particle Physics. Michael E. Peskin.
© Michael E. Peskin 2019. Published in 2019 by Oxford University Press.
DOI: 10.1093/oso/9780198812180.001.0001

The symmetries of this Lagrangian include Lorentz invariance, P, C, and T, global charge conservation (equivalent to quark or baryon number conservation), and the $SU(3)$ color symmetry of QCD.

Other symmetries might be present depending on the values of the quark masses. Consider for the moment a model containing only the two quark species u and d. If $m_u = m_d$, the Lagrangian of this model would be invariant under the continuous group of isospin rotations

$$\Psi_i \to \left(e^{i\vec{\alpha}\cdot\vec{\sigma}/2}\right)_{ij}\Psi_j\,, \tag{14.3}$$

where i, j run over the values u, d. More generally, let m be the 2×2 quark mass matrix,

$$m = \begin{pmatrix} m_u & 0 \\ 0 & m_d \end{pmatrix}\,. \tag{14.4}$$

An isospin rotation changes the Lagrangian according to

$$\overline{\Psi}(i\gamma^\mu D_\mu - m)\Psi \to \overline{\Psi}e^{-i\vec{\alpha}\cdot\vec{\sigma}/2}(i\gamma^\mu D_\mu - m)e^{i\vec{\alpha}\cdot\vec{\sigma}/2}\Psi$$
$$= \overline{\Psi}(i\gamma^\mu D_\mu)\Psi - \overline{\Psi}\left(e^{-i\vec{\alpha}\cdot\vec{\sigma}/2}me^{i\vec{\alpha}\cdot\vec{\sigma}/2}\right)\Psi. \tag{14.5}$$

Then if $[m, \sigma^a] = 0$ for $a = 1, 2, 3$, the Lagrangian is invariant under isospin. The criterion for this is $m_u = m_d$.

In the special case $m_u = m_d = 0$, there is an extension of the group of symmetries. For each of the two flavors, write

$$\Psi_f = \begin{pmatrix} \psi_{fL} \\ \psi_{fR} \end{pmatrix}\,. \tag{14.6}$$

Definition of the Dirac matrix γ^5.

We will find it useful to define the 4×4 matrix

$$\gamma^5 = \begin{pmatrix} -1 & 0 \\ 0 & 1 \end{pmatrix}\,. \tag{14.7}$$

This matrix anticommutes will all four Dirac matrices γ^μ, as one can see by explicit computation or by noting that

$$\gamma^5 = i\gamma^0\gamma^1\gamma^2\gamma^3\,. \tag{14.8}$$

The components ψ_{fL} and ψ_{fR} can then be identified as the eigenstates of γ^5 with eigenvalues -1, $+1$, respectively. Parity reverses the three space dimensions, so when γ^5 appears in a Dirac bilinear, it will acquire a factor -1 in a parity transformation.

Recall from our discussion in Section 8.2 that, if there is no mass term, the two pieces of the Dirac fermion do not couple directly in the Lagrangian. Using this idea, we can rewrite the QCD Lagrangian with zero mass u and d quarks as

$$\mathcal{L} = -\frac{1}{4}(F_{\mu\nu}{}^a)^2 + \sum_{f=u,d}\left\{\psi_{fL}^\dagger i\vec{\sigma}\cdot D\psi_{fL} + \psi_{fR}^\dagger i\sigma\cdot D\psi_{fR}\right\}\,. \tag{14.9}$$

This expression is invariant under two separate isospin symmetries

$$\psi_L \to e^{i\vec{\gamma}\cdot\vec{\sigma}/2}\psi_L\,, \qquad \psi_R \to e^{i\vec{\delta}\cdot\vec{\sigma}/2}\psi_R\,, \tag{14.10}$$

called $SU(2)_L$ and $SU(2)_R$. Alternatively, we can take linear combinations of these and write their actions on the Dirac fields

$$\Psi \to e^{i\vec{\alpha}\cdot\vec{\sigma}\ 2}\Psi \qquad \Psi \to e^{i\vec{\beta}\cdot(\vec{\sigma}\ 2)\ \gamma^5}\Psi \ , \tag{14.11}$$

The first of these symmetries is isospin. The second, which rotates the L and R fields in opposite directions, is called *chiral SU(2)*.

Corresponding to the two isospin symmetries, there are two sets of currents

$$j^{\mu a} = \overline{\Psi}\gamma^\mu \frac{\sigma^a}{2}\Psi \qquad j^{\mu 5 a} = \overline{\Psi}\gamma^\mu\gamma^5\frac{\sigma^a}{2}\Psi \tag{14.12}$$

The currents $j^{\mu a}$ should be conserved when $m_u = m_d$. The Lagrangian (14.2) leads to the two Dirac equations

$$(i\gamma^\mu D_\mu - m)\Psi = 0 \qquad -iD_\mu\overline{\Psi}\gamma^\mu - \overline{\Psi}m = 0 \ , \tag{14.13}$$

with $D_\mu\overline{\Psi} = \partial_\mu\overline{\Psi} + ig\overline{\Psi}A^a_\mu t^a$. Then

$$\partial_\mu j^{\mu a} = \overline{\Psi}\frac{\sigma^a}{2}\gamma^\mu D_\mu\Psi + D_\mu\overline{\Psi}\gamma^\mu\frac{\sigma^a}{2}\Psi$$
$$= -i\overline{\Psi}\frac{\sigma^a}{2}m\Psi + i\overline{\Psi}m\frac{\sigma^a}{2}\Psi$$
$$= i\overline{\Psi}\,[m, \frac{\sigma^a}{2}]\Psi \ . \tag{14.14}$$

This implies that the current is conserved when $[m,\sigma^a] = 0$, in accord with (14.5). We can also check the conservation of $j^{\mu 5 a}$, the *chiral isospin current*. There is an extra (-1) in the first term from anticommuting γ^μ through γ^5. We find

$$\partial_\mu j^{\mu 5 a} = -\overline{\Psi}\frac{\sigma^a}{2}\gamma^5\gamma^\mu D_\mu\Psi + D_\mu\overline{\Psi}\gamma^\mu\gamma^5\frac{\sigma^a}{2}\Psi$$
$$= i\overline{\Psi}\,\{m, \frac{\sigma^a}{2}\}\gamma^5\Psi \ . \tag{14.15}$$

So now the σ^a must anticommute with m if the current is to be conserved. This is true only when $m_u = m_d = 0$.

One might also imagine a symmetry

$$\Psi \to e^{i\phi\gamma^5}\Psi \ , \tag{14.16}$$

called *chiral baryon number*. It can be shown that this is not actually a symmetry of QCD. This is not obvious; in fact, the symmetry is broken only when subtle quantum effects in QCD are taken into account. The symmetry is broken by a strong-interaction effect involving the gluon fields ('t Hooft 1976).

14.2 Spontaneous symmetry breaking

QCD with two massless flavors is thus invariant under the symmetry

$$U(1) \times SU(2) \times SU(2) \ , \tag{14.17}$$

Definition of the *chiral SU(2)* symmetry of QCD with massless u and d quarks.

that is, baryon number, $SU(2)_L$ and $SU(2)_R$. Real QCD might or might not be close to this idealized limit.

We immediately see that, if this symmetry is fully realized, there is a problem. All hadron states would have to be assigned quantum numbers under the full symmetry group. This would disturb the quark model phenomenology. For example, we might assign the left- and right-handed components of the nucleon the baryons number and $SU(2)$ spin quantum numbers

Chiral $SU(2)$ symmetry cannot be a manifest symmetry of QCD.

$$N_L \ : \ (1, \frac{1}{2}, 0) \qquad N_R \ : \ (1, 0, \frac{1}{2}) \qquad (14.18)$$

But then a nucleon mass term, which mixes N_L and N_R, would be forbidden. An alternative strategy would be to assign the full nucleon field to $(1, \frac{1}{2}, 0)$. But then, by parity, there must be another, degenerate, nucleon with the quantum numbers $(1, 0, \frac{1}{2})$. Similar considerations hold for the mesons. A theory with $SU(2)_L \times SU(2)_R$ and parity symmetry requires doubling the number of mesons beyond those expected in the quark model.

However, there is another option. It is possible for $SU(2)_L \times SU(2)_R$ to be a symmetry, in the sense that its generators commute with the Hamiltonian H, but one that is not respected by the states of the theory. In quantum mechanics with a finite number of coordinates, it can be shown that, if Q generates a symmetry of the theory, then the ground state of the theory $|0\rangle$ must obey

$$Q |0\rangle = 0 \qquad (14.19)$$

However, in a system with an infinite number of degrees of freedom, (14.19) can be violated. There can be several ground states of H, all with the same energy, such that any one of these states has an orientation with respect to the symmetry transformations. This situation is called *spontaneous symmetry breaking*. In any particular ground state, the symmetry is not obvious as a relation between the energy levels or the particle interactions. However, there can be other observable consequence of a spontaneously broken symmetry, as we will discuss in the next section.

Definition of *spontaneous symmetry breaking*.

Quantum field theory has an infinite number of quantum degrees of freedom, since it allows the creation of an infinite number of particles from the vacuum state. If the Hamiltonian of a quantum field theory possesses a symmetry, it is possible that there are multiple vacuum states, in each one of which the theory appears asymmetric.

Condensed matter systems, in the thermodynamic limit, have an infinite number of degrees of freedom, and they furnish many illustrations of spontaneous symmetry breaking (Sethna 2006). For example, in a magnet, the Hamiltonian may be invariant under global rotations of the electron spins. However, the state of lowest energy may have the majority of electron spins preferentially aligned in some direction. Then this state does not display spin rotation invariance, even though the rotation generators $\vec{S}$ commute with the Hamiltonian. In this situation, there

Examples of spontaneously broken symmetry in condensed matter physics.

will be several degenerate states of lowest energy, each of which has the electron spins aligned in a different orientation.

Another example occurs in the theory of superconductivity (Tinkham 1966). Electrons near the Fermi surface of a metal bind into pairs which then form a Bose condensate within the metal. This Bose condensate contains an indefinite number of electron pairs. Let $\Phi(x)$ be a field operator that has the quantum numbers of 2 electrons and can therefore annihilate an electron pair. The ground state of a superconductor $|G\rangle$ has

$$\langle G|\, \Phi(x)\, |G\rangle \neq 0 \,. \tag{14.20}$$

The operator Φ has electric charge 2, and so this expectation value would be forbidden if the total electric charge of the ground state were zero (or any other definite value). The fact that the ground state contains a reservoir with an indefinite number of electron pairs is the reason that superconductors have perfect conductivity. The condensate can adjust itself to create a current flow in response to any electrostatic perturbation. Similarly, in the superfluid state of He^4, a condensate forms that contains an infinite number of He atoms. This condensate forms a separate fluid that flows frictionlessly.

If the expectation value (14.20) is a nonzero number, that number may be complex with a definite phase. The degenerate vacuum states of a superconductor or superfluid are characterized by different phases of the expectation value (14.20) (Yang 1962).

A very similar condensate can appear in QCD. Massless quarks and antiquarks cost zero energy to produce. On the other hand, quark-antiquark pairs are bound by QCD forces, which become strong at distances of 1 fm or 1/200 MeV. So it may be energetically favorable for the QCD vacuum state to contain $q\bar{q}$ pairs. Consider, in particular, the state $u_R \bar{u}_R$

Physical origin of a chiral symmetry breaking $q\bar{q}$ condensate in QCD.

$$\tag{14.21}$$

This state is color-singlet and a Lorentz scalar, so the creation of such a $q\bar{q}$ pair leaves a color-singlet, Lorentz-invariant vacuum state. The analogy to superconductivity tells us that the vacuum should fill with a condensate containing an indefinite number of these pairs (and the corresponding states for $u_L \bar{u}_L$, $d_R \bar{d}_R$, and $d_L \bar{d}_L$.

$$\tag{14.22}$$

Now recall that $\bar{u}_R$ is the antiparticle of u_L. If the vacuum is full of $u_R \bar{u}_R$ pairs, a u_L can interact with the vacuum condensate and turn

into a u_R

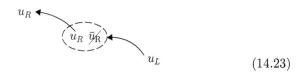

$$(14.23)$$

This a mixing of the massless u_L and u_R. We have seen that this helicity mixing is precisely the effect of a u quark mass term. So, on top of the vacuum state with condensates, the u quark, and also the d quark, obtains a dynamical mass. A similar effect is seen is superconductivity. An energy gap opens in the electron spectrum at the Fermi surface, and an electron needs a finite energy, even at rest, to go above the Fermi energy. The u and d quark dynamical masses should have a size similar to the QCD energy scale of a few hundred MeV. Because of this effect, QCD with zero quark masses in the Lagrangian predicts that valence quarks inside hadrons will apparently have masses of about 300 MeV. The vacuum condensates also allow the proton and neutron to be massive.

Just as superconductivity is characterized by a nonzero operator expectation value in the vacuum, we can characterize the formation of the quark condensates by an operator vacuum expectation value. Call the vacuum state with condensates $|0\rangle$. The state $u_L \bar{u}_L$ is annihilated by the operator

$$\psi_{uR}^\dagger \psi_{uL} \ . \tag{14.24}$$

Thus, a nonzero vacuum expectation value of this operator

$$\langle 0| \, \psi_{uR}^\dagger \psi_{uL} \, |0\rangle \neq 0 \tag{14.25}$$

indicates the presence of a condensate. If all four condensates $u_L \bar{u}_L$, $d_L \bar{d}_L$, $u_R \bar{u}_R$, $d_R \bar{d}_R$ are present in equal amounts, we would have

$$\langle 0| \, \psi_{iR}^\dagger \psi_{jL} \, |0\rangle = \langle 0| \, \psi_{iL}^\dagger \psi_{jR} \, |0\rangle = -\Delta \delta_{ji} \ , \tag{14.26}$$

A nonzero vacuum expectation value that we can use to characterize the broken symmetry vacuum of QCD.

where Δ is a value with the dimensions of $(\text{GeV})^3$. The state $|0\rangle$ is then isospin-invariant and parity-invariant. However, $|0\rangle$ is not invariant under chiral $SU(2)$. To see this, act on $|0\rangle$ with $SU(2)_L$. This is equivalent to an $SU(2)_L$ rotation of the operator ψ_{jL}. We then find a new vacuum state $|\vec{\alpha}\rangle$ with

$$\langle \vec{\alpha}| \, \psi_{iR}^\dagger \psi_{jL} \, |\vec{\alpha}\rangle = -\Delta \big(e^{i\vec{\alpha}\cdot\vec{\sigma}/2} \big)_{ji} \ . \tag{14.27}$$

The state $|\alpha\rangle$ has the same energy as $|0\rangle$, because the $SU(2)_L$ charge commutes with the Hamiltonian. However, it is no longer either isospin or parity symmetric. The parameter $\vec{\alpha}$ can take any value. This gives an infinite number of degenerate vacuum states, in one-to-one correspondence with the elements of the group $SU(2)$.

The logic of the previous paragraph implies that the ground states of the QCD Hamiltonian with two massless quarks form a manifold

isomorphic to the $SU(2)$ group,

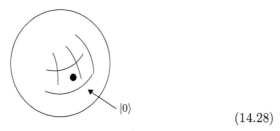

$$(14.28)$$

All of these ground states have the same energy, but none of them fully respects the chiral $SU(2)$ symmetry. In each of these vacuum states, the chiral $SU(2)$ symmetry is spontaneously broken.

14.3   Goldstone bosons

In the situation that the spontaneously broken symmetry is a continuous symmetry such as $U(1)$ or $SU(2)$, the symmetry of the Hamiltonian implies that there are new particles with special properties.

To investigate this statement in QCD, we start with the conserved current associated with the broken $SU(2)$ symmetry. Consider the state created by this current from the broken symmetry vacuum

$$\int d^3x \, j^{05a}(x) \, |0\rangle \ . \qquad (14.29)$$

The operator acting on the vacuum state in (14.29), $\int d^3x \, j^{05a}(x) = Q^{5a}$, is the charge that generates the global chiral $SU(2)$ symmetry. Since Q^{5a} commutes with the Hamiltonian, the energy of (14.29) must be equal to that of $|0\rangle$. The action of Q^{5a} converts the vacuum state $|0\rangle$ into a combination of the other degenerate ground states described in the previous section.

Now consider the state

$$\int d^3x \, e^{-i\vec{p}\cdot\vec{x}} \, j^{05a}(x) \, |0\rangle \ . \qquad (14.30)$$

This is a state with definite nonzero momentum $\vec{p}$. As $\vec{p} \to 0$, the energy of this state (above the energy of the ground state) goes to zero. Thus, this state must be a particle with rest energy zero, that is a particle with zero mass. This observation is *Goldstone's theorem*: For every spontaneously broken continuous symmetry, there is a massless particle created by the symmetry current (Goldstone 1961). This particle is called a *Goldstone boson*.

Goldstone's theorem: Every spontaneously broken continuous symmetry leads to a massless particle, called the *Goldstone boson*.

The annihilation of a Goldstone boson π by the corresponding current j^μ is described by a matrix element

$$\langle 0| \, j^{\mu 5}(x) \, |\pi(\vec{p})\rangle = i \, f \, p^\mu e^{-ip\cdot x} \ , \qquad (14.31)$$

where f is a parameter with the dimensions $(\text{GeV})^1$. This structure (14.31) is the only form allowed by Lorentz symmetry. The parameter

f could in principle be a function of the Lorentz invariant p^2, but for a particle of definite mass, $p^2 = m^2$ is fixed.

Taking the divergence and using current conservation

$$0 = \langle 0| \, \partial_\mu j^{\mu 5}(x) \, |\pi(\vec{p})\rangle = f p^2 e^{-ip \cdot x} \ . \tag{14.32}$$

This implies

$$p^2 = 0 \ . \tag{14.33}$$

This is another way to see that a Goldstone boson must have zero mass.

In QCD with massless u and d quarks, there are three chiral $SU(2)$ currents. These must be spontaneously broken, as just explained, and therefore we must find three Goldstone bosons. The annihilation equation reads

$$\langle 0| \, j^{\mu 5 a}(x) \, |\pi^b(\vec{p})\rangle = i f_\pi p^\mu \delta^{ab} e^{-ip \cdot x} \ , \tag{14.34}$$

where $a, b = 1, 2, 3$. The right-hand side must be proportional to δ^{ab} by isospin invariance. The three chiral currents form an isospin triplet, $I = 1$. The operators j^{05a} have $P = -1$. Thus, the three Goldstone bosons are spin 0, $P = -1$, and $I = 1$. We saw in Section 5.3 that the three hadrons of lowest mass are the pions, which have exactly these properties. Yoichiro Nambu and Giovanni Jona-Lasinio developed this picture and identified the pions as the Goldstone bosons of spontaneously broken chiral $SU(2)$ (Nambu and Jona-Lasinio 1961).

The parameter f_π in the above equation is called the *pion decay constant*. Its value is

Definition of the *pion decay constant*.

$$f_\pi = 93 \text{ MeV} \ . \tag{14.35}$$

I will explain how this value is determined in Section 15.3. Despite the name, there is no intrinsic connection between f_π and the fact that pions decay. On the contrary, f_π is a parameter of the strong interactions that can be calculated (numerically) by solving QCD in the region of strong coupling. The constant f_π plays an important role in the low-energy dynamics of QCD.

14.4   Properties of π mesons as Goldstone bosons

The assumption that the pions are the Goldstone bosons of spontaneously broken chiral $SU(2)$ turns out to contain much information about the behavior of pions at low energy, and about other aspects of strong interaction physics. For example, this assumption leads to specific expressions for the low-energy limit of the pion-pion scattering amplitude and the pion-nucleon scattering amplitude, in both cases, in good agreement with experiment. These developments are described in some detail in (Donoghue *et al.* 1992).

As an example, I will discuss the matrix element of the chiral isospin current in a nucleon state, a quantity that is important in computing the rate of neutron beta decay. To begin, consider the matrix element of the vector isospin current. This matrix element is given by

$$\langle N(p')| j^{\mu a} |N(p)\rangle = 1 \cdot \bar{u}(p')\gamma^\mu u(p) \cdot \frac{\sigma^a}{2} + \cdots , \qquad (14.36)$$

where the omitted terms are less important at low momentum transfer. The coefficient 1 reflects the fact that the nucleons have definite charge $I^3 = \pm\frac{1}{2}$ under isospin. Since the chiral isospin current is not respected by the strong interactions, there is no similar argument for that current, and so its nucleon matrix element can only be written

$$\langle N(p')| j^{\mu 5 a} |N(p)\rangle = g_A \cdot \bar{u}(p')\gamma^\mu \gamma^5 u(p) \cdot \frac{\sigma^a}{2} + \cdots , \qquad (14.37)$$

where g_A is a dimensionless constant. However, it can be shown that the assumption that the chiral $SU(2)$ current is conserved and that the pion is its Goldstone boson leads to the relation

$$g_A = \frac{f_\pi}{m_N} g_{\pi NN} , \qquad (14.38)$$

where $g_{\pi NN}$ is the dimensionless pion-nucleon interaction strength. This formula is called the *Goldberger-Treiman relation* (Goldberger and Treiman 1958). This quite nontrivial relation is reasonably well satisfied. The value of g_A measured from the β decay of the neutron is 1.25, while the measured value $g_{\pi NN} = 13.$ gives a value for the right-hand side equal to 1.31.

If there are small u and d quark masses, the pions will also obtain small masses. One way to see this is to add to the Hamiltonian of the theory with $m_u = m_d = 0$ the mass term

$$\begin{aligned}
\Delta H &= m_u \bar{\Psi}_u \Psi_u + m_d \bar{\Psi}_d \Psi_d \\
&= m_u \psi_{uR}^\dagger \psi_{uL} + m_d \psi_{dR}^\dagger \psi_{dL} + (R \leftrightarrow L) \qquad (14.39)
\end{aligned}$$

Taking the expectation value in the state $|0\rangle$ and using (14.26), we find

$$\langle 0| \Delta H |0\rangle = -2\Delta(m_u + m_d) . \qquad (14.40)$$

This is actually the minimum value of the energy among the set of possible vacuum states $|\alpha\rangle$. This effect will give the Goldstone bosons masses proportional to

$$m_\pi^2 \sim (m_u + m_d) . \qquad (14.41)$$

To make this result more precise, go back to the equation (14.15) for conservation of the chiral currents

$$\partial_\mu j^{\mu 5 a} = i\bar{\Psi}\{m, \frac{\sigma^a}{2}\}\gamma^5 \Psi \qquad (14.42)$$

and put in

$$m = \begin{pmatrix} m_u & 0 \\ 0 & m_d \end{pmatrix} = \frac{1}{2}(m_u + m_d)\mathbf{1} + \frac{1}{2}(m_u - m_d)\sigma^3 . \qquad (14.43)$$

The idea of pions as Goldstone bosons of spontaneously broken chiral $SU(2)$ gives us a way to evaluate the masses of the light quarks taking account of the effects of the strong interaction.

We find

$$\partial_\mu j^{\mu 5a} = i(m_u + m_d)\overline{\Psi}\frac{\sigma^a}{2}\gamma^5\Psi \ , \tag{14.44}$$

plus, for the case $a = 3$, an extra $I = 0$ term that will drop out at the next step. Now take the matrix element between $\langle 0|$ and a 1-pion state

$$\langle 0| \partial_\mu j^{\mu 5a} |\pi^b\rangle = i(m_u + m_d) \langle 0| \overline{\Psi}\frac{\sigma^a}{2}\gamma^5\Psi |\pi^b\rangle \ . \tag{14.45}$$

The left-hand side can be evaluated as in (14.32). By isospin invariance, the right-hand side must be proportional to δ^{ab} (and the additional term from $a = 3$ must give zero). Then, (14.45) implies that

$$f_\pi p^2 \delta^{ab} = (m_u + m_d)\Delta'\delta^{ab} \ , \tag{14.46}$$

where Δ' is a strong interaction quantity with the dimensions $(\text{GeV})^2$. Setting $p^2 = m_\pi^2$, the pion mass satisfies

$$m_\pi^2 = \frac{(m_u + m_d)}{f_\pi}\Delta' \ . \tag{14.47}$$

The size of the parameter Δ' should be set by the QCD energy scale. If we estimate

$$\Delta' = (500 \text{ MeV})^2 \ , \tag{14.48}$$

and use the value $f_\pi = 93$ MeV, we find

$$m_u + m_d \approx 7 \text{ MeV} \ . \tag{14.49}$$

A modern numerical evaluation of the matrix element in QCD confirms that this estimate for the values of the u and d quark masses is about right (Manohar *et al.* 2016).

The picture that emerges is that real QCD is quite close to the limit in which the u and d quark masses in the Lagrangian are zero. The closeness to that limit is measured by the smallness of the ratio

$$\frac{m_\pi^2}{m_\rho^2} = 0.03 \ . \tag{14.50}$$

Note that, because of the spontaneous symmetry breaking, the u and d quarks inside hadrons will move as if they have masses of the order of the QCD scale, about 300 MeV, acquired from their interaction with the quark-antiquark condensate. Thus, the successes of the quark model are quite compatible with the idea that the fundamental masses of the u and d quarks given in the Lagrangian are small.

We can take a further step by considering the model in which the u, d, and s quark masses are all set to zero. In this case, the symmetry of QCD analogous to (14.17) is

$$U(1) \times SU(3) \times SU(3) \ . \tag{14.51}$$

The first $SU(3)$ rotates the left-handed components of the three flavors; the second separately rotates the right-handed components. Similarly to the previous case, we can define vector and axial vector $SU(3)$ symmetries. By the same logic as above, the axial vector symmetry should be spontaneously broken. This leads to a number of Goldstone bosons equal to the number of generators of $SU(3)$, that is, 8. These should all be spin 0, $P = -1$ mesons. The natural candidates are

$$\pi^+ , \ \pi^0 , \ \pi^- , \ K^+ , \ K^0 , \ \overline{K}^0 , \ K^- , \ \eta^0 . \tag{14.52}$$

Repeating the argument above for the pion masses, one can derive the mass formulae

$$m_\pi^2 = (m_u + m_d) \cdot \Delta'/f_\pi ,$$
$$m_{K^+}^2 = (m_u + m_s) \cdot \Delta'/f_\pi ,$$
$$m_{K^0}^2 = (m_d + m_s) \cdot \Delta'/f_\pi ,$$
$$m_\eta^2 = (4m_s + m_u + m_d)/3 \cdot \Delta'/f_\pi . \tag{14.53}$$

where the parameters Δ' all have the same value in the limit of small quark masses. We should add to these expressions small contributions from electromagnetism. For example, the electromagnetic interactions of the pion raise the mass of the $\pi^\pm$ above the mass of the π^0 by about 5 MeV.

These formulae then give us further information about the quark masses. First

$$\frac{m_s}{(m_d + m_u)/2} = \frac{2m_K^2}{m_\pi^2} = 27 . \tag{14.54}$$

Also, the fact that the K^0 is heavier than the K^+ (and the fact that the neutron is heavier than the proton) tells us that the d quark is heavier than the u quark. As noted above, there are also contributions to the π^+-π^0 and K^+-K^0 mass differences from electromagnetism. However, those electromagnetic mass shifts can be shown to cancel in the expression

$$\frac{m_{K^0}^2 - m_{K^+}^2 - m_{\pi^0}^2 + m_{\pi^+}^2}{m_\pi^2} = \frac{m_d - m_u}{m_d + m_u} = 0.3 . \tag{14.55}$$

Using this formula, we find

$$\frac{m_u}{m_d} = 0.6 . \tag{14.56}$$

Through this analysis, we learn that the u and d masses are not close to being equal. From this, we learn that isospin symmetry is not a fundamental symmetry of nature.

The relations in (14.53) give a formula for the mass of the η in terms of the masses of the other pseudoscalar mesons

$$m_\eta^2 = \frac{1}{3}\left[2m_{K^0}^2 + 2m_{K^+}^2 - m_\pi^2\right] . \tag{14.57}$$

This relation, called the *Gell-Mann-Okubo formula*, is satisfied reasonably well,

$$m_\eta^2 = (548 \text{ MeV})^2 \approx \text{RHS} = (567 \text{ MeV})^2 , \tag{14.58}$$

In our discussion of the nonrelativistic quark model in Section 5.3, we were puzzled by the mass pattern of the pseudoscalar mesons. We now see that this pattern is understood in terms of the interpretation of these particles as Goldstone bosons.

This analysis gives the values for the masses of the light quark masses that appear in the QCD Lagrangian (Manohar *et al.* 2016):

Final values of the light quark masses obtained from the analysis of the masses of the pseudoscalar mesons.

$$m_u = 2.2 \text{ MeV} , \qquad m_d = 4.7 \text{ MeV} , \qquad m_s = 96. \text{ MeV} . \qquad (14.59)$$

These values are quite surprising. First of all, these values are much smaller that we might have expected from the meson and baryon masses. Most of the mass of quarks inside hadrons comes from spontaneous chiral symmetry breaking, not from the more fundamental masses that appear in the Lagrangian. But, further, the masses of the u and d quarks are completely different. We might have expected, by isospin symmetry, that $m_u \approx m_d$. We now see that these values differ by a factor of 2. Isospin is not a fundamental symmetry of nature; rather, it is an accident due to the fact that the u and d masses are small. I will have more to say about the quark masses in Chapter 18.

Exercises

(14.1) A field theory with spontaneous symmetry breaking can be constructed as follows: First, write the Klein-Gordon Lagrangian for n fields ϕ_i,

$$\mathcal{L} = \sum_i \left[\frac{1}{2}(\partial_\mu \phi_i)^2 - \frac{1}{2}m_i^2 \phi_i^2 \right] , \qquad (14.60)$$

then replace the mass term by $V(\phi)$, a general nonlinear function of the ϕ_i, to form

$$\mathcal{L} = \sum_i \left[\frac{1}{2}(\partial_\mu \phi_i)^2 \right] - V(\phi) . \qquad (14.61)$$

The function $V(\phi)$ is the potential energy associated with the scalar field value. In a system with spontaneous symmetry breaking, $V(\phi)$ has its minimum at a value $\phi = \Phi$ that does not respect the symmetry of $V(\phi)$. This problem will explore some properties of this theory.

(a) Show that the equations of motion of this theory are

$$\partial_\mu \partial^\mu \phi_i + \frac{\partial}{\partial \phi_i} V(\phi) = 0 \qquad (14.62)$$

(b) The minimum of the potential energy is a constant vector Φ_i satisfying

$$\frac{\partial}{\partial \phi_i} V(\phi)|_{\phi=\Phi} = 0 \qquad (14.63)$$

Writing

$$\phi_i(x) = \Phi_i + \eta_i(x) , \qquad (14.64)$$

expand the equations of motion up to terms of first order in $\eta_i(x)$. Show that the n eigenvalues of the matrix

$$M_{ij}^2 = \frac{\partial^2}{\partial \phi_i \partial \phi_j} V(\phi)|_{\phi=\Phi} \qquad (14.65)$$

give n values of $(\text{mass})^2$ corresponding to n scalar particles in the theory.

(c) Consider the potential

$$V(\phi) = +\frac{1}{2}\mu^2(\phi^2) + \frac{1}{4}\lambda(\phi^2)^2 , \qquad (14.66)$$

where $\phi^2 = \sum_i (\phi_i)^2$ and μ and λ are constants. Sketch this potential. Find the minimum of the potential. Find the masses of the n particles.

(d) Consider the potential

$$V(\phi) = -\frac{1}{2}\mu^2(\phi^2) + \frac{1}{4}\lambda(\phi^2)^2 , \qquad (14.67)$$

where everything is as before, except that I have changed the sign in front of μ^2. Sketch this potential. Find the minimum of V among constant fields of the form

$$\phi = (0, 0, 0, \cdots, 0, v) \qquad (14.68)$$

Show that the minimum occurs for $v \neq 0$, and find the value of v at the minimum. This is spontaneous symmetry breaking. Show that this potential V has an $(n-1)$-dimensional sphere of degenerate minima.

(e) Find the masses of particles in this theory. Show that $(n-1)$ of these masses are zero. This illustrates Goldstone's theorem.

(14.2) The quark masses given by the Particle Data Group (Patrignani *et al.* 2016) are "running masses in the $\overline{MS}$ scheme". Without going into too much detail about the definition, I note that (1) like all other quantities in QCD, the quark masses evolve as functions of the momentum scale Q, and (2) therefore, quark masses must be quoted at a particular value of Q. The PDG chooses to evaluate the masses of the light quarks u, d, s at a common value of the scale $Q_0 = 2$ GeV and the masses of the heavy quarks c, b, t at different values of Q_0 such that $m_q(Q_0) = Q_0$ for each quark. The PDG values are:

flavor	$m_f(Q_0)$	Q_0	flavor	$m_f(Q_0)$	Q_0
u	0.0022	2	c	1.28	1.28
d	0.0047	2	b	4.18	4.18
s	0.096	2	t	164.	164.

$$(14.69)$$

with all mass values in GeV. In this problem, we will compare the quark masses in a more invariant

way. Our analysis will be as simple as possible, to leading order in QCD only.

(a) In (11.68), we found the following expression for α_s:

$$\alpha_s(Q) = \frac{\alpha_s(Q_0)}{1 + (b_0 \alpha_s(Q_0)/2\pi) \log(Q/Q_0)}, \qquad (14.70)$$

where $b_0 = 11 - \frac{2}{3} n_f$ and n_f is the number of quark flavors with $m_f < Q$. Starting from the value of α_s quoted in (11.73), $\alpha_s(91.) = 0.118$, evaluate $\alpha_s(Q)$ at the Q values: $m_b = 4.18$ GeV, 2 GeV, and $m_c = 1.28$ GeV. Note that you will need to use different values of b_0 for $Q > m_b$ and $Q < m_b$ to convert the α_s values.

(b) QCD gives the following equation for the Q-dependence of a quark mass parameter,

$$\frac{d}{d \log Q} m_f(Q) = -8 \frac{\alpha_s(Q)}{4\pi} m_f(Q) \qquad (14.71)$$

Using the formula for α_s in (a), find the solution of this equation that gives $m_f(Q)$ in terms of a reference value $m_f(Q_0)$.

(c) Compute the value of the charm quark mass m_c at $Q = 2$ GeV.

(d) Compute the values of the four lightest quark masses at $Q = m_b$.

(e) Compute the values of all quark masses at $Q = m_t$ given above. You will need to find $\alpha_s(m_t)$ from $\alpha_s(91.)$ using (14.70).

(f) Compute the true ratios of quark masses compared at this common value of Q.

Part III

The Weak Interaction

The Current-Current Model of the Weak Interaction

15

Now we turn to the other subnuclear interaction, the *weak interaction*.

QCD leads to a large spectrum of mesons and baryons. Most of these are unstable, with decay rates of the order of 100 MeV, corresponding to lifetimes of the order of 10^{-23} sec. However, the lightest particles of each type are more stable. For example,

$$\tau(\pi^+) = 2.6 \times 10^{-8} \text{ sec}, \qquad \tau(\Lambda^0) = 2.6 \times 10^{-10} \text{ sec},$$
$$\tau(K^+) = 1.2 \times 10^{-8} \text{ sec}, \qquad \tau(B^0) = 1.5 \times 10^{-12} \text{ sec}. \quad (15.1)$$

Most familiarly, the neutron is unstable by β decay,

$$n \to p\, e^- \, \bar{\nu}_e , \qquad (15.2)$$

though it is very long-lived

$$\tau(n) = 880 \text{ sec}. \qquad (15.3)$$

The great difference between typical hadronic lifetimes and the lifetimes just listed suggests that those particle decays are due to a completely different subnuclear interaction. Now that we understand QCD, this idea is even more compelling. The equations of motion of QCD conserve the number of each flavor of quark. So, any process that changes a quark of one flavor into another—as would be required for all of the decay processes just listed—must necessary require an interaction outside QCD.

This new set of forces is called the *weak interaction*. In this chapter and the next few, I will build up the structure of this interaction from basic properties of weak-interaction decays. Remarkably, the current-current interaction that served as the starting point for our understanding of QED and QCD also plays a central role in this story. In this chapter I will argue that the structure of certain weak-interaction decays requires a special type of current-current interaction, called the

Concepts of Elementary Particle Physics. Michael E. Peskin.
© Michael E. Peskin 2019. Published in 2019 by Oxford University Press.
DOI: 10.1093/oso/9780198812180.001.0001

V−A interaction. Because this coupling has a current-current form, it is natural to suggest that the weak interaction is mediated by a set of spin 1 particles. Our pursuit of this hypothesis will lead us to new theoretical aspects of non-Abelian gauge theories. However, once we understand these, we will be able to predict the properties of the new spin 1 bosons, learn that these bosons actually exist, and test their predicted properties against experiment.

15.1   Development of the V−A theory of the weak interaction

Historically, it took some time to understand that β decay and related processes required a new fundamental interaction. The first guess about β decay was that it corresponded to the ejection of electrons from an atomic nucleus. It was a mystery why the energy spectrum of electrons seemed to be continous rather than a set of discrete lines, as one finds for gamma ray emission from nuclei. In 1930, Wolfgang Pauli explained this by postulating the existence of an invisible particle emitted along with the electron (Pauli 1930). Enrico Fermi called this particle the *neutrino* and gave a unified description of the β decays of nuclei using a general 4-fermion interaction (Fermi 1934)

$$(15.4)$$

In the 1950's, the discovery of strange particles added more elements to the theory of the weak interaction. Strangeness was apparently conserved in the strong interaction production of strange particles, but it must be violated in their decays. This violation could be ascribed to the weak interaction. In addition, it was found that the K^0 could decay by both of the processes

$$K^0 \to \pi^+\pi^- , \qquad K^0 \to \pi^+\pi^-\pi^0 , \qquad (15.5)$$

to final states with $P = +1$ and $P = -1$, respectively. It seemed impossible that these decays could belong to the same particle, since parity was known to be an almost perfect symmetry of atomic physics and nuclear physics. In 1956, Tsung-Dao Lee and Chen-Ning Yang formally proposed the weak interaction as a distinct fundamental force (Lee and Yang 1956). They pointed out that parity conservation had never been tested for this force, and that the weak interaction might indeed violate parity. Very soon after, parity conservation in β decay was tested by Wu, Ambler, Hayward, Hoppes, and Hudson, in the decay of polarized Co^{60} nuclei, and by Garwin, Lederman, and Weinrich and Friedman and Telegdi, in the decay of muons (Wu *et al.* 1957, Garwin *et al.* 1957,

Violation of parity invariance in particle decays showed that the weak interaction was a distinct new force of nature and gave important clues to its structure.

Friedman and Telegdi 1957). Parity violation was not only nonzero, it was seen to be a large effect. In 1958, Feynman and Gell-Mann and Robert Marshak and George Sudarshan proposed a model of the weak interaction based on the idea that the weak interaction violates parity *maximally* (Feynman and Gell-Mann 1958, Marshak and Sudarshan 1958). This model, called the *V−A theory*, proposed that all weak interaction matrix elements could be derived from a current-current interaction of the form (in modern notation)

$$\mathcal{M} = \left\langle \frac{4G_F}{\sqrt{2}} \, j_L^{\mu+} j_{\mu L}^{-} \right\rangle , \tag{15.6}$$

The matrix element of the V−A description of the weak interaction.

where

$$\begin{aligned} j_L^{\mu+} &= \nu_L^\dagger \bar{\sigma}^\mu e_L + u_L^\dagger \bar{\sigma}^\mu d_L + \cdots \\ j_L^{\mu-} &= e_L^\dagger \bar{\sigma}^\mu \nu_L + d_L^\dagger \bar{\sigma}^\mu u_L + \cdots \end{aligned} \tag{15.7}$$

In this equation, and henceforth, I will use the flavor labels e, μ, u, d, *etc.*, to represent the lepton and quark fields. It is a crucial property that only the left-handed components of the Dirac field appear in (15.7) The name V−A ("V minus A") comes from rewriting

$$u_L^\dagger \bar{\sigma}^\mu d_L = \bar{u}\gamma^\mu \left(\frac{1-\gamma^5}{2}\right)d = \frac{1}{2}[\bar{u}\gamma^\mu d - \bar{u}\gamma^\mu\gamma^5 d] , \tag{15.8}$$

a difference of the vector and axial vector currents.

The V−A theory manifestly violates parity (P), because the weak interaction couples to left- but not right-handed electrons. It also manifestly violates charge conjugation (C), because the weak interaction couples to left-handed electrons but not left-handed positrons. The joint operation CP sends a left-handed electron into a right-handed positron, and similarly for other fermions. The symmetry operation CP is then preserved by the V−A theory. However, we will see in Chapter 19 that the full description of the weak interaction requires elements that violate CP.

The parameter G_F is called the *Fermi constant*. It has the dimensions of $(\text{GeV})^{-2}$. Its value is

The *Fermi constant* that gives the strength of the V−A interaction.

$$G_F = 1.166 \times 10^{-5} \, (\text{GeV})^{-2} . \tag{15.9}$$

The most accurate determination of this value comes from the measurement of the muon lifetime, for which the theory will be discussed in the next section. The factor of $\sqrt{2}$ in the definition of G_F is a relic of Fermi's original proposal, which assumed parity conservation.

15.2 Predictions of the V−A theory for leptons

The V−A theory of the weak interaction is very simple, but it is surprisingly rich. It makes a number of detailed and rather unexpected predictions for weak interaction processes that are confirmed by experiment. In the rest of this chapter, I will describe four of these.

First, the theory predicts that electron emitted in the β decay of a nucleus should be preferentially left-handed polarized. For extremely relativistic particles, the field $e_L^\dagger$ creates only left-handed, and not right-handed, electrons. In fact, though, electrons are emitted in nuclear β decay over a wide range of energies. To understand the polarization for a more slowly moving electron, we need to look at the form of the corresponding Dirac spinors. It will still be useful to use the basis (8.7) to represent the Dirac matrices. The matrix γ^5 is diagonal in this basis,

$$\gamma^5 = \begin{pmatrix} -1 & 0 \\ 0 & 1 \end{pmatrix} . \tag{15.10}$$

so projected spinors such as e_L, μ_L, q_L correspond to the top two components of 4-component Dirac spinors.

For an electron moving in the $\hat{3}$ direction, with momentum $p^\mu = (E, 0, 0, p)^\mu$, the Dirac equation takes the form

$$\begin{pmatrix} -m & E - p\sigma^3 \\ E + p\sigma^3 & -m \end{pmatrix} U(p) = 0 . \tag{15.11}$$

The solutions to this equation are the spinors

$$U(e_+) = \begin{pmatrix} \sqrt{E-p} \begin{pmatrix} 1 \\ 0 \end{pmatrix} \\ \sqrt{E+p} \begin{pmatrix} 1 \\ 0 \end{pmatrix} \end{pmatrix} , \qquad U(e_-) = \begin{pmatrix} \sqrt{E+p} \begin{pmatrix} 0 \\ 1 \end{pmatrix} \\ \sqrt{E-p} \begin{pmatrix} 0 \\ 1 \end{pmatrix} \end{pmatrix} . \tag{15.12}$$

Since we have already used the subscripts L, R to denote chirality states, we use here the subscripts $+, -$ to denote the states of definite helicity.

The polarization of an electron measures the probability of finding a left-handed helicity versus a right-handed helicity electron,

$$\text{Pol} = \frac{\text{Prob}(e_-^-) - \text{Prob}(e_+^-)}{\text{Prob}(e_-^-) + \text{Prob}(e_+^-)} . \tag{15.13}$$

The operator $e_L^\dagger$ sees only the top components of these spinors, so only those quantities appear as factors in the β decay amplitudes. The probabilities are proportional to the squares of these factors. So, electrons created by the V−A current (15.7) have the polarization

$$\text{Pol} = \frac{(E+p) - (E-p)}{(E+p) + (E-p)} = \frac{p}{E} . \tag{15.14}$$

That is,

$$\text{Pol} = \frac{v}{c} . \tag{15.15}$$

Figure 15.1 shows a compilation of data from β decay on a variety of nuclei. Indeed, the prediction holds quite accurately. The highest energy electrons emitted in β decay are almost perfectly left-handed polarized.

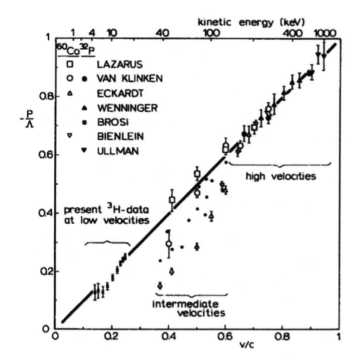

Fig. 15.1 Polarization of electrons emitted in β decay, in units of $\hbar/2$, as a function of the velocity v/c of the emitted electron, from (Koks and van Klinken 1976).

Next, we study the weak interaction decay of the muon. The muon has its own neutrino ν_μ. It appears in the   V−A   theory as a separate term in the currents

$$j_L^{\mu+} = \cdots + \nu_\mu^\dagger \bar\sigma^\mu \mu_L + \cdots \qquad j_L^{\mu-} = \cdots + \mu_L^\dagger \bar\sigma^\mu \nu_\mu + \cdots \qquad (15.16)$$

We will see in Chapter 20 that neutrinos have small nonzero masses. However, for the considerations of this chapter, these masses can be ignored. Then helicity conservation implies that neutrinos produced by the   V−A   interaction are always left-handed, and antineutrinos produced by this interaction are always right-handed.

The muon decays through the process

$$\mu^- \to \nu_\mu e^- \bar\nu_e \,, \qquad (15.17)$$

with invariant matrix element

$$\mathcal{M} = \langle \nu_\mu e^- \bar\nu_e | \frac{4G_F}{\sqrt{2}} \nu_\mu^\dagger \bar\sigma^\mu \mu_L \; e_L^\dagger \bar\sigma_\mu \nu_e | \mu \rangle \qquad (15.18)$$

This somewhat technical section derives the   V−A   prediction for the energy-momentum distribution of electrons emitted in muon decay.

The Feynman diagram for this process has the form

$$(15.19)$$

Using the various fermion fields to destroy and create initial and final particles, the matrix element in (15.18) gives

$$\mathcal{M} = \frac{4G_F}{\sqrt{2}} \, u_L^\dagger(p_\nu)\overline{\sigma}^\mu u_L(p_\mu) \; u_L^\dagger(p_e)\overline{\sigma}_\mu v_L(p_{\overline{\nu}}) \; . \tag{15.20}$$

Now we need to reduce this to an explicit expression in terms of particle 4-vectors.

The matrix element (15.20) is very similar to one that we encountered in our discussion of $eq \to eq$, which is also mediated by a current-current interaction. For that process, in the high-energy limit where we can ignore all masses, we needed the value of the matrix element

Evaluation of the V−A matrix element for muon decay.

$$\mathcal{M}(e_L^- q_L \to e_L^- q_L) \sim u_L^\dagger(p_e')\overline{\sigma}^\mu u_L(p_e) \; u_L^\dagger(p_q')\overline{\sigma}^\mu u_L(p_q) \; .$$

$$(15.21)$$

Evaluating the spinors, we found in (9.48)

$$\left| u_L^\dagger(p_e')\overline{\sigma}^\mu u_L(p_e) \; u_L^\dagger(p_q')\overline{\sigma}^\mu u_L(p_q) \right|^2 = 4s^2 = 4(2p_e \cdot p_\mu)(2p_e' \cdot p_\mu') \; . \tag{15.22}$$

A guess at the evaluation of the matrix element for μ decay, based on results from Section 9.4.

For this current-current matrix element, the answer is similar,

$$\left| u_L^\dagger(p_\nu)\overline{\sigma}^\mu u_L(p_\mu) \; u_L^\dagger(p_e)\overline{\sigma}^\mu v_L(p_{\overline{\nu}}) \right|^2 = \frac{1}{2} \cdot 4(2p_e \cdot p_\nu)(2p_\mu \cdot p_{\overline{\nu}}) \; , \tag{15.23}$$

Derivation of (15.23). I apologize that this derivation uses a number of special tricks. The equation can be derived transparently using the more standard methods for evaluating Feynman diagrams that you will find in textbooks of quantum field theory.

in the limit in which we ignore the masses of e^-, ν_μ and $\overline{\nu}_e$. This is a decay, so the mass of the muon must be retained, and the expression (15.23) does depend correctly on m_μ. The average over the spin of the muon (at rest) gives the factor of $\frac{1}{2}$. In next three paragraphs, I give the derivation of (15.23).

The easiest way to evaluate the matrix element (15.20) is to use the *Fierz identity*, an identity of the σ^μ matrices,

$$(\overline{\sigma}^\mu)_{\alpha\beta}(\overline{\sigma}_\mu)_{\gamma\delta} = 2\epsilon_{\alpha\gamma}\epsilon_{\beta\delta} \tag{15.24}$$

where $\alpha,\beta,\gamma,\delta = 1,2$ are spinor indices and ϵ is the antisymmetric symbol with $\epsilon_{12} = 1$. There are 16 possible values of α, β, γ, δ, so you can check this identity by verifying it for each set of values. This rewrites the product of spinors as

$$2\big(u_\alpha^\dagger(p_\nu)\epsilon_{\alpha\gamma}u_\gamma^\dagger(p_e)\big)\big(u_\beta(p_\mu)\epsilon_{\beta\delta}v_\delta(p_{\overline{\nu}})\big) \; . \tag{15.25}$$

Each term in parentheses is Lorentz invariant. This means that we can evaluate the two terms in different frames and still obtain the correct result for (15.23).

To evaluate the first product, go to the CM frame of e^- and ν_μ. Both particles are massless. Their momenta are back-to-back and can be taken to be along the $\hat{3}$ axis,

$$e^- \longleftarrow \qquad \longrightarrow \nu_\mu \qquad\qquad (15.26)$$

In this frame, the spinors are

$$u_L(p_\nu) = \sqrt{2E_\nu} \begin{pmatrix} 0 \\ 1 \end{pmatrix}, \qquad u_L(p_e) = \sqrt{2E_e} \begin{pmatrix} 1 \\ 0 \end{pmatrix}. \qquad (15.27)$$

Then

$$\left| \left(u^\dagger(p_\nu)_\alpha \epsilon_{\alpha\gamma} u_\gamma^\dagger(p_e) \right) \right|^2 = 4E_\nu E_e = 2p_\nu \cdot p_e . \qquad (15.28)$$

To evaluate the second product in (15.25), work in the frame where the μ^- is at rest. The four-component spinor of the μ at rest is

$$U(p_\mu) = \sqrt{m_\mu} \begin{pmatrix} \xi \\ \xi \end{pmatrix}, \qquad\qquad (15.29)$$

where ξ is the 2-component spinor representing the muon spin orientation. The V−A current sees only the top two components of (15.29). The electron antineutrino can be taken to move in the $+\hat{3}$ direction; then its 2-component spinor is

$$v_L(p_{\bar\nu}) = \sqrt{2E_{\bar\nu}} \begin{pmatrix} 0 \\ 1 \end{pmatrix}. \qquad\qquad (15.30)$$

The product is then

$$\left| \left(u(p_\mu)_\beta \epsilon_{\beta\delta} v_\delta(p_{\bar\nu}) \right) \right|^2 = 2m_\mu E_{\bar\nu} \left| \xi^\dagger \begin{pmatrix} 0 \\ 1 \end{pmatrix} \right|^2. \qquad (15.31)$$

Averaging over the two possible spin directions for ξ, this becomes

$$\left| \left(u(p_\mu)_\beta \epsilon_{\beta\delta} v_\delta(p_{\bar\nu}) \right) \right|^2 = m_\mu E_{\bar\nu} = p_\mu \cdot p_{\bar\nu} . \qquad (15.32)$$

Assembling the results (15.28) and (15.32), and squaring the 2 in (15.25), we find (15.23).

It is convenient to express this result in terms of variables similar to those that we used in Chapter 10 to analyze 3-body phase space. As in (10.45), let

$$x_e = \frac{2E_e}{m_\mu}, \qquad x_{\bar\nu} = \frac{2E_{\bar\nu}}{m_\mu}, \qquad\qquad (15.33)$$

The quantities x_e and $x_{\bar\nu}$ satisfy

$$0 < x_e , \ x_{\bar\nu} < 1 . \qquad\qquad (15.34)$$

In these coordinates, the expression (15.23) for the square of the matrix element is

$$\left| u_L^\dagger(p_\nu)\overline{\sigma}^\mu u_L(p_\mu)\ u_L^\dagger(p_e)\overline{\sigma}_\mu v_L(p_{\overline{\nu}}) \right|^2 = 2m_\mu^4(1 - x_{\overline{\nu}})x_{\overline{\nu}}\ , \qquad (15.35)$$

where I have used

$$(p_e + p_\nu)^2 = (p_\mu - p_{\overline{\nu}})^2 = p_\mu^2 - 2p_\mu \cdot p_{\overline{\nu}} = m_\mu^2(1 - x_{\overline{\nu}})\ . \qquad (15.36)$$

Integration of the muon decay matrix element over phase space.

We can now evaluate the rate of muon decay by integrating this quantity over phase space. The variables x_i are just those used in (7.34), so we can use the formula (7.36). We are ignoring the masses of the three final particles, so, as in (10.47), the x_i are to be integrated over the triangle

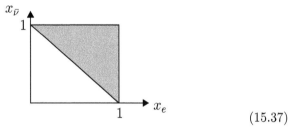

$$(15.37)$$

Then the decay rate is

$$\Gamma = \frac{1}{2m_\mu}\frac{m_\mu^2}{128\pi^3}\int dx_e dx_{\overline{\nu}}\ 16G_F^2 m_\mu^3 x_{\overline{\nu}}(1 - x_{\overline{\nu}})^2$$

$$= \frac{G_F^2 m_\mu^5}{16\pi^3}\int_0^1 dx_e \int_{1-x_e}^1 dx_{\overline{\nu}}\ x_{\overline{\nu}}(1 - x_{\overline{\nu}})\ . \qquad (15.38)$$

The integrand is given in terms of the energy fraction of the $\overline{\nu}_e$, which is unobservable. However, we can integrate over this variable to obtain an expression that only involves the observable electron energy distribution. The integral is

The final result for the V−A prediction of the electron spectrum in muon decay is a simple function that gives a good description of the experimental data.

$$\int_{1-x_e}^1 dx_{\overline{\nu}}\ x_{\overline{\nu}}(1 - x_{\overline{\nu}}) = \int_0^{x_e} dy(1 - y)y = \left(\frac{x_e^2}{2} - \frac{x_e^3}{3}\right)\ . \qquad (15.39)$$

Our final expresssion for the muon decay rate is

$$\Gamma = \frac{G_F^2 m_\mu^5}{32\pi^3}\int_0^1 dx_e\ x_e^2(1 - \tfrac{2}{3}x_e) = \frac{G_F^2 m_\mu^5}{192\pi^3}\ . \qquad (15.40)$$

The shape of the electron energy distribution then is predicted to be

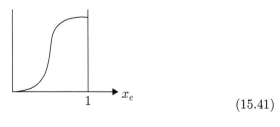

$$(15.41)$$

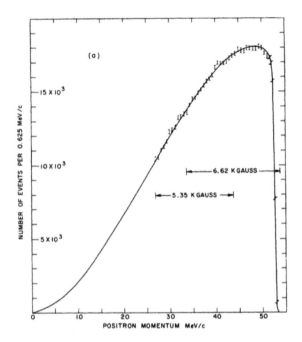

Fig. 15.2 Energy spectrum of positrons emitted in muon decay $\mu^+ \rightarrow e^+ \bar{\nu}_\mu \nu_e$, and comparison to the V−A prediction, from (Bardon, *et al.* 1965).

This is a function with a very characteristic shape. It is quadratic in x_e for small x_e and has a maximum at the endpoint $x_e = 1$.

Figure 15.2 shows the experimental data on the electron energy distribution in the muon decay, which is in very good agreement with this prediction. The slight deviations from the ideal form (15.40) are due to the fact that the outgoing electron can radiate a photon, losing a small fraction of its energy. The theoretical curve shown in the figure takes account of this effect.

The comparison of the total rate formula (including QED corrections) with the measured value of the muon lifetime gives a very accurate value of G_F,

$$G_F = (1.1663787 \pm 0.0000006) \times 10^{-5} \ (\text{GeV})^{-2} \ . \qquad (15.42)$$

Values of G_F obtained from nuclear β decay are consistent with this value (with one subtlety that I will discuss in Section 18.1). The V−A interaction seems to have the constant G_F as a universal strength.

There is one more interesting aspect of the prediction for muon decay. At the endpoint $x_e = 1$, the configuration of the electron and the neutrinos is

$$\qquad (15.43)$$

The ν_μ must be left-handed, the $\bar{\nu}_e$ must be right-handed, and the electron must be left-handed. So the angular momenta of the neutrinos

A special property of muon decay — the complete polarization of electrons with energies at the kinematic endpoint.

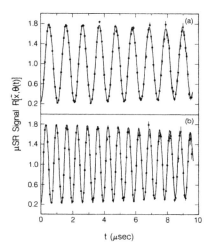

Fig. 15.3 Signal rates as a function of time, as the muon spin is precessed in a magnetic field, in the TRIUMF measurement of the correlation of the positron direction with the muon spin, from (Stoker *et al.* 1985).

cancel, and the total angular momentum in the final state is that carried by the electron spin. This implies that the electron must be emitted in a direction *opposite* to the spin of the muon. The predicted angular distribution for electrons at the endpoint is

$$\frac{d\Gamma}{d\cos\theta} \sim (1 - \cos\theta) , \tag{15.44}$$

with a maximum when the electron is moving opposite to the muon spin and a *zero* when the electron is parallel to the muon spin. This prediction was checked explicitly in an experiment at the TRIUMF laboratory in Vancouver, Canada, in which μ^+s from pion decay were stopped in an absorber and then allowed to decay (Stoker *et al.* 1985). Muons from pion decay are perfectly polarized, for a reason that I will discuss in the next section. A magnetic field was used to precess the spins of the stopped muons, and the decay electrons were counted as a function of time.

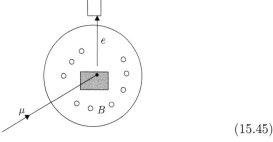

$$\tag{15.45}$$

The signal was seen to oscillate as the muons precess. The data is shown in Fig. 15.3. There is no suppressed zero; the observed extinction when the muon spin points to the detector is almost complete. There is some small depolarization of the muon as it stops in the absorbing medium.

When this is taken into account, the result is consistent with complete left-handed electron polarization in μ decay at the endpoint.

15.3   Predictions of the V−A theory for pion decay

The third example I will discuss is pion decay. The charged pion decays through the weak interaction, by the processes

$$\pi^- \to \mu^- \bar{\nu}_\mu , \quad \pi^- \to e^- \bar{\nu}_e . \tag{15.46}$$

According to the V−A theory, the electron and the muon have identical weak interactions.   However, the ratio of branching ratios for these processes is observed to be

$$\frac{BR(\pi^- \to e^- \bar{\nu})}{BR(\pi^- \to \mu^- \bar{\nu})} = 1.23 \times 10^{-4} . \tag{15.47}$$

A mystery: If the weak interaction has universal strength, why do pions decay much more frequently to muons than to electrons?

How can this be consistent with the V−A theory?

For definiteness, analyze the case of decay to a muon. The V−A interaction mediating the decay is

$$\mathcal{M} = \left\langle \frac{4G_F}{\sqrt{2}} \mu_L^\dagger \bar{\sigma}^\mu \nu_\mu \, u_L^\dagger \bar{\sigma}_\mu d_L \right\rangle . \tag{15.48}$$

We can evaluate the matrix element of the quark current between the pion and the vacuum by casting it into the form of (14.34). Let $|\pi^a\rangle$, $a = 1, 2, 3$, be the pion states with definite isospin indices. Then

$$|\pi^-\rangle = \frac{1}{\sqrt{2}}(|\pi^1\rangle - i\,|\pi^2\rangle) . \tag{15.49}$$

This allows us to evaluate

$$\langle 0|\, u_L^\dagger \bar{\sigma}^\mu d_L \, |\pi^-(p)\rangle = \langle 0|\, \overline{\Psi}_u \gamma^\mu \frac{(1-\gamma^5)}{2} \Psi_d \, \frac{1}{\sqrt{2}}(|\pi^1(p)\rangle - i\,|\pi^2(p)\rangle) \tag{15.50}$$

Write $\Psi = (\Psi_u, \Psi_d)$ and introduce the generators $\sigma^a/2$ of isospin, Then this becomes

$$\langle 0|\, u_L^\dagger \bar{\sigma}^\mu d_L \, |\pi^-(p)\rangle == = -\frac{1}{2\sqrt{2}} \langle 0|\, \overline{\Psi} \gamma^\mu \gamma^5 \left(\frac{\sigma^1 + i\sigma^2}{2}\right) \Psi (|\pi^1(p)\rangle - i\,|\pi^2(p)\rangle)$$

$$= -\frac{1}{2\sqrt{2}} \langle 0|\, (j^{\mu 51} + ij^{\mu 52})(|\pi^1(p)\rangle - i\,|\pi^2(p)\rangle)$$

$$= -\frac{1}{\sqrt{2}} i f_\pi p^\mu . \tag{15.51}$$

The factor of p^μ dots into the lepton current and gives the divergence of this current, which we can evaluate using the Dirac equation as in (14.15),

$$\partial_\mu(\mu_L^\dagger \gamma^\mu \nu_\mu) = im_\mu(\mu_R^\dagger \nu_\mu) . \tag{15.52}$$

Then the lepton matrix element is explicitly proportional to the mass of the lepton.

An easier way to evalute the matrix element is to work in the rest frame of the pion, where $p^\mu = (m_\pi, 0, 0, 0)^\mu$. We find

$$ip_\mu \langle \mu^-(p_\mu)\nu_\mu(p_\nu)| \mu_L^\dagger \sigma^\mu \nu_\mu |0\rangle = m_\pi u_R^\dagger(p_\mu)v_L(p_\nu) . \tag{15.53}$$

The matrix element (15.48) then evaluates to

$$\mathcal{M} = \frac{4G_F}{\sqrt{2}} \frac{1}{\sqrt{2}} f_\pi m_\pi \, u_R^\dagger(p_\mu)v_L(p_{\bar\nu}) . \tag{15.54}$$

Let the muon 4-vector be $p_\mu = (E, 0, 0, k)^\mu$. Then the neutrino 4-vector is $p_\nu = (k, 0, 0, -k)^\mu$, with $E + k = m_\pi$. We have seen above in (15.12) that the two top components of the spinor for a right-handed muon are

$$u_R(p_\mu) = \sqrt{E - k} \begin{pmatrix} 1 \\ 0 \end{pmatrix} . \tag{15.55}$$

With

$$v_L(p_{\bar\nu}) = \sqrt{2E_{\bar\nu}} \begin{pmatrix} 1 \\ 0 \end{pmatrix} , \tag{15.56}$$

the spinor matrix element is

$$\left| u_R^\dagger(p_\mu)v_L(p_{\bar\nu}) \right|^2 = 2k(E - k) . \tag{15.57}$$

The square of the complete decay matrix element is

$$|\mathcal{M}|^2 = 4G_F^2 f_\pi^2 m_\pi^2 \cdot 2k(E - k) . \tag{15.58}$$

In a 2-body decay to one massive and one massless particle, the energies and momenta take the form found in (2.19). Here

$$E = \frac{m_\pi^2 + m_\mu^2}{2m_\pi} , \qquad k = \frac{m_\pi^2 - m_\mu^2}{2m_\pi} . \tag{15.59}$$

Then

$$\Gamma = \frac{1}{2m_\pi} \frac{1}{8\pi} \frac{2k}{m_\pi} \cdot 8G_F^2 f_\pi^2 m_\mu^2 m_\pi \frac{m_\pi^2 - m_\mu^2}{2m_\pi} . \tag{15.60}$$

Finally, we find

$$\Gamma(\pi^- \to \mu^-\nu) = \frac{G_F^2 f_\pi^2 m_\pi^3}{4\pi} \frac{m_\mu^2}{m_\pi^2} \left(1 - \frac{m_\mu^2}{m_\pi^2}\right)^2 . \tag{15.61}$$

From this formula and the measured value of the pion decay rate, assuming that the value of G_F is universal, we obtain the value $f_\pi = 93$ MeV quoted in (14.35).

Using either method of evaluation, the final formula for the decay amplitude is proportional to the mass of the muon. It is easy to understand this by drawing the spins of the muon and neutrino resulting from the pion decay. The pion has spin 0. By V−A , the antineutrino must be right-handed. Then we must have

$$\bar{\nu}_\mu \longleftarrow \qquad \longrightarrow \mu^- \qquad (15.62)$$

To conserve angular momentum, the muon must also be right-handed. This violates helicity conservation, and also the preference of the V−A interaction that the muon be left-handed. To flip the helicity of the muon, we must invoke the muon mass.

Resolution of the mystery: Pion decay requires a violation of helicity conservation. Then the rate of pion decay to a lepton ℓ is proportional to m_ℓ^2/m_π^2.

The decay rate formula is then proportional to m_μ^2. Thus, the V−A interaction naturally predicts a much larger branching ratio for the pion decay to muons rather than electrons. The ratio of these decay rates is predicted to be

$$\frac{BR(\pi^- \to e^- \bar{\nu})}{BR(\pi^- \to \mu^- \bar{\nu})} = \frac{m_e^2}{m_\mu^2}\left(\frac{m_\pi^2 - m_e^2}{m_\pi^2 - m_\mu^2}\right)^2 = 1.28 \times 10^{-4} , \qquad (15.63)$$

in good agreement with the measured value quoted in (15.47).

15.4 Predictions of the V−A theory for neutrino scattering

The final test of the V−A theory that I will discuss comes in deep inelastic neutrino scattering. It is possible to create a neutrino beam using a proton beam from a high-energy accelerator. The method is to shoot the proton beam into a target, produce pions, allow the pions to pass through an empty volume in which they can decay, and then absorb all of the decay products except for the neutrinos, which interact only through the weak interactions and are thus highly penetrating. At Fermilab, the neutrino beam was created by shooting the pion beam horizontally underground. After the pion decay region, the decay products and undecayed pions and other hadrons passed through the earth. The experimenters then dug a a pit 1 km downstream to house the neutrino detector (Benvenuti *et al.* 1973).

The V−A theory predicts neutrino and antineutrino reactions with quarks,

$$\nu_L d_L \to \mu_L^- u_L \qquad \bar{\nu}_R u_L \to \mu_R^+ d_L \qquad (15.64)$$

and antiquarks,

$$\nu_L \bar{u}_R \to \mu_L^- \bar{d}_R \qquad \bar{\nu}_R \bar{d}_R \to \mu_R^+ \bar{u}_R \qquad (15.65)$$

and similar reactions on the s and c quarks and antiquarks in the parton sea. These reactions should be seen as events with hadronic energy deposition and an outgoing muon, called *charged-current events*. An event display for such an event, recorded by the NuTeV experiment at Fermilab (Goncharov *et al.* 2001), is shown in the upper part of Fig. 15.4. The particle going out to the right is a muon, whose momentum is measured

Neutrino deep inelastic scattering.

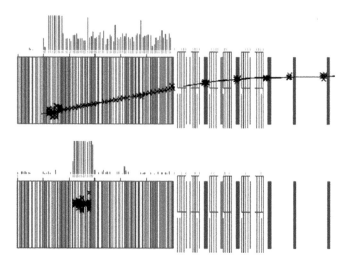

Fig. 15.4 Event displays of charged-current (top) and neutral-current (bottom) neutrino deep inelastic scattering events recorded by the NuTeV experiment at Fermilab (figures courtesy of Kevin McFarland and the NuTeV collaboration).

using a magnetized-iron spectrometer. The experiments are thus very similar in spirit to the classic electron deep inelastic scattering experiments. The outgoing lepton is measured, and the hadronic final states are not discriminated. The neutrino experiments also observe *neutral-current events*, with a neutrino in the final state, as shown in the lower event display in Fig. 15.4. I will discuss these events in Section 16.4.

To predict the cross section for deep-inelastic neutrino scattering, we can follow the derivation that we used earlier for deep inelastic electron scattering. That derivation was based on the formula for electron-quark scattering (9.51),

$$\frac{d\sigma}{d\cos\theta} = \frac{\pi Q_f^2 \alpha^2}{s} \frac{s^2 + u^2}{t^2} \,, \tag{15.66}$$

derived from the formulae for the electromagnetic scattering matrix elements

$$|\mathcal{M}(e_L^- q_L \to e_L^- q_L)|^2 = 4Q_f^2 e^4 \frac{s^2}{t^2} \,,$$

$$|\mathcal{M}(e_L^- q_R \to e_L^- q_R)|^2 = 4Q_f^2 e^4 \frac{u^2}{t^2} \,. \tag{15.67}$$

Notice that we have the factor s^2 in the numerator for the scattering of like-helicity fermions and the factor u^2 in the numerator for the scattering of opposite-helicity fermions. The latter factor appears because the backward scattering of fermions with opposite helicity

$$\tag{15.68}$$

(Calculation of cross sections for deep inelastic neutrino and antineutrino scattering.)

is forbidden by angular momentum conservation. When we transform to the variables x and y of deep inelastic scattering,

$$s^2 \to 1 \qquad u^2 \to (1-y)^2 \,, \qquad (15.69)$$

as we discussed in Section 9.5.

In neutrino scattering, the V−A interaction fixes the helicity to be left-handed for neutrinos and quarks and right-handed for antineutrinos and antiquarks. Changing the prefactors appropriately, the cross sections for neutrino and antineutrino scattering on u and d quarks are

$$\frac{d\sigma}{d\cos\theta}(\nu_L d_L \to \mu_L^- u_L) = \frac{G_F^2}{2\pi s} \cdot s^2 \,,$$

$$\frac{d\sigma}{d\cos\theta}(\bar\nu_R d_L \to \mu_R^+ u_L) = \frac{G_F^2}{2\pi s} \cdot u^2 \,. \qquad (15.70)$$

Similarly, the cross sections for neutrino and antineutrino scattering from antiquarks, through the reactions (15.65), are proportional to u^2 and s^2, respectively.

To derive the formulae for deep inelastic scattering, we integrate these with the pdfs, remembering to average over the initial quark spins. We do not average over the neutrino or antineutrino spin, because the neutrinos are produced completely polarized from π decay. We then find, for neutrino scattering,

$$\frac{d^2\sigma}{dxdy}(\nu p \to \mu^- X) = \frac{G_F^2 s}{\pi}\left[x f_d(x) + x f_{\bar u}(x) \cdot (1-y)^2 \right], \qquad (15.71)$$

and for antineutrino scattering

Characteristic distributions in y for deep inelastic ν and $\bar\nu$ scattering from quarks and antiquarks.

$$\frac{d^2\sigma}{dxdy}(\bar\nu p \to \mu^+ X) = \frac{G_F^2 s}{\pi}\left[x f_u(x) \cdot (1-y)^2 + x f_{\bar d}(x) \right], \qquad (15.72)$$

plus small contributions from heavier sea quarks and antiquarks.

If we concentrated only on the contribution of valence quarks in the proton, we would expect a distribution

$$\frac{d\sigma}{dy}(\nu p) \sim 1 \qquad (15.73)$$

for neutrinos, but a distribution

$$\frac{d\sigma}{dy}(\bar\nu p) \sim (1-y)^2 \qquad (15.74)$$

for antineutrinos. For neutrino scattering from nuclear targets with approximately equal numbers of protons and neutrons, the same regularities should hold. Figure 15.5 shows the distribution in y of neutrino and antineutrino scattering events from the CDHS experiment at CERN (de Groot *et al.* 1979). The prediction is verified in a quite striking way, though there are small deviations from the ideal result due to the scattering from antiquarks.

The V−A theory is thus dramatically successful at describing the weak interactions of quarks and leptons. In the next chapter, I will explain how to obtain the V−A interaction from deeper principles.

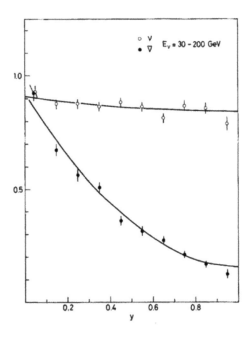

Fig. 15.5 Distribution of neutrino and antineutrino deep inelastic scattering events in y, as measured by the CDHS experiment at CERN, from (de Groot *et al.* 1979).

Exercises

(15.1) The Fierz identity (15.24) is useful in many contexts, so it is worth understanding it in some detail.

  (a) Evalute the left- and right-hand sides of the formula (15.24) for each of the 16 possible values of the indices α, β, γ, δ, and verify that the results match in every case.

  (b) Rearrange the expression

$$u_L^\dagger(p_e')\overline{\sigma}^\mu u_L(p_e)\, u_L^\dagger(p_q')\overline{\sigma}_\mu u_L(p_q) \quad (15.75)$$

  using the Fierz identity. Notice that the result is a Lorentz-invariant product of the spinors of p_e' and p_q' and a second, disconnected Lorentz-invariant product of the spinors of p_e and p_q. Compare this result to (15.22) (or (9.48)).

  (c) We can apply the Fierz identity to products of fermion field operators rather than products of spinors. Fermion field operators create states that obey Fermi statistics, so interchanging the order of two fermion field operators must give a factor (-1) to reflect this.

Including this minus sign, show that, if ψ_L, χ_L are fermion field operators,

$$\psi_L^\dagger\overline{\sigma}^\mu\psi_L \ \chi_L^\dagger\overline{\sigma}_\mu\chi_L = +\psi_L^\dagger\overline{\sigma}^\mu\chi_L \ \chi_L^\dagger\overline{\sigma}_\mu\psi_L \ . \tag{15.76}$$

(15.2) This problem studies weak interaction decays of the τ lepton. The τ is a heavy lepton. The τ and its neutrino ν_τ couple to the weak interaction in exactly the same way as the electron and the muon. The mass of the τ is 1777 MeV.

  (a) The V−A theory predicts that the τ will decay by $\tau^- \to \nu_\tau e^- \overline{\nu}_e$ and $\tau^- \to \nu_\tau \mu^- \overline{\nu}_\mu$. These processes are very similar to muon decay. Compute the partial widths for these decays, using the formulae derived in Section 15.2. (You may ignore the muon mass.)

  (b) Next consider the partial width for the τ to decay to quarks: $\tau \to \nu_\tau d\overline{u}$. Assume that the τ mass is large enough that we can ignore QCD and all quark masses. Then the calculation is just parallel to that for $\tau^- \to \nu_\tau e^- \overline{\nu}_e$.

QCD color must be included. We saw in (11.72) that the first QCD correction is obtained by mutiplying the zeroth order result by

$$(1 + \frac{\alpha_s(m_\tau)}{\pi}) \qquad (15.77)$$

Combining this factor into the zeroth order computation, compute the partial width for $\tau \to \nu_\tau d\bar{u}$.

(c) From the results of parts (a) and (b), compute the τ lifetime and the branching ratio of the τ to leptonic modes. How do these numbers compare with the measured values reported by the Particle Data Group?

(d) A specific hadronic decay of the τ is $\tau^- \to \nu_\tau \pi^-$. Work out the kinematics of this reaction in the frame where the τ is at rest. Let the τ have its spin parallel to the $\hat{3}$ axis, and let the π^- go offat an angle θ with respect to the $\hat{3}$ axis. Write the momentum vectors of the π^- and the ν_τ. Write the spinors $u(p)$ for the τ and the ν_τ. The 2-component spinor in $u_L(\nu_\tau)$ should be left-handed with respect to

the ν_τ direction of motion.

(e) Compute the matrix element for the decay $\tau^- \to \nu_\tau \pi^-$. The calculation is similar to that for π decay to $\mu\nu$. For the hadronic half of the amplitude, you will need the identity related to (15.51)

$$\langle \pi^-(p) | j_L^{-\mu} | \Omega \rangle = -i f_\pi p^\mu / \sqrt{2} \qquad (15.78)$$

where j_L^- is the charge-changing weak interaction current. For the leptonic half of the amplitude, use the explicit spinors for the τ and the ν_τ derived in (d).

(f) Compute the partial width for $\tau^- \to \nu_\tau \pi^-$, using $f_\pi = 93$ MeV. Predict the branching fraction for this decay model, and compare to the Particle Data Group value.

(g) Work out the angular distribution of the pion in $\tau^- \to \nu_\tau \pi^-$ relative to the τ spin direction. Notice that the pion direction is correlated with the τ spin, so measurement of pion momenta in this decay gives an indication of the τ spin direction.

Gauge Theories with Spontaneous Symmetry Breaking

16

The V−A theory, with its current-current interaction, strongly suggests that the weak interaction is generated by the exchange of a spin 1 boson. The current-current interaction would arise from the Feynman diagram

$$(16.1)$$

The new boson is called the W^-. It must have an antiparticle W^+. And, it must be massive. In the diagram, the W^- appears as a resonance, with the Breit-Wigner denominator

The universal V−A interaction can be obtained from Feynman diagrams that include a massive spin 1 boson, the W boson.

$$\frac{1}{q^2 - m_W^2} \, . \qquad (16.2)$$

But, there was no sign of the q^2-dependence in the data that I showed in the previous chapter. This implies that the W^- boson is heavier than about 30 GeV. When we discuss the W boson as a particle, in the next chapter, we will see that its mass is about 80 GeV.

16.1 Field equations for a massive photon

Our need for a massive spin 1 boson forces us to face a problem that we have avoided up to now: What is the wave equation for the associated massive spin 1 field? As we discussed in Section 3.3, it is not straightforward to write a quantum theory for a spin 1 field that is positive and Lorentz-invariant. Maxwell's equations provide such a quantum theory, but Maxwell's equations also require that the associated particle, the photon, is massless.

Concepts of Elementary Particle Physics. Michael E. Peskin.
© Michael E. Peskin 2019. Published in 2019 by Oxford University Press.
DOI: 10.1093/oso/9780198812180.001.0001

We might try simply to add a mass term to Maxwell's equations, but there is a problem. If we have a massive spin 1 particle, we can boost to its rest frame. In this frame, the polarization vector $\vec{\epsilon}$ can point in any of the three space directions. Then the particle must have three independent quantum states. But a photon has only two quantum states. So we not only need to add a mass term for the photon; we also need to supply a new degree of freedom.

There is only one way known to solve this problem. That is to mix the two concepts of gauge invariance and spontaneous symmetry breaking. In this chapter, I will give three examples of gauge theories with spontaneous breaking of the gauge symmetry. In steps, we will build up to the correct theory of the weak interaction.

First, consider a $U(1)$ gauge theory that includes a complex scalar field. The Lagrangian is

The simplest illustrative example of a gauge theory with spontaneous symmetry breaking: a $U(1)$ gauge field (electromagnetism) coupled to a complex-valued scalar field.

$$\mathcal{L} = -\frac{1}{4}(F_{\mu\nu})^2 + \left|D_\mu\phi\right|^2 - V(\phi) \ . \tag{16.3}$$

The covariant derivative on ϕ is

$$D_\mu\phi = (\partial_\mu - ieQA_\mu)\phi \ , \tag{16.4}$$

where Q is the charge of the field ϕ in units of e, and $V(\phi)$ is a potential energy that depends on the value of ϕ.

Lev Landau and Vitaly Ginzburg wrote down this model as a phenomenological description of the electrodynamics of a superconductor (Ginzburg and Landau 1950). We reviewed part of the field-theoretic description of a superconductor in Section 14.2. In a superconductor, e^-e^- pairs form, and these pairs form a Bose condensate in the ground state. The ground state contains an indefinite number of these pairs. This is signalled by the fact that a field $\phi(x)$ that can destroy pairs has a nonzero expectation value in the ground state $|0\rangle$,

This model was originally introduced to model superconductivity in metals at extremely low termperature.

$$\langle 0|\,\phi(x)\,|0\rangle = \frac{v}{\sqrt{2}} \ . \tag{16.5}$$

This expectation value would correspond to the minimum of $V(\phi)$. Because this system has $U(1)$ symmetry, there must actually be a manifold of degenerate ground states, parametrized by

$$\langle \gamma|\,\phi(x)\,|\gamma\rangle = \frac{v}{\sqrt{2}}e^{i\gamma} \ . \tag{16.6}$$

For definiteness, I will expand about the state $|0\rangle$ in which $\langle\phi\rangle$ is real. The expansion of $\phi(x)$ about this ground state has the form

$$\phi(x) = \frac{1}{\sqrt{2}}\big(v + \chi(x) + i\eta(x)\big) \ . \tag{16.7}$$

A constant value of η shifts the vacuum state to one with the phase $\delta\gamma = \eta$. Thus, the field η is the Goldstone boson associated with this symmetry breaking and must have zero mass.

The field ϕ has the quantum numbers of $e^- e^-$, and therefore it is electrically charged, as reflected in the form of the covariant derivative above. To describe superconductivity, $Q = -2$. We can use a local gauge transformation

$$\phi(x) \to e^{-iQ\alpha(x)} \phi(x) \tag{16.8}$$

to remove $\eta(x)$. After doing this, the kinetic term of the ϕ field becomes

$$\left| D_\mu \phi \right|^2 = \left| (\partial_\mu - ieQA_\mu) \frac{1}{\sqrt{2}} (v + \chi(x)) \right|^2$$

$$= \left| -ieQA_\mu \frac{v}{\sqrt{2}} + \cdots \right|^2 . \tag{16.9}$$

This is a mass term for the A_μ field,

Generation of a mass for the photon in the Landau-Ginzburg model

$$\left| D_\mu \phi \right|^2 = \frac{1}{2} e^2 Q^2 v^2 A_\mu A^\mu = \frac{1}{2} m_A^2 A_\mu A^\mu . \tag{16.10}$$

In a superconductor, the quantum state with energy m_A is a quantized oscillation at the plasma frequency. The fact the the photon obtains a mass is manifested experimentally as the *Meissner effect*, the property that a superconductor expels magnetic fields (Tinkham 1966).

It is instructive to count the degrees of freedom. The field $A_\mu(x)$ has 2 degrees of freedom. The Goldstone boson $\eta(x)$ contributes one more degree of freedom. We can eliminate the field $\eta(x)$ by a choice of gauge, but this returns one degree of freedom to $A_\mu(x)$, giving exactly the 3 degrees of freedom required for a massive scalar field. It is often said that *the vector field eats the Goldstone boson and becomes massive*.

The polarization sum for a massive vector boson A_μ is

$$\sum_{i=1,2,3} \epsilon_i^\mu(p) \epsilon_i^{*\nu}(p) = -\left(\eta^{\mu\nu} - \frac{p^\mu p^\nu}{m_A^2} \right) . \tag{16.11}$$

We can check this in the rest frame of the vector boson. In that frame, the right-hand side is the projection onto three spacelike polarization vectors. Since this expression is Lorentz-covariant, it must then be correct in any frame.

This complex of ideas for generating a massive spin 1 field was introduced almost simultaneously in papers by Higgs (1964), Englert and Brout (1964), and Guralnik, Hagen, and Kibble (1964). Parts of the structure were anticipated by Nambu (1960) and Anderson (1963). For brevity, it is called the *Higgs mechanism*. The field ϕ is called the *Higgs field*. The physical quantum state created by the leftover scalar field $\chi(x)$ is called the *Higgs boson*. We will see that these elements have analogs in the realistic theory of the weak interaction.

16.2 Model field equations with a non-Abelian gauge symmetry

Before going to a realistic model, I will consider another illustrative example, this time with a non-Abelian symmetry group. Consider a gauge theory with the gauge group $SO(3)$. There are 3 gauge bosons

$$A_\mu^1 , \quad A_\mu^2 , \quad A_\mu^3 , \tag{16.12}$$

corresponding to rotations about the $\hat{1}$, $\hat{2}$, $\hat{3}$ axes. Introduce a real-valued scalar field Φ^a in the 3-vector representation. This is the adjoint representation of $SO(3)$, and so the covariant derivative on Φ^a is

$$D_\mu \Phi^a = \partial_\mu \Phi^a + g\epsilon^{abc} A_\mu^b \Phi^c . \tag{16.13}$$

We can easily write a potential that is rotationally invariant in the Φ^a space and is minimized when

$$|\langle \Phi^a \rangle| = v \tag{16.14}$$

The minima cover a manifold that has the form of a sphere in 3 dimensions.

The Georgi-Glashow model—an $SO(3)$ gauge theory coupled to a scalar field Φ^a.

For definiteness, I will choose to analyze the vacuum state that points in the $\hat{3}$ direction in the Φ^a space,

$$\langle 0| \Phi^a |0\rangle = v\delta^{a3} . \tag{16.15}$$

We can expand Φ^a about that vacuum,

$$\Phi(x) = (\pi^1(x), \pi^2(x), v + h(x)) . \tag{16.16}$$

The fields π^1, π^2 are Goldstone bosons. Again, we can use a gauge transformation to remove those fields. Then

$$\Phi(x) = (0, 0, v + h(x)) . \tag{16.17}$$

The kinetic term of Φ^a is

$$\frac{1}{2}(D_\mu \Phi^a)^2 = \frac{1}{2}(g\epsilon^{ab3} A_\mu^b (v + h(x)) + \partial_\mu h(x)\delta^{a3})^2 . \tag{16.18}$$

Expanding about the chosen vacuum state, we find

$$\frac{1}{2}(D_\mu \Phi^a)^2 = \frac{g^2}{2}\epsilon^{ab3} A_\mu^b \epsilon^{ac3} A^{\mu c} v^2 + \cdots$$
$$= \frac{g^2 v^2}{2}\left(A_\mu^b A^{\mu b} - A_\mu^3 A^{\mu 3}\right) . \tag{16.19}$$

The fields A_μ^1 and A_μ^2 obtain the mass

$$m_W^2 = g^2 v^2 \tag{16.20}$$

and A_μ^3 remains massless.

Physical explanation of the mass pattern of the Georgi-Glashow model, in which A^1, A^2 become massive but A^3 remains massless.

It is not difficult to understand why the A_μ^3 boson stays at zero mass. If $\langle \vec{\Phi} \rangle$ points in the $\hat{3}$ direction, the symmetry associated with rotation about the $\hat{3}$ axis is not broken,

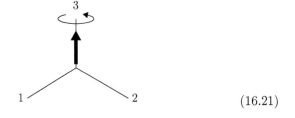

$$\tag{16.21}$$

The unbroken $U(1)$ gauge symmetry protects A^3 from obtaining mass.

The fields A_μ^1, A_μ^2 can be combined into eigenstates of the rotation about the $\hat{3}$ axis,

$$W^\pm = \frac{1}{\sqrt{2}}(A_\mu^1 \mp iA_\mu^2) . \tag{16.22}$$

We are now tempted to identify A_μ^3 as the photon and $W_\mu^\pm$ as the W bosons responsible for creating the V–A interaction. This is a unified model of weak and electromagnetic interactions. It is called the *Georgi-Glashow model* (Georgi and Glashow 1972). Notice that, in this model, the coupling constant g of the weak interaction bosons is equal to the electric charge e.

This model is very attractive, but it is not correct. It identifies electric charge with I^3, the generation of rotations about $\hat{3}$. The neutrino has zero electric charge, but it must be in an isospin multiplet with the electron so that it can be transformed into the electron by a weak interaction. The minimal size multiplet for the neutrino and the electron is an $I = 1$ multiplet

$$\begin{pmatrix} E^+ \\ \nu \\ e^- \end{pmatrix} . \tag{16.23}$$

Then there must be a heavy electron E^+. The model predicts that both of the fermions e^- and E^+ are produced in deep inelastic neutrino scattering,

$$\tag{16.24}$$

Production of the E^+ has not been observed. Searches for the E^+ put a lower bound on the mass of this particle at about 400 GeV.

16.3 The Glashow-Salam-Weinberg electroweak model

Sheldon Glashow suggested another way to construct a unified model of weak and electromagnetic interactions (Glashow 1961). Choose the gauge group

$$SU(2) \times U(1) \tag{16.25}$$

The $SU(2) \times U(1)$ model of unified weak and electromagnetic interactions

There are now 4 vector bosons,

$$A^1 , \quad A^2 , \quad A^3 , \quad B . \tag{16.26}$$

In this structure, we can keep the neutrino-electron system as an $I = \frac{1}{2}$ multiplet $(\nu, e^-)_L$ which transforms under the $SU(2) \times U(1)$ symmetry as

$$\begin{pmatrix} \nu_L \\ e_L^- \end{pmatrix} \to e^{i\vec{\alpha} \cdot \vec{\sigma}/2} e^{-i\beta/2} \begin{pmatrix} \nu_L \\ e_L^- \end{pmatrix} . \tag{16.27}$$

As the Higgs mechanism became understood, Steven Weinberg and Abdus Salam showed that the required mass generation could be accomplished by a Higgs field in the $I = \frac{1}{2}$ representation (Weinberg 1967, Salam 1968). This field transforms under $SU(2) \times U(1)$ as

$$\varphi = \begin{pmatrix} \varphi^+ \\ \varphi^0 \end{pmatrix} \rightarrow e^{i\vec{\alpha}\cdot\vec{\sigma}/2} e^{i\beta/2} \begin{pmatrix} \varphi^+ \\ \varphi^0 \end{pmatrix} . \tag{16.28}$$

Looking at the β terms in these transformation, the lepton doublet and the Higgs field transform under the $U(1)$ symmetry with charges $-\frac{1}{2}$ and $+\frac{1}{2}$, respectively.

Let the potential for the Higgs field be

$$V(\varphi) = -\mu^2 |\varphi|^2 + \lambda(|\varphi|^2)^2 . \tag{16.29}$$

The minimum of the potential satisfies

$$0 = -2\mu^2 \varphi + 4\lambda\varphi |\varphi|^2 \tag{16.30}$$

so, at the minimum,

$$|\varphi|^2 = |\varphi^+|^2 + |\varphi^0|^2 = \frac{\mu^2}{2\lambda} . \tag{16.31}$$

I will define

$$v = \sqrt{2} \langle |\varphi| \rangle = \mu/\sqrt{\lambda} . \tag{16.32}$$

This Higgs field vacuum expectation value spontaneously breaks the $SU(2) \times U(1)$ gauge symmetry.

The Higgs field has 4 degrees of freedom. The minima of $V(\phi)$ form a sphere in this 4-dimensional space. All of these minima are equivalent by $SU(2)$ transformations. For definiteness, I will analyze the vacuum state $|0\rangle$ where

$$\langle \varphi \rangle = \begin{pmatrix} 0 \\ v/\sqrt{2} \end{pmatrix} . \tag{16.33}$$

The Higgs mechanism leaves one remaining physical degree of freedom in the Higgs field φ. The corresponding particle is the *Higgs boson*.

Expanding around this state

$$\varphi(x) = \begin{pmatrix} \pi^+(x) \\ (v + h(x) + i\pi^3(x))/\sqrt{2} \end{pmatrix} . \tag{16.34}$$

The fields $\pi^+ = (\pi^1 + i\pi^2)/\sqrt{2}$ and π^3 are Goldstone bosons. We can set these fields to zero by an $SU(2)$ gauge transformation. This leaves over one real-valued scalar field in φ. This remaining field $h(x)$ is the field of the *Higgs boson*.

The coupling of the gauge fields A_μ^a and B_μ to any fermions and scalars is specified by the covariant derivative

$$D_\mu \Psi = (\partial_\mu - ig A_\mu^a I^a - ig' B_\mu Y)\Psi . \tag{16.35}$$

Here I^a is a generator of the $SU(2)$ gauge symmetry in the appropriate representation. I will refer to the quantum number I^a as *weak isospin*. It is an $SU(2)$ quantum number, but it is important to understand that this $SU(2)$ is an exact symmetry, distinct from the approximate isospin symmetry of the strong interaction. Y is the charge under the $U(1)$ symmetry, which is called the *hypercharge*. The theory has two coupling constants g and g' corresponding to the two independent gauge groups. These coupling constants need not be equal. The ratio g'/g is a an important free parameter in the theory.

The covariant derivative on the Higgs field is

$$D_\mu \varphi = (\partial_\mu - ig A_\mu^a I^a - ig' B_\mu Y) \begin{pmatrix} 0 \\ v/\sqrt{2} \end{pmatrix} + \cdots . \qquad (16.36)$$

The kinetic term of the Higgs field is then

$$|D_\mu \varphi|^2 = \frac{1}{2} (0 \quad v) \left(g A_\mu^a \frac{\sigma^a}{2} + g' B_\mu \frac{1}{2} \right) \left(g A^{\mu b} \frac{\sigma^b}{2} + g' B^\mu \frac{1}{2} \right) \begin{pmatrix} 0 \\ v \end{pmatrix} . \qquad (16.37)$$

Evaluating this expression, the terms with A_μ^1 and A_μ^2 give

$$\frac{1}{2} v^2 g^2 (\frac{1}{2})^2 \left((A_\mu^1)^2 + (A_\mu^2)^2 \right) . \qquad (16.38)$$

In terms of the W fields defined in (16.22), this is

$$(\frac{gv}{2})^2 W_\mu^+ W^{\mu-} = m_W^2 W_\mu^+ W^{\mu-} . \qquad (16.39)$$

The terms involving A_μ^3 and B_μ give

$$\frac{1}{2} v^2 (\frac{1}{2})^2 \left(-g A_\mu^3 + g' B_\mu \right)^2 . \qquad (16.40)$$

This expression is the mass term for one gauge field, which is a linear combination of A_μ^3 and B_μ. It is convenient to define the *weak mixing angle* θ_w by the relation

$$\tan \theta_w = \frac{g'}{g} , \qquad (16.41)$$

and to define parameters c_w, s_w

$$c_w = \cos \theta_w = \frac{g}{\sqrt{g^2 + g'^2}} , \qquad s_w = \sin \theta_w = \frac{g'}{\sqrt{g^2 + g'^2}} . \qquad (16.42)$$

Then we can write the two orthogonal combinations of the fields A_μ^3 and B_μ as

$$Z_\mu = c_w A_\mu^3 - s_w B_\mu$$
$$A_\mu = s_w A_\mu^3 + c_w B_\mu \qquad (16.43)$$

The boson Z_μ receives mass

$$m_Z^2 = \frac{(g^2 + g'^2) v^2}{4} , \qquad (16.44)$$

Introduction of the *weak mixing angle* θ_w. In the remainder of this book, I will abbreviate $\cos \theta_w = c_w$, $\sin \theta_w = s_w$.

and the boson A_μ remains massless.

In the next part of this section, we will see in detail that the massless spin 1 boson A_μ should be identified with the photon. Then the $SU(2) \times U(1)$ model is a unified model of weak and electromagentic interactions. We call this unified force the *electroweak interaction*.

The $SU(2) \times U(1)$ model leaves one exactly massless vector boson, which I will identify with the photon. The photon and Z are linear combinations of the original bosons A^3 and B. Thus, in this model, the weak and electromagnetic interactions are different facets of the same underlying structure. We call these forces collectively the *electroweak interaction*.

It is not difficult to see that there must be a massless spin 1 boson left after the symmetry breaking. The transformation of the Higgs vacuum expectation value is

$$\varphi = \begin{pmatrix} 0 \\ v/\sqrt{2} \end{pmatrix} \rightarrow e^{i\vec{\alpha}\cdot\vec{\sigma}/2} e^{i\beta/2} \begin{pmatrix} 0 \\ v/\sqrt{2} \end{pmatrix} . \tag{16.45}$$

Then a transformation with $\alpha^3 = \beta$ leaves the vacuum expectation value unchanged. This gauge symmetry is not broken. The corresponding gauge boson—A_μ above—remains massless. Any realistic theory must have a massless vector boson that can be identified with the photon. This symmetry principle tells us how to insure that such a massless particle is present.

The masses of the W and Z bosons of the $SU(2) \times U(1)$ model.

The masses of the W and Z bosons are

$$m_W = \frac{gv}{2} , \qquad m_Z = \frac{\sqrt{g^2 + g'^2} v}{2} . \tag{16.46}$$

The predicted mass relation $m_W = m_Z c_w$ provides an important test of the $SU(2) \times U(1)$ model.

These obey the relation

$$m_W = m_Z \cdot c_w . \tag{16.47}$$

If we can measure s_w^2 in another way, this relation is testable experimentally. We will see in the next chapter that, when higher order corrections are included, the relation is obeyed to better than 1% accuracy.

To determine the couplings of the W and Z to quarks and leptons, we need to rewrite the general expression for the covariant derivative in terms of the mass eigenstates. Using the inverse of (16.43)

The couplings of W, Z, A to quarks and leptons are now predicted in terms of the $SU(2) \times U(1)$ quantum numbers of these particles. To find the precise forms of the couplings, we simplify the covariant derivative.

$$A_\mu^3 = c_w Z_\mu + s_w A_\mu ,$$
$$B_\mu = -s_w Z_\mu + c_w A_\mu , \tag{16.48}$$

and the expression (16.22) for $W_\mu^\pm$, we can write the covariant derivative (16.35) as

$$D_\mu \Psi = \left[\partial_\mu - i\frac{g}{\sqrt{2}}(W_\mu^+ \sigma^+ + W^- \sigma^-) \right.$$
$$\left. -ig(c_w Z_\mu + s_w A_\mu)I^3 - ig'(-s_w Z_\mu + c_w A_\mu)Y \right] \Psi . \tag{16.49}$$

We can recast this as

$$D_\mu \Psi = \left[\partial_\mu - i\frac{g}{\sqrt{2}}(W_\mu^+ \sigma^+ + W^- \sigma^-) - ieA_\mu Q - i\frac{g}{c_w}Z_\mu Q_Z \right] \Psi . \tag{16.50}$$

In the electroweak theory, the basic electric charge e is derived from g and g'.

where I have set

$$e = gs_w = g'c_w = \frac{gg'}{\sqrt{g^2 + g'^2}} . \tag{16.51}$$

It is appropriate to identify e with the value of the unit electric charge. The electric charge of a fermion or boson is then

$$Q = I^3 + Y \qquad (16.52)$$

The electric charge Q of each particle in units of e is fixed by its quantum numbers.

Similarly, the Z boson charge Q_Z for any boson or fermion is

$$Q_Z = c_w^2 I^3 - s_w^2 Y \qquad (16.53)$$

In a similar way, the Z charge Q_Z of each particle is fixed as a function of its quantum numbers.

or

$$Q_Z = I^3 - s_w^2 Q . \qquad (16.54)$$

To complete the model, we need to assign to all of the quarks and leptons appropriate quantum numbers under $SU(2) \times U(1)$. I will ignore all masses in this discussion. Then we can treat the left- and right-handed parts of the Dirac field as independent fields. Because the left-handed particles couple to the W bosons but the right-handed particles do not, we will need to assign these fields different quantum numbers. This is a mysterious but also absolutely crucial feature of the model. It is the origin of the V−A structure that, as we have seen in the previous chapter, is required by experiment.

The assignment of different quantum numbers to the left- and right-handed fermions is the origin of the parity violation of the V−A interaction.

In the $SU(2) \times U(1)$ model, the left-handed fields will belong to doublets of $SU(2)$ ($I = \frac{1}{2}$), and the right-handed fields will be in singlets ($I = 0$). We then choose the values of Y to give the appropriate electric charges. For the electron neutrino ν_e, the electron e^-, the u quark, and the d quark, the required charges are

(I^3, Y) quantum numbers for the various species of leptons and quarks.

$$\nu_{eL} : \quad I^3 = +\frac{1}{2} , \ Y = -\frac{1}{2} , \ Q = 0 \qquad \nu_{eR} : \quad I^3 = 0 , \ Y = 0 , \ Q = 0$$

$$e_L^- : \quad I^3 = -\frac{1}{2} , \ Y = -\frac{1}{2} , \ Q = -1 \qquad e_R^- : \quad I^3 = 0 , \ Y = -1 , \ Q = -1$$

$$u_L : \quad I^3 = +\frac{1}{2} , \ Y = \frac{1}{6} , \ Q = \frac{2}{3} \qquad u_R : \quad I^3 = 0 , \ Y = \frac{2}{3} , \ Q = \frac{2}{3}$$

$$d_L : \quad I^3 = -\frac{1}{2} , \ Y = \frac{1}{6} , \ Q = -\frac{1}{3} \qquad d_R : \quad I^3 = 0 , \ Y = -\frac{1}{3} , \ Q = -\frac{1}{3}$$

$$(16.55)$$

Note that the right-handed component of the neutrino has zero coupling to the vector fields and could be omitted from the model. I will return to this point in Chapter 20.

The particles in (16.55) are laid out in $SU(2)$ representations

$$\begin{pmatrix} \nu_L \\ e_L^- \end{pmatrix} \qquad e_R^- \qquad \begin{pmatrix} u_L \\ d_L \end{pmatrix} \qquad u_R \qquad d_R . \qquad (16.56)$$

This structure is called a *generation* of quarks and leptons. There are two more generations known, containing, respectively, ν_μ, μ^-, c and s, and ν_τ, τ^-, t, and b. These have $SU(2) \times U(1)$ quantum numbers with the same values as in (16.55).

Definition of a *generation* of quarks and leptons.

16.4   The neutral current weak interaction

Now that we have formulated a specific theory of the weak interaction, we can work out its observational consequences. From the way that we have constructed the model, with the W^+ and W^- fields coupling only to the left-handed quarks and leptons, the Feynman diagram

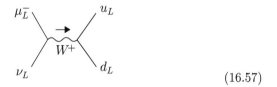

$$(16.57)$$

will produce the current-current interaction

$$\frac{g^2}{2} j^{\mu-} \frac{1}{q^2 - m_W^2} j_\mu^+ \ . \tag{16.58}$$

If $q^2 \ll m_W^2$, we can ignore q^2 in the denominator, and then we find an amplitude with exactly the structure of (15.6). We can identify the coefficient in that formula, in terms of parameters of the $SU(2) \times U(1)$ model as,

$$\frac{4G_F}{\sqrt{2}} = \frac{g^2}{2m_W^2} \tag{16.59}$$

or

$$\frac{G_F}{\sqrt{2}} = \frac{g^2}{8m_W^2} \ , \tag{16.60}$$

where g is the $SU(2)$ gauge coupling in (16.35). The $SU(2) \times U(1)$ weak interaction theory replaces the dimensionful constant G_F of the Fermi theory with a dimensionless coupling strength g and a mass scale set by m_W or v.

The $SU(2) \times U(1)$ theory contains an additional interaction mediated by a virtual Z boson, for example,

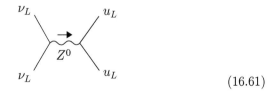

$$(16.61)$$

This diagram leads to the current-current interaction

$$\frac{1}{2} \frac{g^2}{c_w^2} (j_L^{\mu 3} - s_w^2 j_Q^\mu) \frac{1}{q^2 - m_Z^2} (j_{\mu L}^3 - s_w^2 j_{\mu Q}) \ , \tag{16.62}$$

where $j_L^{\mu 3}$ is the left-handed weak isospin current and j_Q^μ is the electric charge current

$$j_Q^\mu = \sum_f Q_f (\overline{f}_L \overline{\sigma}^\mu f_L + \overline{f}_R \sigma^\mu f_R) \ . \tag{16.63}$$

I have included a factor $\frac{1}{2}$ in (16.62) because any given quark or lepton current can appear in either term.

At low energies, we can ignore q^2 relative to m_W^2, m_Z^2. Then, using the relation (16.47), we find a current-current interaction, generalizing (15.6), of the form

$$\mathcal{M} = \left\langle \frac{4G_F}{\sqrt{2}} \left(j_L^{\mu+} j_{\mu L}^- + (j_L^{\mu 3} - s_w^2 j_Q^\mu)^2 \right) \right\rangle . \qquad (16.64)$$

The first term in this current-current interaction, mediated by the W, is called the *charged current interaction*. The second term, mediated by the Z, is called the *neutral current interaction*.

The complete Fermi interaction of the $SU(2) \times U(1)$ model contains the V−A charged current interaction and also a neutral current interaction. Their coefficients have perfect rotational symmetry in the weak interaction $SU(2)$ up to terms proportional to s_w^2.

The neutral current interaction produces a new event type in neutrino scattering, in which a neutrino scatters elastically from a quark or lepton. We have seen an example of an event of this type in the lower part of Fig. 15.4.

For deep inelastic neutrino scattering, we can work out the cross section for neutral current reactions in the same way that we worked out the cross section for charged current reactions. Looking back at (15.71) and (15.72), we see that the formulae for charged current deep inelastic scattering have the form

$$\frac{d^2\sigma}{dxdy} \sim \frac{G_F^2 s}{\pi} \cdot x f(x) , \qquad (16.65)$$

with an extra factor $(1-y)^2$ if the helicities of the beam and target fermion are not matched. The formulae for neutral current deep inelastic scattering will be similar, except that we must include the explicit Z charges from (16.54) or (16.64). These charges are nonzero both for left- and right-handed quarks and antiquarks. For neutrino scattering, the contribution from the quarks is then

$$\frac{d^2\sigma}{dxdy}(\nu p \to \nu X)\Big|_q = \frac{G_F^2 s}{\pi} \left[x f_u(x) \{ (\tfrac{1}{2} - \tfrac{2}{3} s_w^2)^2 + (-\tfrac{2}{3} s_w^2)^2 (1-y)^2 \} \right.$$
$$\left. + x f_d(x) \{ (-\tfrac{1}{2} + \tfrac{1}{3} s_w^2)^2 + (\tfrac{1}{3} s_w^2)^2 (1-y)^2 \} \right] \qquad (16.66)$$

The cross section for deep inelastic neutrino scattering due to the neutral current interaction.

and the contribution from the antiquarks is

$$\frac{d^2\sigma}{dxdy}(\nu p \to \nu X)\Big|_{\bar{q}} = \frac{G_F^2 s}{\pi} \left[x f_{\bar{u}}(x) \{ (\tfrac{1}{2} - \tfrac{2}{3} s_w^2)^2 (1-y)^2 + (-\tfrac{2}{3} s_w^2)^2 \} \right.$$
$$\left. + x f_{\bar{d}}(x) \{ (-\tfrac{1}{2} + \tfrac{1}{3} s_w^2)^2 (1-y)^2 + (\tfrac{1}{3} s_w^2)^2 \} \right] . \qquad (16.67)$$

For an antineutrino beam, the positions of the factors 1 and $(1-y)^2$ reverse.

Neutrino experiments are typically done with very massive targets, made of iron, mineral oil, or another material obtainable in bulk. Then

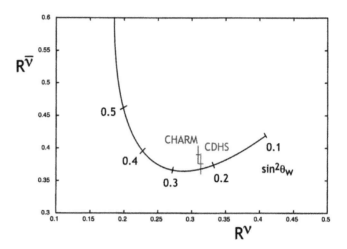

Fig. 16.1 The relation between R^ν and $R^{\bar\nu}$ predicted by (16.73), (16.74), compared to data from the CDHS and CHARM neutrino deep inelastic scattering experiments at CERN. The measured values are taken from (Amaldi *et al.* 1987).

it is relevant to specialize the formulae just given to nuclei with approximately equal numbers of u and d quarks. Let $f_q(x)$ be the pdf for quarks in a nucleus containing a total of A nucleons with equal numbers of protons and neutrons

$$f_q(x) = A(f_u(x) + f_d(x)) . \tag{16.68}$$

The above formulae combine and simplify to

$$\frac{d^2\sigma}{dxdy}(\nu A \rightarrow \nu X) = \frac{G_F^2 s}{\pi}\left[x f_q(x)\{(\frac{1}{2} - s_w^2) + \frac{5}{9}s_w^4[1 + (1-y)^2]\} \right.$$
$$\left. + x f_{\bar q}(x)\{(\frac{1}{2} - s_w^2)(1-y)^2 + \frac{5}{9}s_w^4[1 + (1-y)^2]\} \right] . \tag{16.69}$$

Similarly, the cross section for neutral current scattering of an antineutrino from an isospin singlet target nucleus is

$$\frac{d^2\sigma}{dxdy}(\bar\nu A \rightarrow \bar\nu X) = \frac{G_F^2 s}{\pi}\left[x f_q(x)\{(\frac{1}{2} - s_w^2)(1-y)^2 + \frac{5}{9}s_w^4[1 + (1-y)^2]\} \right.$$
$$\left. + x f_{\bar q}(x)\{(\frac{1}{2} - s_w^2) + +\frac{5}{9}s_w^4[1 + (1-y)^2]\} \right] . \tag{16.70}$$

It is easier to understand these formulae if we divide by the corresponding charged current cross sections

$$\frac{d^2\sigma}{dxdy}(\nu A \rightarrow \mu^- X) = \frac{G_F^2 s}{\pi}\left[x f_q(x) + x f_{\bar q}(x)(1-y)^2 \right] ,$$
$$\frac{d^2\sigma}{dxdy}(\bar\nu A \rightarrow \mu^+ X) = \frac{G_F^2 s}{\pi}\left[x f_q(x)(1-y)^2 + x f_{\bar q}(x) \right] . \tag{16.71}$$

The ratio of the two charged current cross sections can be reduced to

$$r = \frac{\sigma(\bar{\nu}, CC)}{\sigma(\nu, CC)} = \left\langle \frac{x f_q(x)(1-y)^2 + x f_{\bar{q}}(x)}{x f_q(x) + x f_{\bar{q}}(x)(1-y)^2} \right\rangle , \qquad (16.72)$$

where the expectation value indicates that numerator and denominator are integrated over the range of (x, y) covered by the experiment. The quantity r can be measured directly. It depends on the coverage of the detector in x and y, and, typically, it has a value about 0.4. The ratio of neutral to charged current rates for neutrinos and antineutrinos can then be written (Llewellyn Smith 1983).

$$R^{\nu} = \frac{\sigma(\nu, NC)}{\sigma(\nu, CC)} = \frac{1}{2} - s_w^2 + \frac{5}{9} s_w^4 (1 + r) \qquad (16.73)$$

and

$$R^{\bar{\nu}} = \frac{\sigma(\bar{\nu}, NC)}{\sigma(\bar{\nu}, CC)} = \frac{1}{2} - s_w^2 + \frac{5}{9} s_w^4 (1 + \frac{1}{r}) . \qquad (16.74)$$

The ratio of rates for neutral current and charged current deep inelastic scattering, for neutrinos and antineutrinos, provides a simple first test of the $SU(2) \times U(1)$ model.

For a given experiment, using the measured value of r, the values of R^{ν} and $R^{\bar{\nu}}$ lie on a specific curve in the plane of possible values, parametrized by the value of s_w^2. When this curve was introduced, it was popularly known as "Weinberg's nose". Measurements of R^{ν} and $R^{\bar{\nu}}$ thus test the theory and also measure the value of s_w^2. Figure 16.1 shows the curve and the values measured by the two large CERN neutrino experiments of the early 1980's.

The theory passes this test, and the value of s_w^2 is seen to be close to

$$s_w^2 \approx 0.23 . \qquad (16.75)$$

I will present higher-precision tests of the $SU(2) \times U(1)$ weak interaction theory in the next chapter.

Exercises

(16.1) Another illustrative example of spontaneous breaking of a gauge symmetry is given by a theory called *topcolor* (Hill 1995) in which, at very short distances, the (t, b) quarks transform under a different $SU(3)$ color group from the lighter quarks. For simplicity, I ignore the weak interaction in this exercise.

(a) The gauge group of the topcolor theory is $SU(3)_1 \times SU(3)_2$. The theory has two sets of 8 gauge bosons and two independent coupling constants g_1, g_2. The light quarks u, d, s, c

transform only under $SU(3)_1$ according to

$$q_u \to (1 + i\alpha_1^a t^a) q_u , \qquad (16.76)$$

where t^a is a 3×3 representation matrix for $SU(3)$. The b and t quarks transform similarly under $SU(3)_2$,

$$q_t \to (1 + i\alpha_2^a t^a) q_t , \qquad (16.77)$$

The model also contains a complex-valued scalar field Φ which is a (3×3) matrix and transforms as

$$\Phi \to \Phi + (i\alpha_1^a t^a) \Phi + \Phi (-i\alpha_2^a t^a) . \qquad (16.78)$$

Show that the covariant derivatives of the theory are

$$D_\mu q_u = (\partial_\mu - ig_1 A^a_{\mu 1} t^a) q_u ,$$
$$D_\mu q_t = (\partial_\mu - ig_2 A^a_{\mu 2} t^a) q_t ,$$
$$D_\mu \Phi = \partial_\mu \Phi - i(g_1 A^a_{\mu 1} t^a)\Phi + \Phi(+ig_2 A^a_{\mu 2} t^a) .$$
$$(16.79)$$

The Lagrangian of the topcolor theory is

$$\mathcal{L} = -\frac{1}{4}(F^{\mu\nu a}_1)^2 - \frac{1}{4}(F^{\mu\nu a}_2)^2$$
$$+ \sum_{f=u,d,s,c} \bar{q}_f i\gamma \cdot Dq_f + \sum_{f'=b,t} \bar{q}_{f'} i\gamma \cdot Dq_{f'}$$
$$+ \mathrm{tr}[(D_\mu \Phi)^\dagger D^\mu \Phi] - V(\Phi) . \qquad (16.80)$$

(b) Assume that the minimum of the potential $V(\Phi)$ is at the nonzero value

$$\langle\Phi\rangle = V \cdot \mathbf{1} , \qquad (16.81)$$

where $\mathbf{1}$ is the $3{\times}3$ unit matrix and V is a constant with the dimensions of mass. Find the mass terms for $A^a_{\mu 1}$ and $A^a_{\mu 2}$. Show that one linear combination of $A^a_{\mu 1}$ and $A^a_{\mu 2}$ remains massless.

(c) Construct the normalized mass eigenstate fields, by analogy to (16.48). Call the new massless and massive vector fields, respectively, A^a_μ and $\mathbf{A}^a_\mu$. Show that the massive field has the mass

$$m^2 = (g_1^2 + g_2^2)V^2 . \qquad (16.82)$$

(d) Rewrite the covariant deratives on the quark fields in terms of the mass eigenstate vector fields. Show that all quarks now couple to the field A^a_μ with the same coupling constant g, given by

$$\frac{1}{g^2} = \frac{1}{g_1^2} + \frac{1}{g_2^2} . \qquad (16.83)$$

So we find an $SU(3)$ gauge theory just like QCD, with a coupling g smaller than either g_1 or g_2. This property that coupling constants combine like resistors in parallel is often seen in models with spontaneous gauge symmetry breaking.

(16.2) Consider an $SU(3)$ gauge theory coupled to a Hermitian 3×3 matrix scalar field Φ, with Φ transforming under $SU(3)$ as

$$\Phi \to \Phi + (i\alpha^a t^a)\,\Phi + \Phi(-i\alpha^a t^a) . \qquad (16.84)$$

This theory has 8 gauge fields and one coupling constant g. This theory was studied by Weinberg (1972) for reasons that will become clearer as we proceed.

(a) Write out the 8 3×3 matrices that represent the generators of $SU(3)$. These are 3×3 Hermitian matrices orthonormalized such that

$$\mathrm{tr}[t^a t^b] = \frac{1}{2}\delta^{ab} . \qquad (16.85)$$

(b) Write the covariant derivative on Φ.

(c) Assume that the potential for Φ is minimized at a configuration

$$\langle\Phi\rangle = V \begin{pmatrix} a & 0 & 0 \\ 0 & a & 0 \\ 0 & 0 & b \end{pmatrix} , \qquad (16.86)$$

with $a \neq b$. Write the mass matrix for the 8 gauge fields A^a_μ.

(d) Show that 4 of the 8 gauge fields receive zero mass. Show that the other 4 fields obtain masses

$$m^2 = g^2(a-b)^2 V^2 . \qquad (16.87)$$

(e) Let ψ be a fermion field in the $\mathbf{3}$ representation of $SU(3)$. Write the covariant derivative on this field, keeping only the 4 massless vector fields. Show that this is identical to the covariant derivative of an $SU(2) \times U(1)$ gauge theory.

(f) Identify the upper two components of the 3-component field ψ with the $(\nu, e)_L$ doublet. To relate this model to the $SU(2) \times U(1)$ theory of weak interactions, rescale the coupling constant of the $U(1)$ gauge field so that the charge multiplying this coupling constant equals the appropriate hypercharge $Y = \frac{1}{2}$. Show that, after this rescaling, the model has $SU(2)$ and $U(1)$ gauge couplings

$$g = g , \qquad g' = g/\sqrt{3} \qquad (16.88)$$

(g) Compute $\sin^2 \theta_w$ and compare to the value (16.75).

The W and Z Bosons

17

In the previous chapter, I described the $SU(2) \times U(1)$ theory of weak and electromagnetic interactions. In this theory, the V−A weak interaction arises from exchange of the W boson, and there is an additional neutral current interaction due to the exchange of a heavier boson Z. With enough energy in the center of mass, it became possible to produce these bosons directly and study their properties. The W and Z were first seen directly by the UA1 and UA2 experiments at CERN, in a $p\bar{p}$ collider designed for this purpose (Arnison *et al.* 1983a, 1983b, Banner *et al.* 1983a, 1983b). Today, W and Z bosons are produced by the millions at the LHC. Figures 17.1 and 17.2 show beautiful examples of events collected by the ATLAS experiment at the LHC, showing W and Z production with the decays

$$W^\pm \to e^\pm \nu \qquad Z \to e^+ e^- . \qquad (17.1)$$

The $SU(2) \times U(1)$ theory makes detailed predictions for the properties of the W and Z bosons. In this chapter, we will work out those predictions and compare them to experiment.

17.1 Properties of the W boson

To begin, I will work out the major decay rates and production cross sections for the W boson.

From the covariant derivative of the $SU(2) \times U(1)$ model given in (16.50), we read off the matrix element for the leptonic decay $W^+ \to \nu_e e^+$ as

$$\mathcal{M}(W^+ \to \nu_L e_R^+) = \frac{g}{\sqrt{2}} \, u_L^\dagger(p_\nu) \bar{\sigma}^\mu v_L(p_e) \, \epsilon_{W\mu} , \qquad (17.2)$$

Calculation of the partial widths for the decay of the W boson to a pair of leptons or quarks.

where $\epsilon_{W\mu}$ is the polarization vector of the W^+. The product of spinors is the same one that we have seen before in our discussion of $e^+ e^-$ annihilation. From (8.37), we see that, in the $\nu_e e^-$ center of mass frame,

$$u_L^\dagger(p_\nu) \bar{\sigma}^\mu v_L(p_e) \, \epsilon_{W\mu} = 2\sqrt{2} E \, \epsilon_-^* \cdot \epsilon_W . \qquad (17.3)$$

We square the matrix element and average over 3 initial spin directions (or average over angles for the ν_e direction relative to the W polarization

Concepts of Elementary Particle Physics. Michael E. Peskin.
© Michael E. Peskin 2019. Published in 2019 by Oxford University Press.
DOI: 10.1093/oso/9780198812180.001.0001

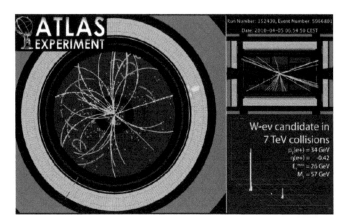

Fig. 17.1 Event display of a $pp \rightarrow W \rightarrow e\nu$ event recorded by the ATLAS experiment at the Large Hadron Collider (figure courtesy of CERN and the ATLAS Collaboration).

Fig. 17.2 Event display of a $pp \rightarrow Z \rightarrow e^{+}e^{-}$ event recorded by the ATLAS experiment at the Large Hadron Collider (figure courtesy of CERN and the ATLAS Collaboration).

vector). This gives

$$\Gamma(W^+ \to \nu e^+) = \frac{1}{2m_W} \frac{1}{8\pi} \frac{g^2}{2} m_W^2 \cdot 2 \cdot \frac{1}{3} \qquad (17.4)$$

or

$$\Gamma(W^+ \to \nu_e e^+) = \frac{\alpha_w}{12} m_W , \qquad (17.5)$$

where

$$\alpha_w = \frac{g^2}{4\pi} = \frac{e^2}{4\pi s_w^2} . \qquad (17.6)$$

Similarly,

$$\Gamma(W^+ \to \nu_\mu \mu^+) = \Gamma(W^+ \to \nu_\tau \tau^+) = \frac{\alpha_w}{12} m_W . \qquad (17.7)$$

For decays to quarks, we must add the color factor of 3. The rate is enhanced by a QCD correction, the same factor that appears in the $e^+e^- \to$ hadrons cross section (11.72). Then

$$\Gamma(W^+ \to u\bar{d}) = \Gamma(W^+ \to c\bar{s}) = \frac{\alpha_w}{12} m_W \cdot 3(1 + \frac{\alpha_s(m_W)}{\pi} + \cdots) . \quad (17.8)$$

The top quark is sufficiently heavy that the decay $W^+ \to t\bar{b}$ is kinematically forbidden.

To evaluate these formulae, we need the value of α_w. This is a good place to pause and collect the values of all of the parameters of the electroweak theory. The two quantities

Values of the coupling constants g and g' and the dimensionful parameter v that characterize the $SU(2) \times U(1)$ theory.

$$\alpha_w = \frac{g^2}{4\pi} , \qquad \alpha' = \frac{g'^2}{4\pi} \qquad (17.9)$$

give the intrinsic strengths of the $SU(2)$ and $U(1)$ interactions. We can evaluate these quantities from the values of α and s_w^2. In Section 17.4, I will point out a number of experimental measurements on the Z bosons that lead to very precise value of s_w^2. The result will be

$$s_w^2 = 0.23116 \pm 0.00012 . \qquad (17.10)$$

This value should be combined with the value of α evaluated at a momentum scale appropriate to the physics of W and Z. This is not $\alpha(Q = 0) = 1/137$ but rather

$$\alpha(m_Z) = 1/129. \qquad (17.11)$$

Making the combination, we find

$$\alpha_w = \frac{\alpha}{s_w^2} = 1/29.8 \qquad \alpha' = \frac{\alpha}{c_w^2} = 1/99.1 \qquad (17.12)$$

The weak interactions are weak, but not exceptionally so. The apparent "weakness" of the weak interactions comes from the small size of G_F, a dimensionful quantity, relative to the mass of the proton. This is due less to the small value of the coupling constant than to the large value of the W boson mass in GeV units. In fact, if we use (16.46) in the relation (16.60) for G_F, the size of the gauge coupling actually cancels out,

$$G_F = \frac{\sqrt{2}g^2}{8m_W^2} = \frac{1}{\sqrt{2}v^2} \ , \tag{17.13}$$

and we see that G_F is completely determined by the Higgs field vacuum expectation value, which we find to be

$$v = 246 \text{ GeV} \ . \tag{17.14}$$

The V$-$A weak interaction is weak because the Higgs vacuum expectation value v is much larger than the proton mass.

Given the values for v, g, and g', we can predict the values of the W and Z masses from (16.46),

Comparison of the predicted W and Z boson masses to the values seen in experiment.

$$m_W = 80.2 \text{ GeV} \qquad m_Z = 91.5 \text{ GeV} \ . \tag{17.15}$$

This is in reasonable agreement with the values found in direct measurement of the particle masses

$$m_W = 80.385 \pm 0.015 \text{ GeV} \qquad m_Z = 91.1876 \pm 0.0021 \text{ GeV} \ . \tag{17.16}$$

It is important to point out that, when comparing numbers at this level of accuracy, we must include the effects of higher order quantum corrections. A particularly important effect for the W and Z masses is the quantum fluctuation of the bosons to quark-antiquark pairs, in particular, to top quarks,

$$\tag{17.17}$$

The value of the top quark mass affects the ratio of the W and Z masses at the 5% level. When the known value of the top quark mass is included, the measured values of the W and Z masses are in very good agreement with the predictions of the electroweak theory, as I will quantity in Section 17.4. Indeed, before the top quark was discovered, precision electroweak measurement of the properties of the Z boson correctly predicted the top quark mass to be in the range 160–180 GeV.

Using the value of α_w above, we find for the total width of the W boson

$$\Gamma_W = \frac{\alpha_w}{12} m_W \cdot \left[3 + 2 \cdot (3.1)\right] = 2.1 \text{ GeV} \ . \tag{17.18}$$

Branching ratios for the W boson decays to lepton and quark pairs.

The branching ratios of the W are predicted to be

$$BR(e\nu_e) = BR(\mu\nu_\mu) = BR(\tau\nu_\tau) = 11\%$$
$$BR(u\bar{d}) = BR(c\bar{s}) = 34\% \ , \tag{17.19}$$

in good agreement with observations.

17.2   *W* production in *pp* collisions

The matrix element for W decay to $u\bar{d}$ can also be used in the opposite direction to compute the production cross section for a W boson in a hadron-hadron collision. At a hadron-hadron collider, the heavy bosons W and Z can be created by quark-antiquark annihilation, for example, $u\bar{d} \to W^+$, $u\bar{u} \to Z$. Such reactions, and reactions such as $u\bar{u} \to \mu^+\mu^-$ involving a virtual photon or Z boson, are called *Drell-Yan processes* (Drell and Yan 1970).

The cross section for Drell-Yan production of a W^+ boson is assembled by combining the parton-level cross section with the pdfs of the colliding hadrons. The parton cross section is

Computation of the Drell-Yan cross section for W boson production in pp collisions.

$$\sigma(u\bar{d} \to W^+) = \frac{1}{2\hat{s}} \int d\Pi \, |\mathcal{M}|^2 \, , \qquad (17.20)$$

where phase space with one particle in the final state is given by

$$\int d\Pi_1 = \int \frac{d^3 p_W}{(2\pi)^3 2E_W} (2\pi)^4 \delta^{(4)}(p_u + p_d - p_W) \, . \qquad (17.21)$$

Comparing to (3.89), we see that

$$\int d\Pi_1 = 2\pi \delta(\hat{s} - m_W^2) \, . \qquad (17.22)$$

We must average the squared matrix element over initial spins and colors and sum over final polarization states. This sum and average is

$$\frac{1}{2 \cdot 2} \frac{1}{3 \cdot 3} \sum_{\text{color, spin}} |\mathcal{M}|^2 = \frac{1}{36} \cdot 3 \cdot \frac{g^2}{2} m_W^2 \cdot 2 = \frac{g^2}{12} m_W^2 \, . \qquad (17.23)$$

Assembling the pieces, we find

$$\sigma(u\bar{d} \to W^+) = \frac{\pi^2 \alpha_w}{3} \delta(\hat{s} - m_W^2) \, . \qquad (17.24)$$

This result must be integrated over the pdfs for the initial state quarks and antiquarks. For pp collisions,

$$\sigma(pp \to W^+) = \int dx_1 f_u(x_1) \int dx_2 f_{\bar{d}}(x_2)$$
$$\cdot \sigma(u(x_1 P_1)\bar{d}(x_2 P_2) \to W^+) + (1 \leftrightarrow 2) \quad (17.25)$$

plus contributions from heavier quarks and antiquarks. To simplify this formula, go to the $p\bar{p}$ CM frame. The parton 4-vectors are

$$p_1 = x_1 P_1 = (x_1 E, 0, 0, x_1 E) \qquad p_2 = x_2 P_2 = (x_2 E, 0, 0, -x_2 E) \, . \qquad (17.26)$$

The total momentum of the W boson is

$$p_W = \big((x_1 + x_2)E, 0, 0, (x_1 - x_2)E\big) \, . \qquad (17.27)$$

This vector is best parametrized by a mass M (eventually to be set equal to m_W) and a rapidity,

$$p_W = \left(M \cosh Y, 0, 0, M \sinh Y \right) . \tag{17.28}$$

The parameters M and Y are related to x_1 and x_2 by

$$x_1 = \frac{M}{\sqrt{s}} e^Y \qquad x_2 = \frac{M}{\sqrt{s}} e^{-Y} . \tag{17.29}$$

To rewrite the integral, we need the Jacobian

$$\frac{\partial(x_1, x_2)}{\partial(M, Y)} = \left| \begin{matrix} e^Y/\sqrt{s} & e^{-Y}/\sqrt{s} \\ Me^Y/\sqrt{s} & -Me^{-Y}/\sqrt{s} \end{matrix} \right| = \frac{2M}{s} . \tag{17.30}$$

Then

$$dx_1 dx_2 \delta(M^2 - m_W^2) = dM \, dY \, \frac{2M}{s} \delta(M^2 - m_W^2)$$
$$= \frac{dY}{s} . \tag{17.31}$$

Finally, we find the simple formula

$$\frac{d\sigma}{dY}(pp \to W^+ + X) = \frac{\pi^2 \alpha_W}{3s} \left[f_u(x_1) f_{\bar{d}}(x_2) + f_{\bar{d}}(x_1) f_u(x_2) \right] , \tag{17.32}$$

where x_1, x_2 are derived from m_W, Y using (17.29). Soon, the measurement of W and Z production cross sections at the LHC will provide the most accurate information on the values of the antiquark pdfs.

17.3   Properties of the Z boson

The properties of the Z boson can be worked out in a similar way. Following the approach of (17.4), we find that the decay width of the Z boson to one chiral species (for example, $e_L^- e_R^+$), is

$$\Gamma(Z \to f\bar{f}) = \frac{1}{2m_Z} \frac{1}{8\pi} \frac{g^2}{c_w^2} m_Z^2 \frac{2}{3} Q_Z^2 , \tag{17.33}$$

where Q_Z is the Z charge given by (16.54). This is written more simply as

$$\Gamma(Z \to f\bar{f}) = \frac{\alpha_w}{6c_w^2} m_Z Q_Z^2 . \tag{17.34}$$

The Q_Z take many values. It is useful to make a table of these for one generation of quarks and leptons. Let Q_{ZL} and Q_{ZR} denote the values of Q_Z for the left- and right-handed fermions, respectively. Then we have

Table of the Z charges for the leptons and quarks of each generation.

species	Q_{ZL}	Q_{ZR}	S_f	A_f
ν	$+\frac{1}{2}$	$-$	0.250	1.00
e	$-\frac{1}{2} + s_w^2$	$+s_w^2$	0.126	0.15
u	$+\frac{1}{2} - \frac{2}{3}s_w^2$	$-\frac{2}{3}s_w^2$	0.144	0.67
d	$-\frac{1}{2} + \frac{1}{3}s_w^2$	$+\frac{1}{3}s_w^2$	0.185	0.94

(17.35)

The quantities S_f and A_f are defined by

$$S_f = Q_{ZL}^2 + Q_{ZR}^2 \quad , \quad A_f = \frac{Q_{ZL}^2 - Q_{ZR}^2}{Q_{ZL}^2 + Q_{ZR}^2} \ . \qquad (17.36)$$

The total rate for Z boson decay to the species f is proportional to S_f. The quantity A_f gives the asymmetry between the production rates for left- and right-handed fermions. Equivalently, it gives the polarization of leptons or quarks emitted in Z boson decay. Notice that left-handed polarization is always preferred, but the size of the polarization varies dramatically among fermions with different quantum numbers.

Adding the partial widths for Z boson decay, to neutrinos, charged leptons, u quarks, and d quarks, we find the total width of the Z to be

$$\Gamma_Z = \frac{\alpha_w m_Z}{6 c_w^2} \Big[3 \cdot 0.250 + 3 \cdot 0.126 + 2 \cdot (3.1) \cdot 0.144 + 2.98 \cdot (3.1) \cdot 0.185 \Big]$$

Prediction of the total width of the Z boson.

$$= 2.49 \text{ GeV} \qquad (17.37)$$

The factor 3.1 is the same one that appears in (17.18); it includes the color factor of 3 for quarks and the QCD correction. I have subtracted 2% from the partial width for $Z \to b\bar{b}$, for a reason to be explained below. The branching ratios of the Z to the various fermions are

$$BR(\nu_e \bar{\nu}_e) = 6.7\% \qquad BR(e^+ e^-) = 3.3\%$$
$$BR(u\bar{u}) = 11.9\% \qquad BR(d\bar{d}) = 15.3\% \qquad (17.38)$$

Branching ratios for the Z boson decays to lepton and quark pairs.

and similarly for the fermions of the second and third generations.

17.4   Precision tests of the electroweak model

In the 1990's, there was a concerted effort to test these predictions by production of the Z boson as a resonance in $e^+ e^-$ annihilation. Figure 17.3 shows measurements by the DELPHI experiment at CERN of the $e^+ e^- \to \mu^+ \mu^-$ and $e^+ e^- \to$ hadrons cross sections at energies up to 200 GeV in the center of mass (Abreu *et al.* 1999). Both cross sections have a huge peak at 91 GeV, increasing the base value by a factor of 100. This is the Z boson resonance

$$e^+ e^- \to Z \to \mu^+ \mu^- \ , \ q\bar{q} \ . \qquad (17.39)$$

Two accelerators, the Large Electron-Positron collider (LEP) at CERN and the Stanford Linear Collider (SLC) at SLAC, were constructed to collect data at this resonance. The experiments at these accelerators systematically tested the values of S_f and A_f in the table above. The complete suite of precision measurements on the properties of the Z resonance is reviewed in (Schael *et al.* 2006).

The key test for S_f is the measurement of the total width of the Z resonance. Figure 17.4 shows a measurement by the OPAL experiment

The line shape of the Z resonance.

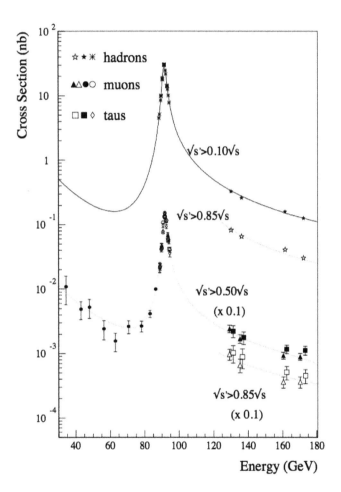

Fig. 17.3 Measurements of the cross section for $e^+e^- \to$ hadrons, $e^+e^- \to \mu^+\mu^-$, and $e^+e^- \to \tau^+\tau^-$, as a function of center of mass energy, by the DELPHI experiment at the LEP collider at CERN (Abreu *et al.* 1999).

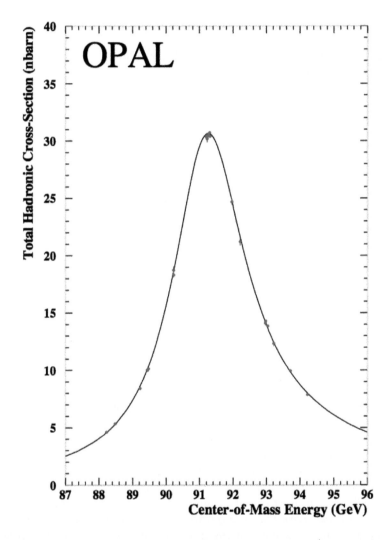

Fig. 17.4 Measurement of the Z boson resonance line shape in e^+e^- annihilation by the OPAL experiment at the LEP collider at CERN; figure courtesy of T. Mori, based on data from (Abbiendi *et al.* 2001).

at LEP of the cross section at steps in energy through the resonance (Abbiendi *et al.* 2001). The experiment is compared to the prediction of the electroweak theory for the best-fit value of s_w^2. The agreement between experiment and theory is quite extraordinary. The theory of the resonance shape begins with a Breit-Wigner resonance

$$\left| \frac{1}{s - m_Z^2 + i m_Z \Gamma_Z} \right|^2 \tag{17.40}$$

and includes the effects of single and multiple collinear photon emission from the colliding electron and positron

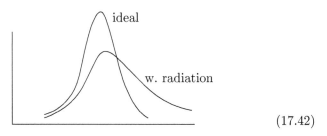

$$\tag{17.41}$$

This radiation decreases the peak height of the resonance and also pushes the resonance to somewhat higher energies. It also gives the resonance a long tail extending to very high energies.

$$\tag{17.42}$$

The effect is shown in Fig. 17.5, along with measured cross sections combined from the four LEP experiments (Schael *et al.* 2006).

The shape distortion of the Z resonance is an effect of QED. The width of the resonance is determined by the weak interaction, with a 4% enhancement of the contribution from decays to quarks due to QCD. Thus, all three of the fundamental interactions of particle physics contribute the excellent agreement of theory and experiment shown in Fig. 17.4.

There are two particularly important outputs from this set of measurements. First, the mass of the Z is measured very precisely. From this, it is possible to determine the value of the weak mixing angle very precisely. Using the $SU(2) \times U(1)$ formulae, we find

From the well-measured observables α, G_F and m_Z, we can construct a very precise reference value of s_w^2.

$$\sin^2 2\theta_w = (2 c_w s_w)^2 = \frac{4\pi\alpha(m_Z)}{\sqrt{2}G_F m_Z^2} \ . \tag{17.43}$$

This translates to a reference value of s_w^2,

$$s_w^2 = 0.231079 \pm 0.000036 \ . \tag{17.44}$$

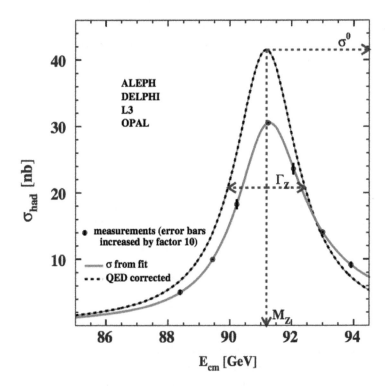

Fig. 17.5 Measurement of the Z boson resonance parameters in e^+e^- annihilation by the LEP experiments (Schael *et al.* 2006). The experimental errors have been inflated by a factor 10 to make them visible. The dotted curve shows the ideal resonance shape, the solid curve shows the predicted resonance shape including the effect of initial-state photon radiation.

Measurement of the number of invisible neutrino species to which the Z boson decays.

Other measurements of quantities depending on s_w^2 can be compared to this standard. The accuracy is such that higher order corrections must be included to make a proper comparison. I will quote some results of this comparison later in this section.

Second, the line-shape of the Z allow us to determine the number of light neutrinos that couple to the Z with the standard $SU(2) \times U(1)$ quantum numbers. Neutrinos are invisible in the Z experiments. Nevertheless, each neutrino contributes to the total width of the Z an amount

$$\Gamma(Z \to \nu_i \bar{\nu}_i) = 170 \text{ MeV} , \qquad (17.45)$$

about 7% of the total width. The presence of one extra neutrino would both increase the width of the resonance and decrease the peak height.

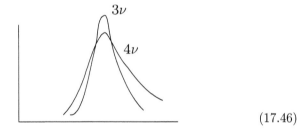

$$(17.46)$$

Careful measurement of the resonance parameters, and fitting to the number of light neutrinos as a continuous variable, gives

$$n_\nu = 2.9840 \pm 0.0082 . \qquad (17.47)$$

Prediction of the decay rate of the Z to $b\bar{b}$, and its comparison to experiment.

Another important measurement related to the S_f is that of

$$R_b = \frac{BR(Z \to b\bar{b})}{BR(Z \to \text{hadrons})} . \qquad (17.48)$$

This ratio can be measured very precisely by selecting $e^+ e^- \to$ hadrons events and then searching within these events for the short-lived B mesons, using a vertex detector such as that described for the BaBar detector at the end of Chapter 6. From our analysis so far, this quantity would be predicted to have a value about 0.220. However, there are higher order corrections involving virtual top quarks that contribute specifically to the partial width for $Z \to b\bar{b}$ through the processes

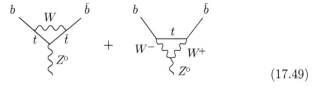

$$(17.49)$$

decreasing the rate for this mode by about 2% for the observed value of the top quark mass. The measurements give

$$R_b = 0.21629 \pm 0.00066 , \qquad (17.50)$$

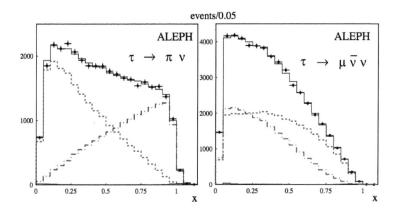

Fig. 17.6 Measurements of the energy distribution of charged particles produced in the decay of τ leptons from $Z \to \tau^+\tau^-$, as a function of $x = E/E_\tau$, by the ALEPH experiment at the LEP collider, from (Heister *et al.* 2001). The dashed and dot-dashed curves show the expectation from τ_L^- and τ_R^-, so the fit to the data shown measures the polarization of τ leptons produced in Z decay: left: $\tau \to \pi\nu$; right: $\tau \to \mu\nu_\tau\bar\nu_\mu$.

in good agreement with this prediction.

The values of A_f can be tested by measurements sensitive to polarization. This is especially interesting because the A_f values are predicted by the $SU(2) \times U(1)$ theory to take very different values for leptons, u quarks, and d quarks.

I will describe two methods for measuring A_f for leptons. The first makes use of the fact that the heavy lepton τ decays by the V−A interaction, which is sensitive to polarization. This is most clearly seen by considering the decay

$$\tau^- \to \nu_\tau \pi^- \ . \tag{17.51}$$

Since the ν_τ is always left-handed and the pion has zero spin, the neutrino must be emitted in the direction opposite to the τ spin direction. In the Z rest frame, the τ^- is highly boosted. Then a τ_R^- will decay to a higher-energy pion and a lower-energy neutrino, and the τ_L^- will decay to a lower-energy pion and a higher-energy neutrino. The actual energy distribution of pions observed at the Z resonance from τ decay, measured by the ALEPH experiment at LEP, is shown in Fig. 17.6 (Heister *et al.* 2001). The fit to the distributions predicted for τ_L^- and τ_R^- shows the expected 15% asymmetry. The similar effect in $\tau \to \mu\nu_\tau\bar\nu_\mu$ is shown on the right-hand side of the figure.

At the SLC, the asymmetry A_e was measured as an asymmetry in the total rate of Z production from e^+e^-. In a circular accelerator, electron beam polarization is typically destroyed as the beams carry out many circuits of the ring. However, linear acceleration naturally preserves the electron polarization. The experiments at SLAC took advantage of this. Using polarized laser light, electrons were produced with preferential left- or right-handed polarization at the front of the accelerator,

Measurements of A_ℓ, the Z polarization asymmetry in decays to leptons.

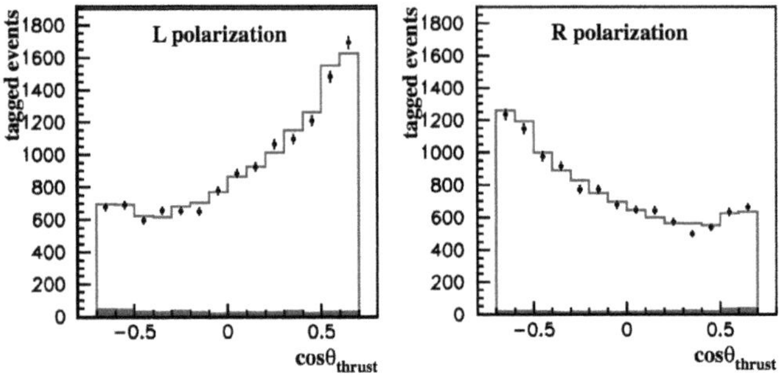

Fig. 17.7 Measurement of the angular distribution of tagged b jets in $Z \to b\bar{b}$ events by the SLD experiment at the SLC collider at SLAC, from (Abe *et al.* 1998).

transported over 4 km to the collider interaction point, and then annihilated with positrons to create Z bosons. The correlation of the laser polarization with the rate for Z production allowed a measurement of the asymmetry in which almost all systematic errors cancelled. The experiment measured (Abe *et al.* 2001a)

$$A_e = 0.1516 \pm 0.0021 \ . \tag{17.52}$$

It is interesting that

$$A_e = \frac{(\frac{1}{2} - s_w^2)^2 - (s_w^2)^2}{(\frac{1}{2} - s_w^2)^2 + (s_w^2)^2} = \frac{\frac{1}{4} - s_w^2}{2s_w^4 + (\frac{1}{4} - s_w^2)} \approx 8 \left(\frac{1}{4} - s_w^2\right) . \tag{17.53}$$

Since the actual value of s_w^2 is close to $\frac{1}{4}$, this very accurate value of A_e turns into an even more accurate value of s_w^2,

$$s_w^2 = 0.23109 \pm 0.00026 \ . \tag{17.54}$$

For b quarks, the polarization asymmetry is expected to be almost maximal. This prediction could be tested at the SLC by using the polarized e^- beam to produce events with b quarks in the final state. Recall that the angular distributions in polarized e^+e^- annihilation depend on the fermion polarizations

Measurement of A_b, the Z polarization asymmetry in decays to b quarks.

$$\frac{d\sigma}{d\cos\theta}(e_L^- e_R^+ \to b_L \bar{b}_R) \sim (1 + \cos\theta)^2 \ ,$$
$$\frac{d\sigma}{d\cos\theta}(e_R^- e_L^+ \to b_L \bar{b}_R) \sim (1 - \cos\theta)^2 \ . \tag{17.55}$$

If the production of b_L dominates, the angular distribution should be highly forward peaked for an e_L^- beam and highly backward peaked for an e_R^- beam. The data from the SLD experiment at the SLC is shown in Fig. 17.7 (Abe *et al.* 1998). The asymmetries are diminished because it is difficult to distinguish the b from the $\bar{b}$ jet, but, nevertheless, the effect is striking. The observed distributions are consistent with the almost maximal asymmetry predicted by the $SU(2) \times U(1)$ theory.

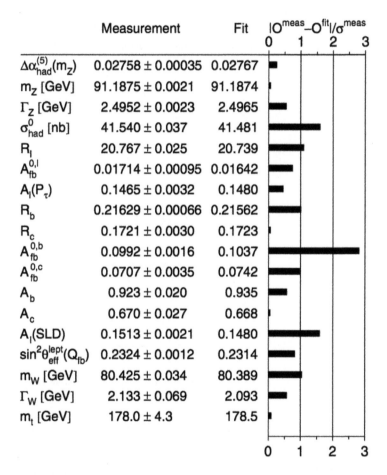

	Measurement	Fit	$|O^{meas} - O^{fit}|/\sigma^{meas}$
$\Delta\alpha_{had}^{(5)}(m_Z)$	0.02758 ± 0.00035	0.02767	
m_Z [GeV]	91.1875 ± 0.0021	91.1874	
Γ_Z [GeV]	2.4952 ± 0.0023	2.4965	
σ_{had}^0 [nb]	41.540 ± 0.037	41.481	
R_l	20.767 ± 0.025	20.739	
$A_{fb}^{0,l}$	0.01714 ± 0.00095	0.01642	
$A_l(P_\tau)$	0.1465 ± 0.0032	0.1480	
R_b	0.21629 ± 0.00066	0.21562	
R_c	0.1721 ± 0.0030	0.1723	
$A_{fb}^{0,b}$	0.0992 ± 0.0016	0.1037	
$A_{fb}^{0,c}$	0.0707 ± 0.0035	0.0742	
A_b	0.923 ± 0.020	0.935	
A_c	0.670 ± 0.027	0.668	
A_l(SLD)	0.1513 ± 0.0021	0.1480	
$\sin^2\theta_{eff}^{lept}(Q_{fb})$	0.2324 ± 0.0012	0.2314	
m_W [GeV]	80.425 ± 0.034	80.389	
Γ_W [GeV]	2.133 ± 0.069	2.093	
m_t [GeV]	178.0 ± 4.3	178.5	

Fig. 17.8 Compilation of precision electroweak measurements, and comparison to the predictions of the $SU(2) \times U(1)$ model using the best-fit parameters, from (Schael *et al.* 2006).

Figure 17.8 gives a compilation of precision measurements on the W and Z by the LEP Electroweak Working Group and the five major contibuting experiments (Schael *et al.* 2006). The figure lists the measurements of a large number of Z decay rates and asymmetries and some other quantities that affect the $SU(2) \times U(1)$ predictions. The second column gives the measured values, averaged among the various experiments. The third column gives the values predicted by the $SU(2) \times U(1)$ theory for the best-fit values of g, g', and m_Z. The fourth column shows graphically the discrepancy between the best-fit theory and experiment, in units of the standard deviation of each masurement. The $SU(2) \times U(1)$ model indeed gives a very accurate explanation for the properties of the W and Z bosons.

Exercises

(17.1) The properties of the Z boson can be evaluated in a similar way to the the the properties of the W boson computed in detail in the text.

(a) Derive the formula (17.34) for the partial width of the Z boson decay to $f\bar{f}$.

(b) Work out the formula corresponding to (17.32) for the cross section for Drell-Yan production of a Z boson in pp collisions.

(c) Estimate numerically the cross sections for Drell-Yan W and Z production at LHC energies. Compute the ratio of these cross sections to the pp total cross section of about 100 mb.

(17.2) The Z can appear as an intermediate state in e^+e^- annihilation. The contributions from the intermediate virtual γ and Z should be added in the amplitude and can interfere.

(a) Considering only an intermediate γ, recall the differential cross sections for $e_L^- e_R^+ \to \mu_L^- \mu_R^+$, $e_L^- e_R^+ \to \mu_R^- \mu_L^+$, $e_R^- e_L^+ \to \mu_L^- \mu_R^+$, $e_R^- e_L^+ \to \mu_R^- \mu_L^+$ computed in (8.47) and (8.48). For example,

$$\frac{d\sigma}{d\cos\theta}(e_L^- e_R^+ \to \mu_R^- \mu_L^+) = \frac{\pi\alpha^2}{2s}(1+\cos\theta)^2 . \tag{17.56}$$

(b) Draw the Feynman diagrams for a virtual γ and a virtual Z and compare them. Show that the cross section for $e_L^- e_R^+ \to \mu_L^- \mu_R^+$ in the full electroweak theory is given by multiplying the

result in (a) by

$$\left| 1 + \frac{1}{c_w^2 s_w^2}\left(\frac{1}{2} - s_w^2\right)^2 \frac{s}{s - m_Z^2 + im_Z\Gamma_Z} \right|^2 . \tag{17.57}$$

(c) In a similar way, compute the cross sections for the other possible helicity states.

(d) The *forward-backward asymmetry* A_{FB} for the reaction $e^+e^- \to \mu^+\mu^-$ is defined by

$$A_{FB} = \frac{\sigma(\cos\theta > 0) - \sigma(\cos\theta < 0)}{\sigma(\cos\theta > 0) + \sigma(\cos\theta < 0)} . \tag{17.58}$$

Compute the forward-backward asymmetry for the polarized reaction in (a) and show that it equals 3/4.

(e) Now consider the A_{FB} for the unpolarized process $e^+e^- \to \mu^+\mu^-$. A_{FB} obtains contributions from each of the 4 possible polarized reactions. Show that $A_{FB} = 0$ for $\sqrt{s} \ll m_Z$. Find A_{FB} just on the Z resonance, where the contribution from the virtual γ can be ignored.

(f) Write the leading term in the expression for the cross section in (b) in the limit $\sqrt{s} \gg m_Z$.

(g) Consider the unbroken $SU(2) \times U(1)$ theory. In this theory, the process $e^+e^- \to \mu^+\mu^-$ is mediated by virtual A^3 and B boson exchange. Compute the cross section for the polarized process in (b). You should find agreement with your answer in (f).

(h) Check this agreement for the other three helicity states. Apparently, spontaneous symmetry breaking only affects cross sections at low energy. In some sense, a spontaneously broken symmetry is restored at sufficiently high energy.

Quark Mixing Angles and Weak Decays

18

The theory of the weak interaction that we have developed so far still omits some of the processes with which we began our discussion of this theory. We still have not proposed a mechanism for the strangeness-changing decays

$$K^0 \rightarrow \pi^- e^+ \nu \qquad \Lambda^0 \rightarrow pe^- \overline{\nu} \ . \qquad (18.1)$$

These decays seem to call for a contribution to the weak charged current of the form

$$u_L^\dagger \overline{\sigma}^\mu s_L \ . \qquad (18.2)$$

However, there is a strong constraint on this modification of the V−A theory described in Chapter 15. Although the charged-current weak interaction exhibits sizable rates for processes that change the quark generation, the neutral-current weak interaction does not. To see this, compare a process based on $s \rightarrow u\mu\nu$,

$$BR(K_L^0 \rightarrow \pi\mu\nu) = 0.27 \ , \qquad (18.3)$$

with one based on $s \rightarrow d\mu^+\mu^-$,

$$BR(K_L^0 \rightarrow \mu^+\mu^-) = 7 \times 10^{-9} \ . \qquad (18.4)$$

Similarly, in B meson decays,

$$BR(b \rightarrow se^+e^-) \approx BR(b \rightarrow s\mu^+\mu^-) = 4 \times 10^{-6} \ . \qquad (18.5)$$

Our theory of weak interactions must provide for flavor-changing charged-current decays while restricting flavor-changing neutral current decays. In this chapter, we will see that both aspects of generation change in the weak interaction are naturally accounted for in the $SU(2) \times U(1)$ model.

Experiment requires additional charged-current interactions that change the quark generation.

Weak decays that change the quark generation can appear in charged-current processes but are highly suppressed in neutral-current processes.

18.1 The Cabibbo mixing angle

To begin, we must work out what interaction strength we need for the $s \rightarrow u$ weak decays. Writing the matrix elements for the weak interaction

Concepts of Elementary Particle Physics. Michael E. Peskin.
© Michael E. Peskin 2019. Published in 2019 by Oxford University Press.
DOI: 10.1093/oso/9780198812180.001.0001

as a V−A interaction with the Fermi constant measured in muon decay, I will write the weak interaction current as

$$j^{\mu+} = \nu^\dagger \bar\sigma^\mu \mu_L + \cdots + V_{us} u_L^\dagger \bar\sigma^\mu s_L + \cdots . \tag{18.6}$$

That is, V_{us} gives the strength of the strangeness changing interaction relative to the strength of the weak interaction in muon decay.

It is possible to determine the value of V_{us} from the rates of Λ^0, Σ^-, and K meson β decay. Look back at the discussion of the normalization of current matrix elements in (14.36) and (14.37). The axial vector current matrix elements may contain new dynamical factors such as g_A, but the vector current matrix elements, at zero momentum transfer, have a fixed normalization given by the flavor charges. The best situation is found for the decay $K \to \pi \ell \nu$. The matrix element

Measurement of the strength V_{us} of the weak interaction transition $s \to u$.

$$\langle \pi | \,\bar u \gamma^\mu (1 - \gamma^5) s \,| K \rangle \tag{18.7}$$

involves only the vector current, because both K and π have $P = -1$. In the limit of zero quark mass, in which K and π are massless Goldstone bosons, the flavor current is conserved and this matrix element contains only one allowed kinematic structure,

$$\langle \pi | \,\bar u \gamma^\mu s \,| K \rangle = i(p_\pi + p_K)^\mu f_+(q) \tag{18.8}$$

where $q = p_K - p_\pi$. At $q = 0$, the value of the matrix element is fixed by the flavor charge,

$$f_+(q = 0) = 1 . \tag{18.9}$$

The corrections to these formulae are proportional to the u, d, and s quark masses and can be worked out systematically. By measuring the rate of $K \to \pi e \nu$ decays, the KLOE experiment at the INFN Frascati laboratory in Frascati, Italy, determined (Ambrosino 2008)

$$V_{us} = 0.2249 \pm 0.0010 . \tag{18.10}$$

This question is coupled to another one. To a first approximation, the strength of the V−A interaction in the β decay of nuclei is equal to that in muon decay. But, is this equality exact? Beginning in the late 1950's, attempts were made to measure the strength of the weak interaction in β decay precisely. To discuss this strength quantitatively, we might parametrize the $d \to u$ term in the V−A charged current as a term in (18.6) of the form

Measurement of the strength V_{ud} of the weak interaction transition $d \to u$ seen in nuclear β decay.

$$j^{\mu+} = \cdots + V_{ud} u_L^\dagger \bar\sigma^\mu d_L + \cdots . \tag{18.11}$$

In the $SU(2) \times U(1)$ theory as we have discussed it so far, gauge invariance would require that the W boson couple to muon, electron, and (u, d) doublets with the same strength. Then we would have $V_{ud} = 1$. However, persistently, the values from experiment were somewhat smaller.

The best experimental determinations come from the rates of *superallowed* β decay transitions between 0^+ nuclei. These use only the vector current. An illustrative example is given by N^{14} and its excited states.

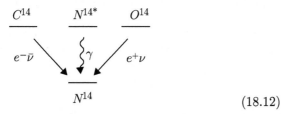

$$(18.12)$$

C^{14} is a β^- emitter, O^{14} is a β^+ emitter. These two states are members of an $I = 1$ triplet with N^{14*}, a state that decays by gamma ray emission. The three vector current matrix elements are related by isospin, so the weak interaction matrix elements can be normalized relative to the measured rate of the electromagnetic decay. Then the normalization factor V_{ud} can be extracted from the rates of the weak interaction decays (Hardy and Towner 2009). The best current value obtained from these measurements is

$$V_{ud} = 0.97425 \pm 0.00022 \ . \qquad (18.13)$$

This value is significantly less than 1.

In 1963, working from the much more uncertain numbers then available, Nicola Cabibbo suggested that these two values fit together through the relation (Cabibbo 1963)

$$|V_{ud}|^2 + |V_{us}|^2 = 1 \ . \qquad (18.14)$$

The reduced strength of the $s \to u$ weak interaction transition is explained by the Cabibbo angle.

That is, we can represent

$$V_{ud} = \cos\theta_C \ , \qquad V_{us} = \sin\theta_C \ , \qquad (18.15)$$

where θ_C is called the *Cabibbo angle*. Evaluating the relation from the numbers above,

$$|V_{ud}|^2 + |V_{us}|^2 = 0.9997 \pm 0.0005 \ . \qquad (18.16)$$

Apparently, the $SU(2)$ gauge interaction does couple with the same strength to quarks as to leptons—as is required by the structure of the gauge theory—but it couples the u quark to a linear combination of d and s.

18.2 Quark and lepton mass terms in the $SU(2) \times U(1)$ model

The structure I have just described can arise in a natural way in the $SU(2) \times U(1)$ model. To understand this, we must first explore how quark and lepton masses arise in that model. A mass term is a term in the Lagrangian

$$\Delta\mathcal{L} = -m_f(f_R^\dagger f_L + f_L^\dagger f_R) \qquad (18.17)$$

In the $SU(2) \times U(1)$ model, quark and lepton masses arise from terms involving fermion interactions with the Higgs field. These terms take the form of mass terms after the Higgs field acquires a vacuum expectation value.

linking the two chiral components of a fermion field. However, we are forbidden to write such a term for any quark or lepton. The $SU(2) \times U(1)$ theory puts the left-handed quarks and leptons into $I = \frac{1}{2}$ doublets but assigns the right-handed quarks and leptons $I = 0$. Thus, any mass term violates the $SU(2)$ gauge symmetry.

Thus, generation of mass for any quark or lepton requires the spontaneous breaking of $SU(2) \times U(1)$. The Higgs field φ has the quantum numbers $I = \frac{1}{2}$, $Y = \frac{1}{2}$. So it is consistent with all symmetries of the theory to add to the Lagrangian the terms

$$\Delta \mathcal{L} = -y_e L_a^\dagger \varphi_a e_R - y_d Q_a^\dagger \varphi_a d_R - y_u Q_a^\dagger \epsilon_{ab} \varphi_b^* u_R + h.c. \qquad (18.18)$$

where $a, b = 1, 2$ and

$$L = \begin{pmatrix} \nu \\ e^- \end{pmatrix}_L , \qquad Q = \begin{pmatrix} u \\ d \end{pmatrix}_L . \qquad (18.19)$$

The coefficients y_f are called *Yukawa couplings*. Each term in (18.18) is invariant under isospin, and each term has the sum of the hypercharges of the fields summing to zero. For example, in the middle term, the hypercharges are

$$-\frac{1}{6} + \frac{1}{2} - \frac{1}{3} = 0 . \qquad (18.20)$$

Note that the Yukawa coupling term for the u quark has a slightly different structure from the others, with φ^* rather tnan φ.

If we replace the Higgs field by its vacuum expectation value

$$\varphi \to \begin{pmatrix} 0 \\ v/\sqrt{2} \end{pmatrix} \qquad (18.21)$$

we find that (18.18) becomes

$$\Delta \mathcal{L} = -\frac{y_e v}{\sqrt{2}} e_L^\dagger e_R - \frac{y_d v}{\sqrt{2}} d_L^\dagger d_R - \frac{y_u v}{\sqrt{2}} u_L^\dagger u_R + h.c. \qquad (18.22)$$

Formula for the masses of quarks and leptons in terms of their Yukawa couplings to the Higgs field.

Comparing this equation to (8.24), we see that it has just the structure of mass terms for the e, d, and u. Then

$$m_f = y_f \frac{v}{\sqrt{2}} \qquad (18.23)$$

for all three species.

In writing (18.18) and (18.22), I have omitted mass terms for the neutrinos. This is an excellent approximation for particle physics at GeV energies. However, the assumption that the neutrino masses are zero has important consequences in the analysis to be presented in Section 18.3. I will return to the question of neutrinos masses in Chapter 20.

The construction I have presented here gives an origin for the quark and lepton mass terms. But, it does not solve the problem of the large range of values of these terms. It only pushes the problem back one level, onto the physics of the fermion couplings to the Higgs field. This does not make the problem of quark and lepton masses any less mysterious.

18.3   Discrete space-time symmetries and generation mixing

In nature, we see three fermions with each type of quantum number, for example, e, μ, and τ for charged leptons. We refer to the three states of each kind as belonging to three *generations*. To give mass to the second and third generations, we could simply repeat the structure above. However, it is instructive to write a more general set of Yukawa couplings, in fact, the most general set of couplings consistent with $SU(2) \times U(1)$ gauge invariance. In this section, I will analyze that quite general theory and derive from it some surprising conclusions.

Gauge invariance requires that the gauge couplings of the fermions of the three generations are absolutely identical. But, gauge invariance puts much weaker constraints on the Yukawa couplings. The most general Yukawa couplings consistent with gauge invariance include arbitrary mixtures of couplings among the three generations. Letting $i, j = 1, 2, 3$ label generations, this most general set of Yukawa couplings is written

$$\Delta \mathcal{L} = -y_e^{ij} L_a^{\dagger i} \varphi_a e_R^j - y_d^{ij} Q_a^{\dagger i} \varphi_a d_R^j - y_u^{ij} Q_a^{\dagger i} \epsilon_{ab} \varphi_b^* u_R^j + h.c. \quad (18.24)$$

where the y_f^{ij} are complex-valued 3×3 matrices of general symmetry.

We can simplify this structure by diagonalizing the y_f matrices and making appropriate changes of variables among the fields. The Yukawa matrices are not Hermitian. But, they can be diagonalized as follows: Construct the matrices

$$y_f y_f^{\dagger} , \qquad y_f^{\dagger} y_f . \quad (18.25)$$

These are Hermitian and positive and have the same eigenvalues. We can represent them as

$$y_f y_f^{\dagger} = U_L^{(f)} \mathbf{Y}_f U_L^{(f)\dagger} , \qquad y_f^{\dagger} y_f = U_R^{(f)} \mathbf{Y}_f U_R^{(f)\dagger} . \quad (18.26)$$

where $U_L^{(f)}$, $U_R^{(f)}$ are (in general, different) unitary matrices and $\mathbf{Y_f}$ is real, positive, and diagonal, and identical in the two formulae. Then if

$$Y_f = \sqrt{\mathbf{Y}_f} = \begin{pmatrix} \sqrt{\mathbf{Y}_{f1}} & & \\ & \sqrt{\mathbf{Y}_{f2}} & \\ & & \sqrt{\mathbf{Y}_{f3}} \end{pmatrix} , \quad (18.27)$$

we have

$$y_f = U_L^{(f)} Y_f U_R^{(f)\dagger} . \quad (18.28)$$

For leptons, we now make the change of variables

$$e_R^i \to U_{Rij}^{(e)} e_R^j , \qquad L^i \to U_{Lij}^{(e)} L^j . \quad (18.29)$$

The matrices $U_L^{(e)}$, $U_R^{(e)}$ disappear from the Yukawa couplings. The lepton mass terms are now diagonal in generation, and the new fields L^i, e_R^i correspond to mass eigenstates. These are now the fields of the familar leptons e, μ, and τ.

The equation (18.24) seems to have much more generality than we require in our theory of the weak interaction. But in this section, we will systematically simplify it, using several changes of variables. You will be surprised by the final result.

Using the representation (18.28) for the Yukawa matrix, we simplify the lepton terms in the Lagrangian.

The change of variables (18.29) moves the matrices $U_L^{(e)}$ and $U_R^{(e)}$ to the lepton kinetic terms, for example,

$$e_R^\dagger(i\sigma \cdot D)e_R \to e_R^\dagger U_R^{(e)\dagger}(i\sigma \cdot D)U_R^{(e)}e_R \ . \qquad (18.30)$$

But here these matrices cancel out completely, because the three generations have the same gauge interactions. The formula (18.30) becomes

$$= e_R^\dagger(i\sigma \cdot D)U_R^{(e)\dagger}U_R^{(e)}e_R = e_R^\dagger(i\sigma \cdot D)e_R \ . \qquad (18.31)$$

Lepton number conservation is automatic in the $SU(2) \times U(1)$ model with zero neutrino masses.

There are no interactions remaining that couple the lepton generations. Thus, lepton number conservation, separately for each generation, is a consequence, not an assumption, of the $SU(2) \times U(1)$ theory.

Please note that, in this argument, I have used the property of our $SU(2) \times U(1)$ model that there are no neutrino mass terms. If we had included a neutrino mass term, the matrices $U_L^{(e)}$, $U_R^{(e)}$ would not have cancelled out of that term, and we would have found very small generation-changing interactions proportional to the neutrino masses. I will discuss this effect in Chapter 20.

The construction for the quarks is somewhat more complicated. We make the change of variables

$$u_R^i \to U_{Rij}^{(u)}u_R^j \qquad u_L^i \to U_{Lij}^{(u)}u_L^j$$
$$d_R^i \to U_{Rij}^{(d)}d_R^j \qquad d_L^i \to U_{Lij}^{(d)}d_L^j \qquad (18.32)$$

Using the representation (18.28) for the Yukawa matrices, we simplify the quark terms in the Lagrangian.

After this change of variables, the matrices U_L, U_R have disappeared from the Yukawa couplings. The new u^i and d^i fields correspond to mass eigenstates—the physical quarks u, c, t and d, s, b. The unitary matrices are transfered to the quark kinetic terms. Then they cancel, just as for the leptons—at least, in the couplings to the gluon, photon, and Z boson. We now see that, for the most general structure of Yukawa couplings, the neutral current interaction mediated by the Z boson is always diagonal in flavor.

In the coupling to the W boson, the unitary matrices do not completely cancel. Instead, we find

$$u_L^\dagger(i\bar\sigma^\mu)d_L \to u_L^\dagger U_L^{(u)\dagger}(i\bar\sigma^\mu)U_L^{(d)}d_L$$
$$= u_L^\dagger(i\bar\sigma^\mu)V_{CKM}d_L \ , \qquad (18.33)$$

where

$$V_{CKM} = U_L^{(u)\dagger}U_L^{(d)} \ . \qquad (18.34)$$

The last vestige of the unitary transformations that diagonalize the quark mass matrices produces precisely the Cabibbo mixing in the weak interaction and its generalization to three generations.

The U_L matrices can thus be combined into a single unitary matrix, V_{CKM}, called the *Cabibbo-Kobayashi-Maskawa matrix*. After the changes of variables, this is the only term in the weak interaction Lagrangian that contains generation-changing interactions. The matrix elements of V_{CKM} are exactly the parameters V_{ud}, V_{us}, etc., that we introduced in (18.6) and (18.11),

$$V_{CKM} = \begin{pmatrix} V_{ud} & V_{us} & V_{ub} \\ V_{cd} & V_{cs} & V_{cb} \\ V_{td} & V_{ts} & V_{tb} \end{pmatrix} \ . \qquad (18.35)$$

Thus, each physical u quark is linked by charged-current interactions to a different linear combination of the d quarks. V_{CKM} is a unitary matrix, and so these linear combinations are orthogonal. At this point, the combinations have complex coefficients. The imaginary parts of the coefficients can be shown to lead to CP- and T-violating interactions.

However, we can simplify the structure even further. A 3×3 unitary matrix has 9 parameters. If this matrix were real-valued, it would be a rotation matrix in 3 dimensions, parametrized by 3 Euler angles. So a 3×3 unitary matrix is parametrized by 3 angles and 6 phases. By a further change of variables to change the phases of the quark fields

$$u_L^j \to e^{i\alpha_j} u_L^j , \qquad d_L^j \to e^{i\beta_j} d_L^j , \qquad (18.36)$$

we can remove 5 phases. The overall phase of the quark fields drops out of the Lagrangian and cannot be used to simplify V_{CKM}. So, finally, V_{CKM} can be written with 4 parameters—3 angles and 1 phase. This phase is a single parameter that produces CP and T violation in the weak interaction.

We will see in the next chapter that certain weak interaction decays do show CP and T violation. This explanation for the origin of CP violation was first put forward by Makoto Kobayashi and Toshihide Maskawa in 1973. Note that, if we had only 2 generations, V_{CKM} would be a 2×2 parametrized by one angle, the Cabibbo angle, and all phases could be removed. Thus, the Kobayashi-Maskawa theory connects CP violation in the weak interaction to the existence of three generations of quarks. Remarkably, Kobayashi and Maskawa proposed the existence of the third generation before the discovery of the τ lepton and even before the discovery of the c quark (Kobayashi and Maskawa 1973).

It turns out that there is one defect in this argument. The same strong-interaction physics of gluons that destroys the possible chiral $U(1)$ symmetry of QCD with massless quarks also allows a possible CP-violating term in QCD, parametrized by an angle θ. This term potentially generates CP- and T- violating effects in the strong interaction, for example, the generation of an electric dipole moment for the neutron. Measurements of the neutron electric dipole moment, which we will discuss in Section 19.2, require that $|\theta| < 10^{-10}$. The θ parameter is shifted by the phase transformation (18.36). Still, it is possible to introduce additional mechanisms, requiring new particles or interactions, that guarantee that θ is sufficiently small. Having called your attention to this problem, I will ignore it from here on. For further discussion of this issue, see (Dine 2000).

At the end of the simplifications, the weak interaction contains one CP violating parameter. This parameter could be transformed away if there were fewer than three generatioms of quarks.

18.4 The Standard Model of particle physics

We have now derived a remarkable result. We wrote down the most general Lagrangian allowed by $SU(3) \times SU(2) \times U(1)$ gauge symmetry. After spontaneous symmetry breaking and some changes of variables, we have reduced that Lagrangian to the following form:

$$\mathcal{L} = -\frac{1}{4}\sum_a (F^a_{\mu\nu})^2 + m_W^2 W^+_\mu W^{-\mu} + \frac{1}{2}m_Z^2 Z_\mu Z^\mu$$

$$+ \sum_f \overline{\Psi}_f (i\gamma \cdot D_f - m_f)\Psi_f + \frac{1}{2}(\partial_\mu h)^2 - V(h) \ . \quad (18.37)$$

where the sum over a runs over the generators of $SU(3) \times SU(2) \times U(1)$ and the sum over f runs over all quark and lepton flavors. The covariant derivatives D_f are of the form

$$D_{\mu f} = \partial_\mu - ieQ_f A_\mu - i\frac{g}{c_w}Q_{Zf}Z_\mu - i\frac{g}{\sqrt{2}}(W^\pm_\mu t^\pm) - ig_s A^a_\mu t^a \ , \quad (18.38)$$

representing electromagnetic, Z, W, and gluon couplings to fermions. For the quarks, the W couplings also generation mixing by V_{CKM}. The interactions of the Higgs boson field $h(x)$ are generated by the replacement $v \to v + h(x)$ in the mass terms for W, Z, quarks, and leptons. This theory is called the *Standard Model* of particle physics.

The Standard Model Lagrangian automatically has many highly accurate approximate symmetries (Weinberg 1973, Nanopoulos 1973):

(1) The Lagrangian conserves overall quark number or baryon number and, separately, overall lepton number. Note that these conservation laws are outputs of the analysis, not assumptions.

(2) All terms except for the couplings of the W and Z bosons to fermions conserve P, C, and T. In particular, it is automatically true that the strong and electromagnetic interactions conserve P, C, and T.

(3) All terms except for the couplings of the W boson preserve the fermion number for each individual fermion species. For the leptons, the weak interaction also connects each charged lepton to one neutrino. This explains the fact that, in experiment, each lepton seems to carry a separate conserved quantum number. For example,

$$BR(\mu^- \to e^-\gamma) < 2.4\times10^{-12} \ , \qquad BR(\tau^- \to \mu^-\gamma) < 4.4\times10^{-8} \ . \quad (18.39)$$

(4) The W and Z couplings violate P and C in a maximal way. However, if these couplings are real-valued, they preserve the joint symmetry CP. Since CPT is a symmetry of any quantum field theory, real-valued couplings also preserve T. There is one possible source of CP and T violation in the Standard Model, and that is the one remaining phase in the CKM matrix. The Standard Model associates CP violation with interactions of the third generation. I will discuss tests of this idea in Chapter 19.

18.5   Quark mixing including heavy quarks

We have already discussed the value of the CKM parameter V_{us}. Since a two-generation theory has only one angle, we must use processes in-

volving the third generation to determine the other two angles. The
angle V_{cb} is extracted from the decay rate of B mesons. Ignoring for
a moment the effects of the strong interaction, we can estimate the rate
of B meson decay from the formula for b quark decay that is analogous
to the formula (15.40) for the rate of muon decay. That is,

$$\Gamma(b \to cf\overline{f}) = |V_{cb}|^2 \, \frac{G_F^2 m_b^5}{192\pi^3}(3 + 2 \cdot 3) \, , \qquad (18.40)$$

Determination of the parameters V_{cb} and V_{ub} that control the rate of b quark decay.

where the decays to $f\overline{f} = e\nu, \mu\nu, \tau\nu, u\,\overline{d}$, and $c\overline{s}$ are included. The decays
to τ and c are substantially reduced by phase space, but also there is
a relatively large enhancement from QCD corrections. Evaluating the
simple formula (18.40), we find

$$\Gamma(b \to cf\overline{f}) = 4 \times 10^{-10} \text{ GeV} \cdot |V_{cb}|^2 \qquad (18.41)$$

or

$$\tau(b) = 1.7 \times 10^{-15} \text{ sec} \cdot |V_{cb}|^2 \, . \qquad (18.42)$$

The measured B meson lifetime is

$$\tau(B) = 1.5 \times 10^{-12} \text{ sec} \, . \qquad (18.43)$$

If we interpret this as the rate of b quark decay, we would estimate
$V_{cb} \approx 0.03$. The best current estimate, which includes the effects of the
strong interaction in the b quark binding and decay, is

$$V_{cb} = (4.09 \pm 0.11) \times 10^{-2} \, . \qquad (18.44)$$

Decays with $b \to u$ are a small fraction of B meson decays and are
somewhat harder to relate to measured quantities. The best current
estimate gives

$$V_{ub} = (4.15 \pm 0.49) \times 10^{-3} \, . \qquad (18.45)$$

A very convenient parametrization of the CKM matrix is that developed by Wolfenstein (1983). This parametrization uses the fact that
V_{us}, V_{cb}, and V_{ub} are successively smaller. From these elements, the
whole unitary matrix can be constructed using the requirement that, in
a unitary matrix, the rows and the column are orthogonal vectors. The
following formula maintains this orthogonality up to terms of order V_{us}^4:

Wolfenstein's useful parametrization of the CKM matrix.

$$V_{CKM} = \begin{pmatrix} 1 - \lambda^2/2 & \lambda A\lambda & \lambda^3(\rho - i\eta) \\ -\lambda & 1 - \lambda^2/2 & A\lambda^2 \\ A\lambda^3(1 - \rho - i\eta) & -A\lambda^2 & 1 \end{pmatrix} . \qquad (18.46)$$

The current best values of the parameters are

$$\lambda = 0.225$$
$$A = 0.81$$
$$|\rho - i\eta| = 0.37 \qquad (18.47)$$

Wolfenstein placed the possible phase of the CKM matrix in the extreme corners. This makes it more explicit that CP violation appears only in processes that involve either the b or the t quark in some way.

To conclude this discussion, I return to the issue of flavor-changing neutral weak interactions. In the Standard Model, leptons have no flavor-changing weak interactions. For quarks, there are no flavor-changing neutral current terms in the Lagrangian, but very small flavor-changing effects can be generated by Feynman diagrams that make use of the charged-current interactions. We can study this in the example of the process $\bar{s}d \to \bar{d}s$, which converts a K^0 meson to a $\overline{K}^0$ meson. I will discuss the physical consequences of this mass mixing in Section 19.1.

The top left submatrix of the CKM matrix is, to a good approximation,

$$\begin{pmatrix} \cos\theta_C & \sin\theta_C \\ -\sin\theta_C & \cos\theta_C \end{pmatrix} . \tag{18.48}$$

This structure was originally proposed by Glashow, Iliopoulos, and Maiani (1970) to explain the smallness of the K^0–$\overline{K}^0$ mixing amplitude. As we will see in a moment, the unitarity of the matrix (18.48) gives rise to what is called a *GIM cancellation* that makes this amplitude much smaller than one might expect.

Early in the study of the weak interaction, it was realized that it is possible to convert K^0 to $\overline{K}^0$ through the process

$$\tag{18.49}$$

This matrix element has the form

$$m^2(\overline{K}^0, K^0) \sim m_K^2 f_K^2 \frac{G_F^2}{(4\pi)^2} \sin^2\theta_C \cos^2\theta_C \cdot m_W^2 . \tag{18.50}$$

In this formula, m_K is the mass of the K^0, and f_K is the K meson analogue of the pion decay constant defined in (14.34). Dimensional analysis requires another factor with the dimensions of $(\text{GeV})^2$. I have written this factor as m_W^2 because, in the computation of the diagram, the momentum of the off-shell W bosons is allowed to run up to m_W. Evaluating this equation, we find

$$m^2(\overline{K}^0, K^0) \approx 10^{-12} \text{ GeV}^2 . \tag{18.51}$$

This result is much larger than the value measured in experiment,

$$m^2(\overline{K}^0, K^0) \approx 3.5 \times 10^{-15} \text{ GeV}^2 . \tag{18.52}$$

GIM suggested—in 1970, before the discovery of the J/ψ—that postulating a c quark with the weak interaction coupling structure above could solve this problem. In the GIM theory, there were three more

Although the Standard Model Lagrangian contains no flavor-changing neutral current couplings, such processes can be induced by higher-order corrections. In this passage, I discuss the flavor-changing amplitude that converts a K^0 to a $\overline{K}^0$ and vice versa.

diagrams,

$$(18.53)$$

The fact that the charge-changing weak interactions arise from a unitary matrix leads to cancellations in the induced amplitudes for neutral flavor-changing processes. These *GIM cancellations* are essential for agreement with the measured values.

Note that the diagrams with both u and c have the opposite sign, from the factors of $(-\sin\theta_C)$. The four diagrams contain integrals over momenta that have the same form for exchanged momenta $q \gg m_c$. So, the four diagrams cancel for large momentum transfers. This cancellation removes the entire region in which the momenta carried by the off-shell W bosons is larger than m_c. The sum of the four diagrams is then of the form of (18.50) but with

$$m_W^2 \to m_c^2 \ . \tag{18.54}$$

This lowers the estimate of Δm_K by a factor of 10^{-4}, to about the correct value. A full QCD analysis with the now known parameters of the Standard Model gives a value in good agreement with experiment. Similar GIM cancellations predict that the D^0–$\overline{D}^0$ and B^0–$\overline{B}^0$ mixing amplitudes also have very small values, again as required by experiment.

Exercises

(18.1) Since most of the degrees of freedom in the quark Yukawa matrices can be transformed away, there is a great deal of freedom to make proposals for the underlying form of these matrices. A simple proposal, due to Fritzsch (1977), is that the u quark Yukawa matrix is diagonal while the d quark Yukawa matrix is symmetric with zeros on the diagonal.

   (a) Fritzsch's original proposal was for two generations (d, s). He proposed

$$Y_d = \begin{pmatrix} 0 & A \\ A & B \end{pmatrix} \ . \tag{18.55}$$

      Diagonalize this matrix, obtaining formulae for the s and d masses and the matrices $U_L^{(d)}$ and $U_R^{(d)}$. Construct the 2×2 CKM matrix for this model.

   (b) Show that the zero element of the original Y_d

implies the relation

$$\tan\theta_C = \sqrt{m_d/m_s} \ . \tag{18.56}$$

(c) Evaluate this formula using (14.59) and (18.10). How well does it work?

(d) A generalization to three generations is

$$Y_d = \begin{pmatrix} 0 & A & 0 \\ A & 0 & B \\ 0 & B & C \end{pmatrix} \ . \tag{18.57}$$

Show that the zero in the central element of this matrix leads to the prediction

$$\frac{V_{ts}}{V_{cs}} \approx \sqrt{\frac{m_s}{m_b}} \ . \tag{18.58}$$

Evaluate this expression using the Wolfenstein parameters (18.47). How well does it work?

(18.2) At the time that the b quark was discovered, there was no direct evidence for the existence of the t

quark. An alternative possibility was that the b_L and b_R were both $SU(2)$ singlets. The b quark would have weak interaction decays if the weak interaction mixing matrix mixed the b_R with the other right-handed, $Y = -\frac{1}{3}$, quark fields.

(a) For simplicity, set $\theta_C = 0$ and ignore the (u, d) quarks. Then the masses of s, c, b can be generated by the Lagrangian terms

$$\Delta \mathcal{L} = -y_s Q_a^\dagger \varphi_a \tilde{s}_R - y_c Q_a^\dagger \epsilon_{ab} \varphi_b^* c_R$$
$$- M b_L^\dagger b_R + h.c. \quad (18.59)$$

where $Q = (c, s)_L$ and

$$\tilde{s}_R = \cos \alpha \, s_R + \sin \alpha \, b_R . \quad (18.60)$$

Show that (18.59) is invariant under $SU(2) \times U(1)$.

(b) Let the Higgs field acquire its vacuum expectation value v. Write the quark mass matrix and diagonalize it. Construct the matrices U_L and U_R needed to diagonalize the (s, b) mass matrix.

(c) Make the change of variables analogous to (18.32). This removes U_L and U_R from the Higgs couplings and introduces these matrices in the weak interaction couplings. Write the term in the Lagrangian that leads to the charged-current weak interaction of b.

(d) Compute the decay rate for $b \to c \mu^- \bar{\nu}$ in the simple approximation used in (18.40).

(e) Show that the flavor-changing neutral current interaction of b is nonzero in this model. Write the term in the Lagrangian that leads to the neutral current weak interaction of b.

(f) Compute the decay rate for $b \to s \mu^+ \mu^-$ in the same approximation as in part (d).

(g) Compute the ratio of branching ratios

$$BR(b \to s\mu^+\mu^-)/BR(b \to c\mu^-\bar{\nu}) \quad (18.61)$$

in this model. Compare your result to the Particle Data Group values

$$BR(b \to \mu^- \nu X) = 0.11$$
$$BR(b \to s\mu^+\mu^-) = 4 \times 10^{-6} . \quad (18.62)$$

(18.3) The fact that, in the Standard Model, the Higgs boson couplings are exactly diagonal in flavor is an important part of the understanding of K^0–$\overline{K}^0$ mixing. To see this, consider the consequences of introducing another scalar particle h_2 that could mediate d-s flavor changes. In particular, write for the h_2 the interaction

$$\Delta \mathcal{L} = i \frac{y_2}{\sqrt{2}} h_2 (\bar{s}\gamma^5 d - \bar{d}\gamma^5 s) . \quad (18.63)$$

(a) To evaluate the K^0–$\overline{K}^0$ mixing amplitude, we need the value of $\langle 0| \bar{s}\gamma^5 d |K^0\rangle$. Using the derivation in Chapter 14 from (14.45) to (14.48), evaluate this amplitude in terms of the parameter Δ' and then numerically.

(b) Draw the Feynman diagram by which an s-channel exchange of the h_2 generates a K^0–$\overline{K}^0$ mixing amplitude. Evaluate the contribution to this amplitude in terms of the coupling constant y_2.

(c) Set y_2 equal to the s quark Yukawa coupling and the mass of the h_2 to 100 GeV to estimate the size of the induced mixing. How does this compare to the measured value (18.52)?

CP Violation

19

We saw in the previous chapter that the 3-generation $SU(2) \times U(1)$ model has room for one phase angle, which would signal violation of CP and T. In this chapter, I will discuss the evidence for CP violation in hadronic weak decays. We will see that CP violation, though it is a very small effect, is clearly observed in specific weak interaction processes. These observations, as I will show, are well explained by the Kobayashi-Maskawa phase in the mixing matrix for charge-changing weak interactions.

The study of CP violation is fascinating from another point of view. CP violation is difficult to observe directly using the observables that we have discussed so far in this book. Typically, CP violation leads to only very small asymmetries in the rates of weak interaction decays between particles and antiparticles. The most compelling evidence for CP violation comes from a different kind of experiment in which we observe the time-dependent evolution of a particle that decays through the weak interaction. In such a system, CP violation can be observed as a nonzero phase in the quantum interference of two components of the wavefunction of the decaying state. In some cases, this quantum interference plays out over macroscopic distances, of the order of meters. In these systems, the experiments on weak interaction decay test not only the details of a particle physics model but also the underlying fundamental principles of quantum mechanics. CP violation experiments are reviewed from this point of view in (Testa 2007).

19.1 CP violation in the K^0–$\overline{K}^0$ system

I will begin by describing the evidence for CP violation in the K^0–$\overline{K}^0$ system. In Section 18.5, I pointed out that there is a very small amplitude that mixes the K^0 and $\overline{K}^0$ states. This observation leads to some unexpected phenomena in K^0 decays even in the case where CP is conserved. In this section, I will first develop this theory assuming CP conservation, and then generalize it to the case in which CP is violated.

The neutral K meson is a 2-state quantum system that evolves according to

$$e^{-i\mathbf{M}\tau} , \tag{19.1}$$

Evolution of an initial K^0 state under the influence of K^0–$\overline{K}^0$ mixing, under the assumption that CP is conserved.

Concepts of Elementary Particle Physics. Michael E. Peskin.
© Michael E. Peskin 2019. Published in 2019 by Oxford University Press.
DOI: 10.1093/oso/9780198812180.001.0001

where τ is the time measured in the rest frame (proper time), and $\mathbf{M}$ is a mass matrix for the two-state system. If CP is conserved, $\mathbf{M}$ has the form

$$\mathbf{M} = \begin{pmatrix} \overline{m} - i\overline{\Gamma}/2 & \delta m - i\delta\Gamma/2 \\ \delta m - i\delta\Gamma/2 & \overline{m} - i\overline{\Gamma}/2 \end{pmatrix} , \tag{19.2}$$

symmetrical between particles and antiparticles. The parameters $\overline{m}$ and δm contribute to the masses of the eigenstate particles. The parameters $\overline{\Gamma}$ and $\delta\Gamma$ contribute to their decay rates; the factor of $(-i)$ turns (19.1) into an exponential decay. CPT requires that the diagonal elements of this matrix are equal. C and P act on $|K^0\rangle$ and $|\overline{K}^0\rangle$ as

$$\begin{aligned} P|K^0\rangle &= -|K^0\rangle & P|\overline{K}^0\rangle &= -|\overline{K}^0\rangle \\ C|K^0\rangle &= +|\overline{K}^0\rangle & C|\overline{K}^0\rangle &= +|K^0\rangle . \end{aligned} \tag{19.3}$$

Thus, CP symmetry implies that the off-diagonal elements of (19.2) are equal. The eigenstates of this mass matrix are CP eigenstates,

$$\begin{aligned} CP = +1 && |K_S^0\rangle &= \frac{1}{\sqrt{2}}\left(|K^0\rangle - |\overline{K}^0\rangle\right) , \\ CP = -1 && |K_L^0\rangle &= \frac{1}{\sqrt{2}}\left(|K^0\rangle + |\overline{K}^0\rangle\right) . \end{aligned} \tag{19.4}$$

The corresponding mass and decay rate eigenvalues are

$$\begin{aligned} M_S &= \overline{m} - \delta m - i(\overline{\Gamma} - \delta\Gamma)/2 \\ M_L &= \overline{m} + \delta m - i(\overline{\Gamma} + \delta\Gamma)/2 \end{aligned} \tag{19.5}$$

A particle produced as a K^0 will propagate as a linear combination of K_S^0 and K_L^0. The two components of the wavefunction will have different decay rates and will oscillate with different frequencies.

The K^0 and $\overline{K}^0$ are stable with respect to the strong interaction but can decay by the weak interaction, through

$$s \to u e^- \overline{\nu}_e , \qquad s \to u \mu^- \overline{\nu}_\mu , \qquad s \to u d \overline{u} . \tag{19.6}$$

Computation of QCD corrections gives a large enhancement for the purely hadronic decay modes. In particular, the decay

$$K^0 , \ \overline{K}^0 \to \pi\pi \tag{19.7}$$

The K^0 splits into two components, K_S^0 and K_L^0, with lifetimes that differ by a factor of almost 1000.

is enhanced by about a factor of 100 relative to other modes. The decay

$$K^0 , \ \overline{K}^0 \to 3\pi \tag{19.8}$$

also has a QCD enhancement, but at the same time it is suppressed by the large denominator in the formula for 3-body phase space and by the fact that $(m_K - 3m_\pi)$ is small. For pions in an S-wave, the dominant final states have the CP quantum numbers

$$CP|\pi\pi\rangle = +|\pi\pi\rangle \qquad CP|\pi\pi\pi\rangle = -|\pi\pi\pi\rangle . \tag{19.9}$$

Then the state called K_S in (19.4) is allowed to decay to 2π, but for the state called K_L this decay is forbidden by CP conservation. This implies that the two mass eigenstates of the K^0–$\overline{K}^0$ system have two very different lifetimes

$$\tau_S = 0.895 \times 10^{-10} \text{ sec} \qquad c\tau_S = 2.68 \text{ cm}$$
$$\tau_L = 5.116 \times 10^{-8} \text{ sec} \qquad c\tau_L = 15.34 \text{ m} \qquad (19.10)$$

The two states are appropriately called "K-short" and "K-long". It is an interesting accident that the K_L^0–K_S^0 mass difference

$$m_L - m_S = 3.48 \times 10^{-15} \text{ GeV} , \quad \frac{\hbar}{2(m_L - m_S)} = 0.95 \times 10^{-10} \text{ sec} ,$$
$$(19.11)$$

corresponds to a time very close to the lifetime of the K_S^0.

The structure of the K_S^0 and K_L^0 states leads to some remarkable physical consequences. If a K^0 is produced, for example, in the reaction

$$\pi^- p \to \Lambda^0 K^0 , \qquad (19.12)$$

the K^0 state resolves itself into the two CP eigenstates. The K_S^0 component decays to $\pi\pi$ in a few cm. This has a probability of 50%. The other part of the K^0 wavefunction decays to 3π and other final states over a distance of tens of meters. This alternative possibility also has a probability of 50%. If we created mutiple K^0s using a beam of π^-s, the decay vertices appear as

$$\text{cm} \qquad\qquad (19.13)$$

If we go meters downstream from the K^0 production target, we have essentially a pure K_L^0 beam. The particles in this beam are coherent mixtures of K^0 and $\overline{K}^0$, as indicated in (19.4). By disturbing the quantum state, it is possible to change the relative amplitudes of K^0 and $\overline{K}^0$ in the wavefunctions. According to the rules of quantum mechanics, this should *regenerate* a K_S^0 component. We can do this in practice by placing an absorber in the path of the kaon beam (Pais and Piccioni 1955). The $\overline{K}^0$ ($s\bar{d}$) state contains a d antiquark and so has a larger inelastic cross section on matter. Thus, after the K^0 state passes through the absorber, the original K_L^0 wavefunction now has a larger K^0 component. We can represent the kaon state that exits the absorber as the quantum state

The K_S^0 component of the wavefunction may be regenerated by an absorber placed in the path of the kaons.

$$a|K^0\rangle + b|\overline{K}^0\rangle = \alpha|K_L^0\rangle + \beta|K_S^0\rangle , \qquad (19.14)$$

where, if $a \neq b$, β will be nonzero. We will then see $K^0 \to \pi\pi$ decays in the few cm behind the absorber with probability $|\beta|^2$,

$$(19.15)$$

There are specific final states, such as $\pi^- e^+ \nu$, to which both K_S^0 and K_L^0 can decay. For these final states, we will see quantum interference of the two decay processes in this same region. Meters behind the regenerator, the state reverts again to a pure K_L^0 state.

So far, I have been analyzing the K^0–$\overline{K}^0$ system under the assumption that CP is conserved. However, in 1964, the picture was made more complicated. In an experiment at the Brookhaven National Laboratory, Christenson, Cronin, Fitch, and Turlay (1964) carefully observed K_L^0 decays in a meters-long decay region filled with helium. They discovered that there is a small component of decays to $\pi^+\pi^-$ with the time dependence of the K_L^0 lifetime. This decay

$$\left|K_L^0\right\rangle \to |\pi\pi\rangle \tag{19.16}$$

Modification of the above analysis in the true case that CP is violated.

cannot proceed unless CP is violated. The branching ratio is

$$BR(K_L^0 \to \pi\pi) = 2.8 \times 10^{-3} , \tag{19.17}$$

so the effect is doubly small, a small effect in comparison to the already small K_L^0 decay rate.

There is a place for this CP violating effect within the Standard Model. The t quark can appear as an intermediate state in the K^0–$\overline{K}^0$ mixing amplitude, and diagrams with the t quark can carry a phase

$$(19.18)$$

The effect on the K^0–$\overline{K}^0$ mass matrix is to change (19.2) to

$$\mathbf{M} = \begin{pmatrix} \overline{m} - i\overline{\Gamma}/2 & \delta m(1+i\zeta) - i\delta\Gamma/2 \\ \delta m(1-i\zeta) - i\delta\Gamma/2 & \overline{m} - i\overline{\Gamma}/2 \end{pmatrix}, \tag{19.19}$$

The eigenstates of this matrix are, to first order in ζ,

$$\left|K_S^0\right\rangle = \frac{1}{\sqrt{2}}\left((1+\epsilon)|K^0\rangle - (1-\epsilon)|\overline{K}^0\rangle\right) ,$$

$$\left|K_L^0\right\rangle = \frac{1}{\sqrt{2}}\left((1+\epsilon)|K^0\rangle + (1-\epsilon)|\overline{K}^0\rangle\right) . \tag{19.20}$$

where

$$\epsilon = \frac{i\zeta\delta m/2}{\delta m - i\delta\Gamma/2} . \tag{19.21}$$

The states $|K_S^0\rangle$ and $|K_L^0\rangle$ are not orthogonal, but this is permitted because the modified mass matrix is not Hermitian.

The parameter δm is half of the K_L^0–K_S^0 mass difference. The K_S^0 and K_L^0 decay rates are

$$\Gamma_S = \overline{\Gamma} - \delta\Gamma \qquad \Gamma_L = \overline{\Gamma} + \delta\Gamma \tag{19.22}$$

which implies

$$\delta\Gamma \approx -\frac{1}{2}\Gamma_S \ . \tag{19.23}$$

Using these relations, we can write (19.21) as

$$\epsilon = \frac{i\zeta(m_L - m_S)/2}{m_L - m_S + i\Gamma_S/2} \ . \tag{19.24}$$

I have pointed out above that the real and imaginary parts of the denominator are almost equal. This implies that the phase of ϵ is close to $45°$. More precisely,

$$\epsilon = |\epsilon|e^{i\phi} \qquad \text{with} \qquad \phi = 44° \ . \tag{19.25}$$

To describe the effects of this change in the mass matrix, it is useful to write the eigenstates of $\mathbf{M}$, given by (19.20), in terms of the CP eigenstates (19.4), which I will now refer to as $|K_+^0\rangle$ and $|K_-^0\rangle$. We find

$$\begin{aligned}\left|K_S^0\right\rangle &= \left|K_+^0\right\rangle + \epsilon\left|K_-^0\right\rangle \\ \left|K_L^0\right\rangle &= \left|K_-^0\right\rangle + \epsilon\left|K_+^0\right\rangle .\end{aligned} \tag{19.26}$$

It follows from this formula that

$$\frac{\Gamma(K_L^0 \to \pi\pi)}{\Gamma(K_S^0 \to \pi\pi)} = |\epsilon|^2 \ . \tag{19.27}$$

Evaluating this formula, we find

$$|\epsilon| = 2.23 \times 10^{-3} \ . \tag{19.28}$$

Each of the states $\left|K_S^0\right\rangle, \left|K_L^0\right\rangle$ evolves, in its rest frame, according to

$$e^{-im\tau}e^{-\Gamma\tau/2} \ , \tag{19.29}$$

where τ is proper time. For a moving K^0 state, the oscillation plays out as function of position along its path. A coherent state of $\left|K_S^0\right\rangle$ and $\left|K_L^0\right\rangle$ then displays an interference pattern. Since both states can decay to $\pi^+\pi^-$, we can see this interference in the decay rate to $\pi^+\pi^-$. For a K meson state behind a regenerator, with the wavefunction (19.14), the decay rate is proportional to

CP violation is manifested in a characteristic pattern of quantum interference between the K_S^0 and K_L^0 decays to $\pi\pi$.

$$\Gamma(K^0 \to \pi\pi) \sim \left|\epsilon\,\alpha\,e^{-im_L\tau - \Gamma_L\tau/2} + \beta\,e^{-im_S\tau - \Gamma_S\tau/2}\right|^2$$

$$\sim |\beta|^2 \left|e^{-\Gamma_S\tau/2} + \frac{\epsilon\,\alpha}{\beta}e^{-i(m_L - m_S)\tau - \Gamma_L\tau/2}\right|^2 . \tag{19.30}$$

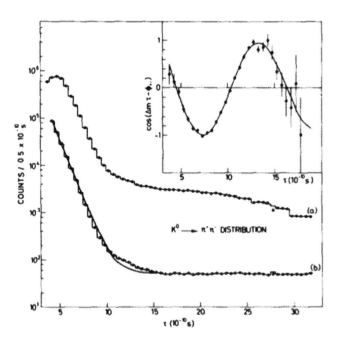

Fig. 19.1 Distribution of $K^0 \to \pi^+\pi^-$ decays behind a regenerator as a function of proper time, from (Geweniger *et al.* 1974).

This function has the form of an oscillation superposed on an exponential decay,

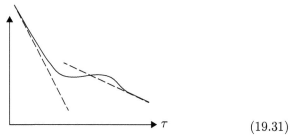

$$(19.31)$$

This is quantum interference over a macroscopic length scale. Some examples of such interference patterns seen in real experiments are shown in Figs. 19.1 and 19.2.

A different interference effect appears in the decays

$$K^0 \to \pi^\pm e^\mp \nu \,, \quad K^0 \to \pi^\pm \mu^\mp \nu \,. \tag{19.32}$$

The K^0 ($\bar{s}d$) decays only to e^+; the $\overline{K}^0$ ($s\bar{d}$) decays only to e^-. A state that is originally K^0 has its time-dependence determined by the resolution into mass eigenstates,

$$\left|K^0\right\rangle = \frac{1}{\sqrt{2}(1+\epsilon)} \left[\left|K^0_S\right\rangle + \left|K^0_L\right\rangle\right] \,. \tag{19.33}$$

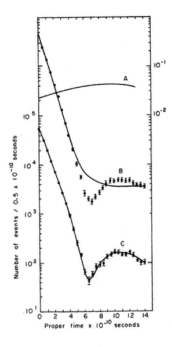

Fig. 19.2 Distribution of $K^0 \to \pi^+\pi^-$ decays behind a regenerator as a function of proper time, from (Carithers *et al.* 1975).

From this formula, we can work out the K^0 and $\overline{K}^0$ components of the original K^0 wavefunction as a function of τ,

$$
\begin{aligned}
|K^0(\tau)\rangle &= \frac{1}{\sqrt{2}(1+\epsilon)}\left[|K_S^0\rangle e^{-im_S\tau - \Gamma_S\tau/2} + |K_L^0\rangle e^{-im_L\tau - \Gamma_L\tau/2}\right] \\
&= \frac{1}{2(1+\epsilon)}\left[\left((1+\epsilon)|K^0\rangle - (1-\epsilon)\big|\overline{K}^0\big\rangle\right)e^{-im_S\tau - \Gamma_S\tau/2}\right. \\
&\quad \left. + \left((1+\epsilon)|K^0\rangle + (1-\epsilon)\big|\overline{K}^0\big\rangle\right)e^{-im_L\tau - \Gamma_L\tau/2}\right].
\end{aligned} \tag{19.34}
$$

Looking at the K^0 and $\overline{K}^0$ content of these eigenstates, we can read off the decay rate to e^+

$$
\Gamma(K^0 \to e^+\pi\nu) \sim \frac{|1+\epsilon|^2}{|1+\epsilon|^2}\left|e^{-im_S\tau - \Gamma_S\tau/2} + e^{-im_L\tau - \Gamma_L\tau/2}\right|^2, \tag{19.35}
$$

and to e^-

$$
\Gamma(K^0 \to e^-\pi\nu) \sim \frac{|1-\epsilon|^2}{|1+\epsilon|^2}\left|e^{-im_S\tau - \Gamma_S\tau/2} - e^{-im_L\tau - \Gamma_L\tau/2}\right|^2, \tag{19.36}
$$

The charge asymmetry

$$
A(\tau) = \frac{N(e^+) - N(e^-)}{N(e^+) + N(e^-)} \tag{19.37}
$$

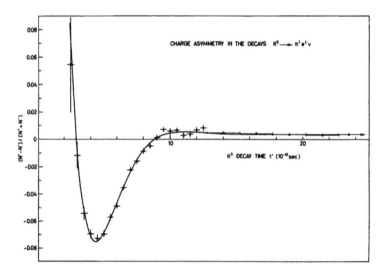

Fig. 19.3 Charge asymmetry in $K^0 \to e^+\pi^-\nu$ / $K^0 \to e^-\pi^+\overline{\nu}$ decays as a function of proper time, from (Gjesdal *et al.* 1974).

goes through an oscillation, as shown in Fig. 19.3. The asymmetry tends to a nonzero constant at large values of τ. This reflects the asymmetry of the K_L^0 component of the state which remains at long times. We find

CP violation in K^0–$\overline{K}^0$ mixing also predicts a small excess of $K^0 \to e^+\nu\pi$ over $K^0 \to e^-\nu\pi$ decays, which is observed.

$$A(\tau) \to \frac{|1+\epsilon|^2 - |1-\epsilon|^2}{|1+\epsilon|^2 + |1-\epsilon|^2} \approx 2 \, \mathrm{Re} \, \epsilon \, . \tag{19.38}$$

The numerical prediction for the asymmetry is

$$A(\tau) \to 3.3 \times 10^{-3} \, , \tag{19.39}$$

in good agreement with the data.

19.2 Electric dipole moments

For a long time, *CP* violation was only seen in the K^0–$\overline{K}^0$ system, and all nonzero *CP*-violating observables were consistent with an origin in the complex phase of the K^0–$\overline{K}^0$ mixing amplitude. In the 1990's, a small *CP*-violating contribution to the $K^0 \to \pi\pi$ decay amplitude was also discovered (Barr *et al.* 1993). Other quantities that might show *CP* and *T* violation are the electric dipole moments of elementary particles. For a spin-$\frac{1}{2}$ particle, the spin indicates an orientation. An electric dipole moment is then a charge polarization in the direction of the spin. *T* reverses the spin but does not reverse the polarization; hence, an electric dipole moment is a *T*-violating effect. Naively, one might expect that the electric dipole moment of the neutron might be as large as

CP violation can potentially lead to nonzero electric dipole moments for elementary particles. In the Standard Model, electric dipole moments are predicted to be extremely small, in agreement with experiment.

$$d_n \sim e \cdot 1 \, \mathrm{fm} \sim 10^{-13} e\text{-cm} \, . \tag{19.40}$$

In fact, the neutron electric dipole moment is known to be much, much smaller. The current limit is (Pendlebury *et al.* 2015, Serebrov *et al.* 2015)

$$d_n < 0.3 \times 10^{-25} e\text{-cm} . \tag{19.41}$$

The limit on the electron electric dipole moment is (Baron *et al.* 2014)

$$d_e < 8.7 \times 10^{-29} e\text{-cm} . \tag{19.42}$$

These values turn out to be consistent with the expectations for these quantities in the Standard Model. Since the neutron and the electron contain, to high degree of approximation, only particles of the first generation, the *CP*-violating effects predicted for these systems are extremely small.

19.3 *CP* violation in the B^0–$\overline{B}^0$ system

There is a more tantalizing way to search for additional quantities exhibiting *CP* violation. In the Standard Model, *CP* violation is expected to come from an order-1 phase associated with heavy quarks. If this is true, there must be a heavy quark weak interaction process with order-1 *CP* violation. How do we find it?

Bigi, Carter, and Sanda suggested that one could see order-1 effects of the CKM phase in the time-dependence of decays of B mesons to exclusive final states with definite *CP* (Carter and Sanda 1980, Bigi and Sanda 1981). The simplest example is

$$B^0 , \overline{B}^0 \to J/\psi K^0_S . \tag{19.43}$$

Consider, for definiteness, the decay of the $\overline{B}^0$ ($b\bar{d}$). The $\overline{B}^0$ can reach the $J/\psi K^0_S$ final state in two ways. First, it can decay directly, through the weak interaction process $b \to c\bar{c}s$,

$$\sim V_{cb}V_{cs}^* \tag{19.44}$$

But also, it can decay through B^0–$\overline{B}^0$ mixing, followed by the process $\bar{b} \to \bar{c}cs$. The K^0–$\overline{K}^0$ mixing matrix must also be used to cause the final states to interfere. So the second path follows the Feynman diagram

$$\sim V_{cb}^*V_{cs} . \tag{19.45}$$

In the exclusive process $\overline{B}^0 \to J/\psi K^0_S$ two alternative quantum paths interfere with a phase that displays *CP* violation.

The B^0–$\overline{B}^0$ mixing amplitude is dominated by the process

$$\sim V_{tb}V_{td}^*V_{td}^*V_{tb} \qquad (19.46)$$

and the K^0–$\overline{K}^0$ mixing amplitude is dominated by the process

$$\sim V_{cs}^*V_{cd}V_{cd}V_{cs}^* \ . \qquad (19.47)$$

The two paths then differ by a relative factor proportional to

$$-\left[V_{cb}^*V_{cs}V_{tb}V_{td}^*V_{cs}^*V_{cd}\right]^2 \ , \qquad (19.48)$$

where the extra minus sign is that in the K_S^0 wavefunction (19.4). In the Wolfenstein parametrization of the CKM matrix (18.46), the only factor in this formula that has a phase is V_{td}, which can be represented as

$$V_{td} = A\lambda^3(1 - \rho - i\eta) = \mathcal{C} \ e^{-i\beta} \ . \qquad (19.49)$$

So, the relative phase between the two paths is $-e^{2i\beta}$. Any phases arising from the strong interaction matrix elements are identical along the two paths and factor out of the decay amplitude.

I will now discuss how this phase can be measured experimentally in the simplest situation. To explain this clearly, I will use a number of approximations that are accurate for the particular process $B^0/\overline{B}^0 \to J/\psi K_S^0$. For a complete discussion of this and other time-dependent B decay processes, see (Bevan *et al.* 2014).

The B^0–$\overline{B}^0$ system is somewhat simpler than the K^0–$\overline{K}^0$ system, in that the hadronic decays of the B meson are decays to complex multiparticle final states with both possible values of CP. Hence, the decay rates of the two mass eigenstates are nearly equal, so that $\delta\Gamma$ can be neglected. The B^0–$\overline{B}^0$ mass matrix is then well approximated by

$$\mathbf{M} = \begin{pmatrix} \overline{m} - i\Gamma/2 & -e^{2i\beta}\delta m/2 \\ -e^{-2i\beta}\delta m/2 & \overline{m} - i\Gamma/2 \end{pmatrix} \ . \qquad (19.50)$$

In writing (19.50), I have used the result in (19.46) that the $\overline{B}^0 \to B^0$ mixing amplitude has the phase of $(V_{td}^*)^2 \sim e^{2i\beta}$. The parameter δm is real-valued, and it turns out to be positive. The lifetime of the B^0 mesons is

$$\tau = 1.52 \times 10^{-12} \text{ sec} \qquad \Gamma = 4.3 \times 10^{-13} \text{ GeV} \ . \qquad (19.51)$$

The eigenstates of the matrix (19.50) are

$$|B_L^0\rangle = \frac{1}{\sqrt{2}}\left(|B^0\rangle + e^{-2i\beta}|\overline{B}^0\rangle\right) \ ,$$

$$|B_H^0\rangle = \frac{1}{\sqrt{2}}\left(|B^0\rangle - e^{-2i\beta}|\overline{B}^0\rangle\right) \ , \qquad (19.52)$$

with eigenvalues

$$\overline{m} - \delta m/2 - i\Gamma/2 \,, \qquad \overline{m} + \delta m/2 - i\Gamma/2 \qquad (19.53)$$

for B_L^0 and B_H^0, respectively. The mass difference of the two states is

$$m_H - m_L = \delta m = 3.3 \times 10^{-13} \text{ GeV} \,, \qquad (19.54)$$

The value of $(m_H - m_L)$ is accidentally quite close to the decay rate Γ. This means that the time-dependent interference terms in B^0 decay might be observable.

The states $|B_L^0\rangle$ and $|B_H^0\rangle$ have simple time-dependence, for example,

$$|B_L^0(\tau)\rangle = \exp[-i(\overline{m} - \delta m/2 - i\Gamma/2)\tau]|B_L^0\rangle \,. \qquad (19.55)$$

Then we can use (19.52) to compute the time-dependence of the $|B^0\rangle$ and $|\overline{B}^0\rangle$ states. For $|B^0\rangle$,

$$
\begin{aligned}
|B^0(\tau)\rangle &= \frac{1}{\sqrt{2}}\left[|B_L^0(\tau)\rangle + |B_H^0(\tau)\rangle\right] \\
&= \frac{1}{2}e^{-i\overline{m}\tau - \Gamma\tau/2}\left[|B^0\rangle(e^{i\delta m\,\tau/2} + e^{-i\delta m\,\tau/2}) \right. \\
&\qquad \left. + |\overline{B}^0\rangle e^{-2i\beta}(e^{i\delta m\,\tau/2} - e^{-i\delta m\,\tau/2})\right] \\
&= e^{-i\overline{m}\tau - \Gamma\tau/2} \\
&\qquad \left(|B^0\rangle \cos\frac{\delta m\,\tau}{2} + i|\overline{B}^0\rangle\, e^{-2i\beta}\, \sin\frac{\delta m\,\tau}{2}\right) \,.
\end{aligned}
$$
$$(19.56)$$

Similarly, for $|\overline{B}^0\rangle$,

$$
\begin{aligned}
|\overline{B}^0(\tau)\rangle &= e^{-i\overline{m}\tau - \Gamma\tau/2} \\
&\qquad \left(|\overline{B}^0\rangle \cos\frac{\delta m\,\tau}{2} + i|B^0\rangle\, e^{+2i\beta}\, \sin\frac{\delta m\,\tau}{2}\right) \,.
\end{aligned}
$$
$$(19.57)$$

We have now dealt with the B^0–$\overline{B}^0$ mixing, so all that remains is to include the decay the B^0 and $\overline{B}^0$ states directly to $J/\psi\, K_S^0$. Recalling again that there is a minus sign between the $s\overline{d}$ and $d\overline{s}$ components of the K_S^0, the matrix elements for the full process of time evolution and decay have the form

$$
\begin{aligned}
\mathcal{M}(B^0(\tau) \to J/\psi\, K_S^0) &= e^{-i\overline{m}\tau - \Gamma\tau/2}\mathcal{A} \\
&\qquad \left(|B^0\rangle \cos\frac{\delta m\,\tau}{2} - i|\overline{B}^0\rangle\, e^{-2i\beta}\, \sin\frac{\delta m\,\tau}{2}\right) \,. \\
\mathcal{M}(\overline{B}^0(\tau) \to J/\psi\, K_S^0) &= e^{-i\overline{m}\tau - \Gamma\tau/2}\mathcal{A} \\
&\qquad \left(|\overline{B}^0\rangle \cos\frac{\delta m\,\tau}{2} - i|B^0\rangle\, e^{+2i\beta}\, \sin\frac{\delta m\,\tau}{2}\right) \,.
\end{aligned}
$$
$$(19.58)$$

The decay amplitude $\mathcal{A}$ can be complex, with a phase due to the strong interaction, but this factor is the same for B^0 and $\overline{B}^0$ decays due to the CP invariance of the strong interaction.

Squaring the amplitudes (19.58), we find the time-dependence of the decay rates

$$\Gamma(B^0(\tau) \to J/\psi \ K_S^0) \sim e^{-\Gamma\tau}(1 - \sin\delta m \, \tau \ \sin 2\beta)$$
$$\Gamma(\overline{B}^0(\tau) \to J/\psi \ K_S^0) \sim e^{-\Gamma\tau}(1 + \sin\delta m \, \tau \ \sin 2\beta) \quad (19.59)$$

The asymmetry in the rates is

$$\frac{\Gamma(\overline{B}^0 \to J/\psi K_S^0) - \Gamma(B^0 \to J/\psi K_S^0)}{\Gamma(\overline{B}^0 \to J/\psi K_S^0) + \Gamma(B^0 \to J/\psi K_S^0)} = +\sin\delta m \, \tau \ \sin 2\beta \quad (19.60)$$

So, the decay is shifted forward in time for an initial $\overline{B}^0$ and backward in time for an initial B^0. The asymmetry is predicted to have a time-dependence governed by δm with amplitude $\sin 2\beta$. For the process $B^0/\overline{B}^0 \to J/\psi \ K_L^0$, the relative minus sign in the decay amplitudes from B^0 and $\overline{B}^0$ becomes a plus sign, and so the asymmetry becomes $(-\sin\delta m\tau \ \sin 2\beta)$. The angle β in this formula is the phase angle taken directly from the CKM matrix through (19.49), with no corrections due to the strong interaction.

To understand how to measure this time-dependent asymmetry, we must first think about the production of B^0 and $\overline{B}^0$ mesons in e^+e^- annihilation. We have seen that e^+e^- annihilation leads to a state with $J = 1$. For production of a pair of spin 0 mesons, the two mesons are in an $L = 1$ wavefunction, which must then be antisymmetric in the other meson quantum numbers. The B mesons go outward from the production point. After some time, one of the mesons decays. A decay to an e^+ or μ^+ tags this meson—at that time—as a B^0. The other meson must then be a $\overline{B}^0$. This state propagates for an additional time Δt, possibly mixing to B^0 during that time, and then decays to the observed final state.

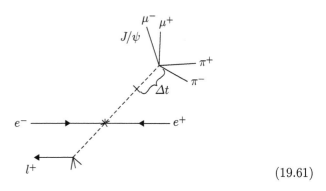

$$(19.61)$$

The relative time Δt, or, rather $\tau = \Delta t/\gamma$, is the time that would appear in the formula above. The relative time Δt can be negative, corresponding to the case in which the selected exclusive decay takes place *before* the tagging leptonic decay.

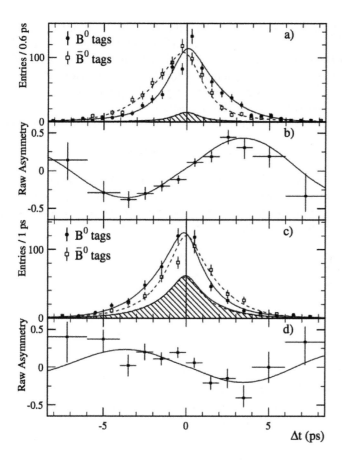

Fig. 19.4 Proper time distribution of $B^0\overline{B}^0 \to J/\psi K^0$ decays at the $\Upsilon(4S)$, measured by the BaBar experiment at the PEP-II collider at SLAC, from (Aubert *et al.* 2002). Panel (a) shows the decay distributions for $B^0\overline{B}^0 \to J/\psi K^0_S$. Panel (b) shows the rate asymmetry (19.60). Panel (c) shows the decay distributions for $B^0\overline{B}^0 \to J/\psi K^0_L$. Panel (d) shows the corresponding rate asymmetry (19.60).

The lifetime of the B meson is about 1.5 ps, so it is difficult to measure the decay time directly. However, Pier Oddone suggested that one might construct an asymmetric colliding beam accelerator, in which the e^+e^- center of mass frame is moving with respect to the lab (Oddone 1989). In the realistic case, the boost of the center of mass is $v/c \sim 0.5$. Then the two B decays would be separated by about 200 microns, a distance that is resolvable using a silicon tracking detector to pinpoint the decay vertices. In the late 1990's, two asymmetric e^+e^- colliders were constructed, one at SLAC (9.0 GeV e^- × 3.1 GeV e^+), for the BaBar experiment, and one at KEK in Tsukuba, Japan (8.0 GeV e^- × 3.5 GeV e^+), for the BELLE experiment. In 2001, both experiments observed the CP-violating asymmetry in $B^0 \to J/\psi K^0_S$ (Abe *et al.* 2001b, Aubert *et al.* 2001).

Figure 19.4 shows the displacements of the decay distributions for

The B^0 decay distributions in time can be converted to observable distributions in space by creating the $B^0\overline{B}^0$ states at an asymmetric e^+e^- collider.

$B^0 \to J/\psi K^0$ and $\overline{B}^0 \to J/\psi K^0$ measured by the BaBar experiment (Aubert *et al.* 2002). Note that the distributions are labelled by the *tagging B* meson, so the points labeled "B^0 tags" indicate $\overline{B}^0(\tau)$ decays, and vice versa. The distributions for B^0 and $\overline{B}^0$ are shifted substantially with respect to one another, in just the directions predicted below (19.60). The shifts are in the opposite directions for K_L^0 instead of K_S^0 in the final state. The current best value of β from this measurement is

$$\sin 2\beta = 0.679 \pm 0.20 , \qquad (19.62)$$

that is, $\beta = 21°$. This is indeed a large *CP*-violating effect.

A useful way to visualize the phase of the CKM matrix is to plot the complex parameter $(\rho + i\eta)$ and use it to define a triangle, called the *unitarity triangle* (Bjorken and Dunietz 1987).

The unitarity triangle is a visualization of the *CP* violation of the Standard Model. The Standard Model has *CP*-violating interactions as long as the angles β and γ are nonzero.

$$(19.63)$$

The internal angles of the triangle are called (α, β, γ) or, alternatively, (ϕ_2, ϕ_1, ϕ_3). The angle γ is the phase of $(\rho + i\eta)$. The angle β is the angle defined in (19.49). There is *CP* violation as long as β and γ are nonzero and the triangle does not collapse to a line.

The left and right sides of this triangle can be expressed more generally as

$$(\rho + i\eta) = -\frac{V_{ud}V_{ub}^*}{V_{cd}V_{cb}^*} , \qquad (\rho + i\eta - 1) = \frac{V_{td}V_{tb}^*}{V_{cd}V_{cb}^*} . \qquad (19.64)$$

It should be noted that these ratios of V_{CKM} matrix elements are invariant to changes of phase of the quark fields. The closure of the triangle,

$$1 - (\rho + i\eta) - (1 - \rho - i\eta) = 0 , \qquad (19.65)$$

is equivalent to the relation

$$V_{ud}V_{ub}^* + V_{cd}V_{cb}^* + V_{td}V_{tb}^* = 0 , \qquad (19.66)$$

which expresses the orthogonality of the first and third columns of the CKM matrix.

The angles α and γ can also be measured by observable parameters of B decays. The angle α is given by time-dependent asymmetries in B decay to light quarks,

$$B^0 \to \pi^+\pi^- , \quad \pi^\pm\rho^\mp , \quad \rho^+\rho^- . \qquad (19.67)$$

The angle γ can be extracted from asymmetries in B decays to DK. These constraints are shown in Fig. 19.5, together with constraints from the value of $|V_{ub}|$, the values of the B^0–$\overline{B}^0$ mixing amplitude, the value of

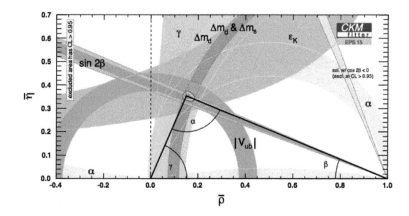

Fig. 19.5 Constraints on the CKM parameters (ρ, η) from measurements of CP violation, showing the fit by the CKMfitter collaboration (Charles *et al.* 2005).

B_s^0–$\overline{B}_s^0$ mixing amplitude, and the value of ϵ from the neutral K system (Charles *et al.* 2005). In the Standard Model, all of these parameters must be consistent with a common value of $(\rho + i\eta)$. You can see that this is the case, and also that the ρ and η parameters are quite well determined.

I have told you earlier that any quantum field theory is invariant under CPT, so CP violation implies T violation. However, it is interesting to ask whether one can directly see T violation in heavy quark decays. The BaBar experiment demonstrated this in the following way: We have seen that, in e^+e^- annihilation, B mesons are produced as pairs in a quantum coherent wavefunction. The decay of one meson breaks the coherence, identifying one meson of the pair as a B^0 or a $\overline{B}^0$, for a leptonic decay, or as a $CP = +$ or $CP = -$ state (B_+ or B_-), for a decay to a CP eigenstate. We can then pick out events in which the leptonic decay happens first, followed by time evolution to a CP eigenstate, and also events in which the CP decay happens first, followed by time evolution to a state with a definite leptonic decay. If the equations of motion of nature were T symmetric, the rates for time evolution in the two directions would be equal. They are not. The asymmetries between the rates for pairs of time-reversed processes (e.g., $B^0 \to B_-$ vs. $B_- \to B^0$) are shown in Fig. 19.6 (Lees *et al.* 2012). Note that the asymmetries reverse when one changes from B^0 to $\overline{B}^0$ and from even to odd CP, consistent with the physics described above. This is the most direct evidence that the equations of nature violate time reversal invariance.

Measurement of the time-dependence of $B^0\overline{B}^0$ decays to exclusive final states shows explicitly that T is violated by the weak interaction.

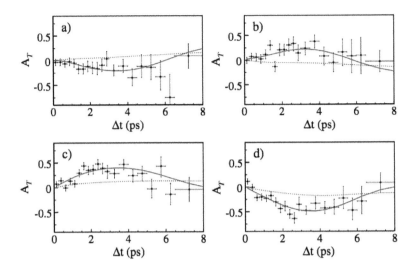

Fig. 19.6 Time reversal violating asymmetries measured as a function of proper time by the BaBar experiment at the PEP-II collider, from (Lees *et al.* 2012). The four panels refer to the transitions: (a). $\overline{B}^0 \to B_-$, (b). $B_+ \to B^0$, (c). $\overline{B}^0 \to B_+$, (d). $B_- \to B^0$.

Exercises

(19.1) Consider the time-dependent B meson decay $B^0 \to \pi^+\pi^-$.

(a) Draw the Feynman diagram similar to (19.44) giving the decay of a B^0 to $\pi^+\pi^-$. Collect the factors of CKM matrix elements that appear in the evaluation of this diagram.

(b) Draw the Feynman diagram similar to (19.45) giving the direct decay of a B^0 to $\pi^+\pi^-$ through B^0–$\overline{B}^0$ mixing. Collect the factors of CKM matrix elements that appear in the evaluation of this diagram.

(c) Show that the quantum interference term between these diagrams is proportional to

$$(V_{ud}V_{ub}^*V_{tb}V_{td}^*)^2 . \qquad (19.68)$$

Show that the phase of this quantity is given by the angle α in (19.63).

(19.2) In the discussion of the K^0–$\overline{K}^0$ system in Section 19.1, I included CP violation in the neutral kaon mass matrix but ignored the possibility of CP violation in decay amplitudes of the various neutral kaon states to pions. This problem will add that

effect. The analysis is straightforward but much more involved than you might have expected.

(a) The neutral K mesons are particles with strong interaction isospin $I = \frac{1}{2}$. A π meson has $I = 1$, so a $\pi\pi$ in the S wave has $I = 0$ or $I = 2$. (Why is $I = 1$ not allowed?) Write down the $I = 0$ state as a linear combination of $|\pi^+(p_1)\pi^-(p_2)\rangle$, $|\pi^0(p_1)\pi^0(p_2)\rangle$, and $|\pi^-(p_1)\pi^+(p_2)\rangle$. Show that, if the final state of a $K^0 \to \pi\pi$ decay is purely $I = 0$, then the decay amplitudes would satisfy

$$\mathcal{M}(K \to \pi^+(p_1)\pi^-(p_2))$$
$$= \mathcal{M}(K \to \pi^0(p_1)\pi^0(p_2)) . \quad (19.69)$$

Remembering that, for identical particles, we integrate phase space over only half of 4π, show that this implies

$$\Gamma(K \to \pi^+\pi^-) = 2\Gamma(K \to \pi^0\pi^0) . \quad (19.70)$$

(b) Assume that the decay amplitude for the neutral K meson leads to $I = 0$ states only. This amplitude could have a CP-violating phase.

It also will have a phase δ_0 resulting from strong final state interactions between the two pions. Include this complex decay amplitude in the analysis leading to (19.27). Show that the complex number is squared, so that the CP-violating phase (and, in fact, any other contribution from the decay amplitude) has no effect on (19.27).

(c) CP violation in the decay amplitude can have an observable effect as an interference between the phases of the decay amplitudes to $I = 0$ and $I = 2$ $\pi\pi$ states. However, the $I = 0$ (or $\Delta I = \frac{1}{2}$) amplitude is found experimentally to be much larger than the $I = 2$ (or $\Delta I = \frac{3}{2}$) amplitude. The evidence for this comes from the following observations: First, the K^+ meson can decay to $\pi^+\pi^0$ only through the $\Delta I = \frac{3}{2}$ amplitude. Second, the decay $K^+ \to \pi^+\pi^0$ is much slower than the decay $K^0 \to \pi\pi$. To verify these statements, first, argue that $\left|\pi^+\pi^0\right\rangle$ in the S-wave is a state with $I = 2$. Then, look up the lifetime of the K^+ (and the branching ratio for $K^+ \to \pi^+\pi^0$) at the Particle Data Group web site, compare to (19.10), and estimate the ratio of the $\Delta I = \frac{1}{2}$ and the $\Delta I = \frac{3}{2}$ decay amplitudes. This large ratio is consistent with the results of QCD numerical calculations.

(d) Construct the $\pi\pi$ state with $I = 2$, $I^3 = 0$, noting that it must be orthogonal to the state with $I = 0$, $I^3 = 0$, and show that

$$\mathcal{M}(K \to \pi^+(p_1)\pi^-(p_2))$$
$$= -\frac{1}{2}\mathcal{M}(K \to \pi^0(p_1)\pi^0(p_2)) .$$

$$(19.71)$$

(e) Add the $I = 0$ and $I = 2$ decay amplitudes with factors a_0 and a_2 representing their magnitudes. Show that the four possible ampli-

tudes are consistently represented as

$$\mathcal{M}(K^0 \to \pi^+\pi^-) = a_0 e^{i\delta_0} + a_2 e^{i\delta_2}$$
$$\mathcal{M}(K^0 \to \pi^+\pi^-) = a_0^* e^{i\delta_0} + a_2^* e^{i\delta_2}$$
$$\mathcal{M}(\overline{K}^0 \to \pi^0\pi^0) = a_0 e^{i\delta_0} - 2a_2 e^{i\delta_2}$$
$$\mathcal{M}(\overline{K}^0 \to \pi^0\pi^0) = a_0^* e^{i\delta_0} - 2a_2^* e^{i\delta_2} .$$

$$(19.72)$$

Note that the strong interaction final-state phases do not change sign when we replace particles by antiparticles. According to part (b), one overall phase is not observable. It is conventional to represent this by taking a_0 to be real.

(f) Using the expressions in (19.72), work through the derivation of (19.27) for the two distinct final states and show that, for $a_0 \gg |a_2|$,

$$\frac{\Gamma(K_L^0 \to \pi^+\pi^-)}{\Gamma(K_S^0 \to \pi^+\pi^-)} ,$$
$$= |\epsilon(1 + i\frac{\mathrm{Im}a_2}{a_0}e^{i(\delta_2 - \delta_0)})|^2 ,$$
$$\frac{\Gamma(K_L^0 \to \pi^0\pi^0)}{\Gamma(K_S^0 \to \pi^0\pi^0)} ,$$
$$= |\epsilon(1 - 2i\frac{\mathrm{Im}a_2}{a_0}e^{i(\delta_2 - \delta_0)})|^2 .$$

$$(19.73)$$

These equations are conventionally written

$$\frac{\Gamma(K_L^0 \to \pi^+\pi^-)}{\Gamma(K_S^0 \to \pi^+\pi^-)} = |\epsilon + \epsilon'|^2 ,$$
$$\frac{\Gamma(K_L^0 \to \pi^0\pi^0)}{\Gamma(K_S^0 \to \pi^0\pi^0)} = |\epsilon - 2\epsilon'|^2 . \quad (19.74)$$

From experiment,

$$\mathrm{Re}(\epsilon'/\epsilon) = (1.66 \pm 0.23) \times 10^{-3} . \quad (19.75)$$

Neutrino Masses and Mixings

20

In the Standard Model as I have presented it in the previous two chapters, the neutrinos are assumed to have zero mass. This was a good approximation for all of the processes that we have discussed so far. However, it is straightforward to include the possibility that neutrinos are massive.

The masses of neutrinos turn out to be very small on the scale of other elementary particle masses. This makes it difficult to observe these masses experimentally. We will see that the evidence for neutrino masses is tied to the existence of another effect, the conversion of neutrinos from one flavor to another in flight. This latter phenomenon is observable due to quantum interference of the sort that we saw in the K^0–$\overline{K}^0$ and B^0–$\overline{B}^0$ system, but now playing out over larger distances from km to thousands of km.

20.1   Neutrino mass and β decay

Studies of β decay require that the mass of the electron neutrino, at least, is very small. A bound on the mass of the ν_e can be obtained by studying the endpoint of the electron energy distribution in β decay. The rate for β decay of a nucleus A to B has the form

$$\Gamma(A \to B e^- \overline{\nu}) = \frac{1}{2m_A} \int \frac{d^3p_B d^3p_e d^3p_\nu}{(2\pi)^9 2E_B 2E_e 2E_\nu}$$
$$(2\pi)^4 \delta^{(4)}(p_A - p_B - p_e - p_\nu)|\mathcal{M}|^2 . \quad (20.1)$$

Since A and B are very heavy compared to their mass difference, which is typically a few MeV, it is a good approximation to assume that the final nucleus B takes up the recoil momentum, so that the directions of the electron and neutrino are uncorrelated. In this limit, the energies of the final electron and neutrino sum to

Kinematics and phase space for β decay.

$$E_e + E_\nu = m(A) - m(B) = \Delta m_{AB} . \quad (20.2)$$

Concepts of Elementary Particle Physics. Michael E. Peskin.
© Michael E. Peskin 2019. Published in 2019 by Oxford University Press.
DOI: 10.1093/oso/9780198812180.001.0001

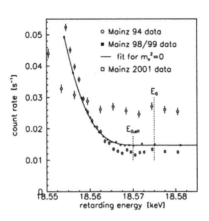

Fig. 20.1 Measurement of the endpoint of the electron energy spectrum in tritium β decay, from (Kraus *et al.* 2005).

Then

$$\Gamma = \frac{1}{2m_A} \frac{1}{(2\pi)^5 2m_B} \int \frac{dp_e p_e^2}{2E_e} \int \frac{dp_\nu p_\nu^2}{2E_\nu} \delta(\Delta m_{AB} - E_e - E_\nu) |\mathcal{M}|^2 \; .$$
(20.3)

At the endpoint of the electron energy spectrum, we can approximate the matrix element by a constant. Then, using

$$dp_e p_e = dE_e E_e \; , \qquad dp_\nu p_\nu = dE_\nu E_\nu \; ,$$
(20.4)

we can write the decay rate as

$$\Gamma \sim \int_{m_e}^{\Delta m_{AB}} dE_e E_e (\Delta m_{AB} - E_e)^2$$
(20.5)

The *Kurie plot*, which visualizes the distortion in the electron spectrum in β decay due to the presence of a neutrino mass.

Assuming that the neutrino has zero mass, this gives

$$\frac{d\Gamma}{dE_e} \sim (\Delta m_{AB} - E_e)^2 \; .$$
(20.6)

This energy distribution is conventionally represented by a *Kurie plot*, plotting the square root of the event rate as a function of the electron energy. This should be a straight line for a zero mass neutrino (Kurie *et al.* 1936). If the neutrino is massive, the plot falls off at the kinematic endpoint $E_e = \Delta m_{AB} - m_\nu$,

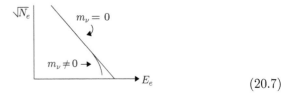

(20.7)

Measurements of β decay exclude ν_e masses of more than a few eV. Unfortunately, for neutrino masses as small as eV, there are extra complications. The β electron can lose an energy of order eV when it exits the atom, and it loses eV/mm in traversing material. These energy losses must be accounted for in the interpretation of the electron energy distribution. The most careful experiments, done with cryogenic tritium films at Mainz and Troitsk (Kraus *et al.* 2005, Aseev *et al.* 2011), give a limit

$$m_{\nu_e} < 2.05 \text{ eV} . \qquad (20.8)$$

The endpoint of the Mainz spectrum is shown in Fig. 20.1. The large open circles show data from an earlier version of the experiment in which the energy loss of the electron was incorrectly estimated because of *roughening* of the surface of the tritium film, an effect of having insufficiently low temperature.

The direct limits on the masses of the ν_μ and ν_τ are weaker. However, I will argue later that the three neutrino masses are all within 1 eV of one another. Given this, there is another class of constraints on the neutrino masses. Massive neutrinos moving relativistically in the early universe would transfer energy and smear out cosmic structure, giving an observably different distribution of clusters of galaxies if the neutrino masses are sufficiently large. The absence of this effect gives a bound currently estimated to be (Ade *et al.* 2016)

$$\sum_{i=1,2,3} m_{\nu_i} < 0.23 \text{ eV} . \qquad (20.9)$$

20.2   Adding neutrino mass to the Standard Model

Neutrino masses are thus very small compared to the weak interaction mass scale, sufficiently small that it is unclear how they can be observed. To understand the evidence for neutrino mass, we need to develop further the theory of neutrino massses within the Standard Model.

If we assume $SU(2) \times U(1)$ symmetry, neutrino masses can arise in one of two ways. The simplest mechanism is to assume that there exist right-handed neutrinos that couple to the left-handed neutrinos through Yukawa couplings. That is, we add to the Standard Model Lagrangian a term

$$\Delta\mathcal{L} = -y_\nu^{ij} L_a^{\dagger i} \epsilon_{ab} \varphi_b^* \nu_R^j + h.c. \qquad (20.10)$$

Consequences of adding a neutrino mass term to the Standard Model Lagrangian.

similar to the u quark mass term in (18.24). In principle, we could treat this term in the same way that we treated the quark and lepton mass terms in Section 18.3. However, this is not appropriate. In elementary particle reactions, neutrinos are typically emitted at MeV or higher energies, at which effects of eV-scale masses are unimportant. Therefore it is most convenient to retain our earlier convention that the left-handed neutrinos are described in the basis that diagonalizes their weak interactions. We then treat the new term by making the change of variables

$$L^i \to U_{Lij}^{(e)} L^j , \qquad (20.11)$$

Definition of the *flavor eigenstates* of neutrinos. The PMNS matrix relates the neutrino flavor eigenstates to the mass eigenstates.

just as we did in (18.29). This transforms

$$y_\nu \to y'_\nu = U_L^{(e)\dagger} y_\nu .\tag{20.12}$$

Notice that this transformation diagonalizes the charged lepton Yukawa matrix but does not necessarily diagonalize the neutrino Yukawa matrix. I will refer to this basis for neutrino states as the basis of *flavor eigenstates*. In this basis, the ν_e is the linear combination of the three neutrino states that is produced in a weak interaction decay together with an e^+, and the ν_μ and ν_τ are defined similarly.

We can now diagonalize y'_ν as before,

$$y'_\nu = U_L^{(\nu)} Y_\nu U_R^{(\nu)} ,\tag{20.13}$$

where Y_ν is real and diagonal. We can transform away $U_R^{(\nu)}$, but we cannot get rid of the matrix $U_L^{(\nu)}$. This is a fixed unitary transformation between the basis of flavor eigenstates and the basis of mass eigenstates. I will refer to the the mass eigenstates as ν_1, ν_2, ν_3, with masses m_1, m_2, m_3. As we did with the quark mixing matrix, we can redefine phases in $U_L^{(\nu)}$ so that $U_L^{(\nu)}$ contains three angles but only one phase. The mixing matrix $U_L^{(\nu)}$ is called the *Pontecorvo-Maki-Nakagawa-Sakata* or *PMNS matrix* and is more commonly notated V or V_{PMNS} (Pontecorvo 1958, Maki, Nakagawa, and Sakata 1962).

I can now describe the physical effect of a neutrino mass term. I choose the process of π^+ decay as an example. The π^+ decays to $\mu^+\nu_\mu$, that is, specifically to the ν_μ weak interaction eigenstate. The ν_μ is a linear combination of the three mass eigenstates. If the π^+ energy is fixed, the three components are emitted with slightly different values of momentum

$$p_i = E - \frac{m_i^2}{2E} + \cdots .\tag{20.14}$$

This is permitted, because the pion decay region is of finite size, allowing the momentum to be uncertain. This uncertainty is small enough that the components of the ν_μ wavefunction are created with quantum coherence.

The outgoing neutrino wavefunction then has the form

$$\sum_{i=1,2,3} V_{\mu i} e^{+i(E-m_i^2/2E)x} |\nu_i\rangle .\tag{20.15}$$

At very large distances x, the components of this wavefunction go out of phase. Then the probability of finding a ν_μ is no longer 1. Instead, it is given by

$$\text{Prob}(\nu_\mu \to \nu_\mu) = \left|\sum_i V_{\mu i} V_{\mu i}^* e^{-i(m_i^2/2E)x}\right|^2 .\tag{20.16}$$

It is easiest to understand this formula if we evaluate it for the case of two-neutrino mixing with mixing angle θ,

$$V = \begin{pmatrix} \cos\theta & -\sin\theta \\ \sin\theta & \cos\theta \end{pmatrix} .\tag{20.17}$$

In that case, the formula becomes

$$\text{Prob}(\nu_\mu \to \nu_\mu) = \left| \cos^2 \theta e^{-i(m_1^2/2E)x} + \sin^2 \theta e^{-i(m_2^2/2E)x} \right|^2 , \quad (20.18)$$

which can be rewritten as

$$\text{Prob}(\nu_\mu \to \nu_\mu) = 1 - \sin^2 2\theta \sin^2 \left[\frac{\Delta m^2}{4E} x \right] , \quad (20.19)$$

where $\Delta m^2 = (m_2^2 - m_1^2)$. There is an oscillation between the flavor eigenstates with an oscillation length

$$L = 4\pi \frac{E}{\Delta m^2} = (2.48 \text{ m}) \frac{E \text{ (MeV)}}{\Delta m^2 \text{ (eV}^2)} . \quad (20.20)$$

The conclusion is quite surprising. We can detect the presence of small neutrino masses if the neutrinos also exhibit flavor mixing. Then the effect of the mass term is to generate a *flavor oscillation* as a function of the distance from the neutrino source. For MeV neutrinos with 10^{-2} eV masses or for GeV neutrinos with 10^{-1} eV masses, the length scale of the oscillation can be km.

The evidence for the masses of neutrinos comes from the observation of oscillation between flavor eigenstates as neutrino travel over macroscopic distances.

This is just the opposite of the way that we determine the masses and weak interaction flavor mixing among quarks. For quarks, we observe the particles as mass eigenstates, inside hadrons of definite mass. Decays through the weak interaction show that the mass eigenstates are linear combinations of weak interaction eigenstates. For neutrinos, the primary way that we observe the particles is through weak interaction decay. Then we characterize the neutrino eigenstates according to their weak interaction properties. It is the flavor mixing as the neutrinos travel that demonstrates that there is a mass eigenstate basis, with different masses for the three neutrinos, that is different from the flavor basis.

There is another way to add neutrino masses to the Standard Model that is consistent with Lorentz invariance and $SU(2) \times U(1)$. We can write

$$\Delta \mathcal{L} = -\frac{1}{2} \mu_{ij} (L_{a\alpha}^i \epsilon_{ab} \varphi_b)(L_{c\beta}^j \epsilon_{cd} \varphi_d) \epsilon_{\alpha\beta} + h.c. , \quad (20.21)$$

where $\alpha, \beta = 1, 2$ are the indices of 2-component spinors. The expression (20.21) is Lorentz-invariant. It does not violate any gauge symmetry of the Standard Model. The expression does violate lepton number, but you might recall from Section 18.4 that lepton number conservation is not a postulate in the description of the Standard Model. When the Higgs field φ acquires an expectation value and breaks $SU(2) \times U(1)$, (20.21) leads to a mixing of the ν_L states with their antiparticles $\bar{\nu}_R$, generating masses given by the eigenvalues of

The *Majorana mass term* for neutrinos.

$$m_{ij} = \mu_{ij} \frac{v^2}{2} . \quad (20.22)$$

This mass term, resulting from particle-antiparticle mixing, is called a *Majorana mass term* (Majorana 1937).

Origin of the Majorana neutrino mass
from the influence of very heavy right-
handed neutrinos.

The quantity μ_{ij} has the dimensions $(\text{GeV})^{-1}$, so we might also write
the mass formula as

$$m_{ij} = \frac{\overline{\mu}_{ij}v^2/2}{M} \ , \qquad (20.23)$$

where $\overline{\mu}$ is dimensionless and M sets the mass scale. For reasons that I
will explain in a moment, the elements of $\overline{\mu}$ might be expected to have
the size of Yukawa couplings. If we estimate $\tilde{\mu} \sim (10^{-2})^2$, then we find
sub-eV neutrino masses for $M \sim 10^{10}$ eV. Yukawa couplings cover a
wide range, so M could be orders of magnitude larger or smaller.

We can obtain this structure naturally by starting from a Lagrangian
with neutrino Yukawa couplings and a lepton-number violating mass
term for the right-handed neutrinos,

$$\Delta\mathcal{L} = -\frac{1}{2}M_{ij}\nu^i_{R\alpha}\nu^j_{R\beta}\epsilon_{\alpha\beta} + h.c. \qquad (20.24)$$

This is a direct Majorana mass term for the right-handed neutrinos.
Note that, because the right-handed neutrinos do not transform under
$SU(2) \times U(1)$, we are free to write this term without violating any symmetry of the Standard Model. Thus, while quark, lepton, and vector
boson masses are restricted to be of the size of the Higgs field expectation value (17.14), there is no reason why the scale of masses of the
right-handed neutrinos cannot be very much larger. When we use (20.24)
together with the neutrino Yukawa coupling (20.10), the diagram

$$(20.25)$$

generates Majorana masses for the left-handed neutrinos of the form
(20.23) with the mass scale M given by the right-handed neutrino mass.
This is called the *seesaw mechanism* for generating small neutrino masses.
It produces small masses by a modification of the theory at very high energies (Minkowski 1977, Gell-Mann, Ramond, and Slansky 1979, Yanagida
1980).

The consequences of the Majorana mass term for neutrinos are almost
the same as those of the Dirac mass term. We can diagonalize the
Majorana neutrino mass as

$$m_{ij} = (V\overline{m}V^T)_{ij} \ , \qquad (20.26)$$

where $\overline{m}$ is complex diagonal and V is complex unitary. The matrix V
is the PMNS matrix, reducible to three angles and one phase. There
are two more possible phases in $\overline{m}$. These have no significant effect on
neutrino flavor oscillations.

Neutrinoless double β decay.

However the Majorana mass term generates a new weak interaction
process, called *neutrinoless double β decay*.   At some points in the
periodic table, ordinary β decay is energetically forbidden, but double
β decay is allowed. For example,

$$m(\text{Cs}^{136}) > m(\text{Xe}^{136}) > m(\text{Ba}^{136}) \qquad (20.27)$$

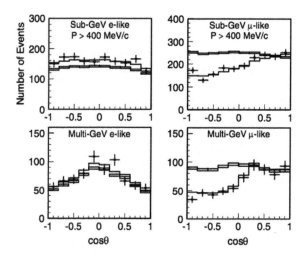

Fig. 20.2 Measurement of the flux of electron- and muon-type neutrinos from atmospheric cosmic ray events, compared to models of neutrino production with and without neutrino mixing, from (Ashie *et al.* 2005).

Then Xe^{136} can decay by

$$Xe^{136} \to Ba^{136} + e^{-}\bar{\nu}_e e^{-}\bar{\nu}_e \ . \tag{20.28}$$

Double β decay processes are some of the rarest physical processes known. For example, the EXO experiment measured (Ackerman *et al.* 2011)

$$\tau(Xe^{136}) = 2 \times 10^{21} \ \text{yr} \ . \tag{20.29}$$

If the neutrino $\bar{\nu}_e$ has a lepton-number violating Majorana mass term, then also a decay process

$$Xe^{136} \to Ba^{136} + e^{-}e^{-} \ . \tag{20.30}$$

is allowed, with no final-state neutrinos. The rate of this decay is expected to be small even in comparison to (20.29). This and similar decays are being intensively searched for, but none has yet been observed.

20.3 Measurements of neutrino flavor mixing

Now that we know how to look for neutrino mass, we can discuss the experimental evidence that the neutrino masses are indeed nonzero.

The first clear evidence for neutrino flavor mixing, and, thus, for neutrino mass, came in the study of the neutrinos produced in cosmic ray interactions in the atmosphere. These were observed in underground water Cherenkov detectors originally built to look for proton decay. Persistently since the 1980's, it was observed that the flux of ν_e from atmospheric interactions was close to the predictions, while the flux of ν_μ

Flavor mixing of atmospheric neutrinos.

was too small by a factor of 2. In 1998, the SuperKamiokande experiment, a very large water Cherenkov detector in the Kamioka mine in Japan, resolved this question by observing the directions of ν_μ's from their conversion to muons in charge-changing interactions (Fukuda *et al.* 1998). The downward-going ν_μ were present with a flux that was essentially unsuppressed, while upward-going ν_μ, created on the other side of the earth, were highly supressed. For ν_e, the ratio of the predicted to the observed flux was independent of direction. The data is shown in Fig. 20.2. This strongly indicated a flavor mixing $\nu_\mu \leftrightarrow \nu_\tau$ on the scale of the earth's diameter. The mixing angle was consistent with a maximal value

$$\sin^2 2\theta = 1 . \tag{20.31}$$

This flavor mixing has since been confirmed by accelerator experiments that create beams of ν_μ at GeV energies and detect the neutrinos over a long path length. The experiment K2K has a baseline of 250 km, from KEK to the Kamioka mine (Ahn *et al.* 2006). The experiment MINOS has a baseline of 750 km, from Fermilab to the Soudan mine in northern Minnesota (Michael *et al.* 2006). The current best values of the oscillation parameters are

$$\Delta m^2 = (2.43 \pm 0.08) \times 10^{-3} \text{ eV}^2 = (5 \times 10^{-2} \text{ eV})^2 ,$$
$$\sin^2 \theta = 0.386 \pm 0.023 . \tag{20.32}$$

The value of $\sin^2 \theta$ seems smaller than (20.31), but it is still consistent within statistics with the maximal mixing value of 0.5.

Production of neutrinos by the sun.

The mass of the ν_e is related to another long-standing anomaly in neutrino physics. In the 1960's, John Bahcall suggested testing the mechanism of energy generation in the sun by observing the flux of neutrinos produced by the sun (Bahcall 1964). Raymond Davis took up the challenge. He designed an experiment with a tank containing 600 tons of CCl_4 underground in the Homestake mine in South Dakota. Solar neutrinos would convert Cl^{37} to Ar^{37} at the rate of atoms/month. The radioactive Ar atoms could then be extracted and counted. The rate of Ar production was observed to be consistently low compared to the solar model prediction (Davis *et al.* 1968).

The production of neutrinos by the sun is quite complex. The dominant process, accounting for 99% of solar neutrinos, is

$$pp \rightarrow D + e^+\nu_e , \tag{20.33}$$

where D is a deuterium nucleus. However, the resulting neutrinos, at 0.5 MeV energy, are of too low energy to be detected in Davis's experiment. Instead, rarer reactions are needed to give neutrinos of energy above the 0.8 MeV threshold for this detection technique. A typical solar neutrino spectrum is shown in Fig. 20.3 (Serenelli *et al.* 2011). Over the decades, solar neutrino experiments were mounted in other energy regions, and eventually experiments with a gallium detection medium observed the neutrinos from the dominant pp process. Always, the rate was smaller than required.

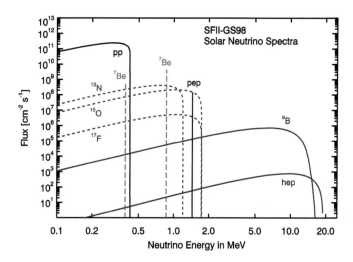

Fig. 20.3 Predicted energy spectrum of neutrinos from the sun (figure courtesy of A. Serenelli, based on the analysis in (Serenelli, Haxton, and Pena-Garay 2011)).

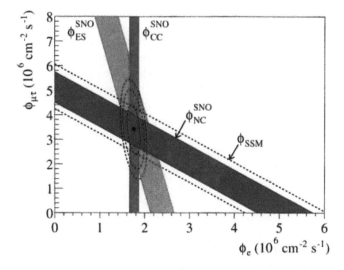

Fig. 20.4 Fluxes of solar neutrinos of the various types, extracted from the data of the SNO experiment, from (Ahmad *et al.* 2002). The estimates of ν_e and ν_μ/ν_τ fluxes from the three processes listed in (20.34) are shown as the red, blue, and green bands, respectively.

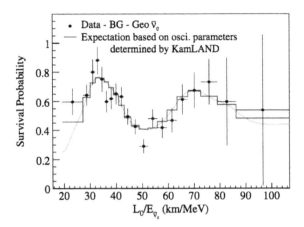

Fig. 20.5 Probability of $\nu_e \to \nu_e$ for neutrinos from nuclear reactors as a function of proper time, as measured by the KamLAND experiment, from (Abe *et al.* 2008).

Flavor mixing of solar neutrinos.

Finally, the situation was resolved by the Sudbury Neutrino Observatory (SNO) experiment, using a heavy water (D_2O) Cherenkov detector located in the Sudbury mine in northern Ontario (Ahmad *et al.* 2002). This experiment was sensitive only to the highest energy solar neutrinos, from $B^8 \to Be^8 e^+ \nu_e$. However, it was able to simultaneously observe three different neutrino reactions,

$$\nu_e D \to p p e^- \ ,$$
$$\nu_i D \to p n \nu_i \ ,$$
$$\nu_i e^- \to \nu_i e^- \ . \tag{20.34}$$

The first reaction in (20.34), charged current neutrino scattering from deuterium, measures the flux of ν_e. The second reaction is the neutral current scattering from deuterium, which has equal cross section for all three neutrino species. Neutrino-electron scattering is sensitive to all neutrino species, but the cross section for ν_e is larger than that for ν_μ, ν_τ by about a factor 6, reflecting contributions from both Z and W exchange processes,

$$\tag{20.35}$$

The flux determinations from SNO are shown in Fig. 20.4. The flux of ν_e is indeed smaller than expected by more than a factor of 2, but the total neutrino flux is in good agreement with the prediction for ν_e production in solar models. Apparently, the solar neutrinos are converting to ν_μ and ν_τ on their way to the earth.

This neutrino flavor oscillation, which requires a small Δm^2, was confirmed by the KamLAND experiment, a scintillator detector in the

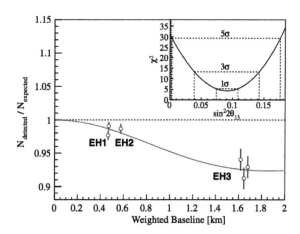

Fig. 20.6 Probability of $\nu_e \rightarrow \nu_e$ for neutrinos from nuclear reactors as a function of distance from the reactor, as measured by the Daya Bay experiment, from (An *et al.* 2012).

Kamioka mine which observed antineutrinos from nuclear reactors in Japan at baselines of order 100 km (Abe *et al.*2008). The oscillation in the $\bar{\nu}_e$ survival probability as a function of neutrino energy is shown in Fig. 20.5. The current best values for the oscillation parameters are

$$\Delta m^2 = (7.54 \pm 0.24) \times 10^{-5} \text{ eV}^2 = (0.9 \times 10^{-2} \text{ eV})^2 \;,$$
$$\sin^2 \theta = 0.307 \pm 0.017 \;. \tag{20.36}$$

So, there are two small neutrino mass differences of rather different scale. The large ratio between the twoΔ m^2 values justifies the use of two-neutrino mixing formula to parametrize each oscillation. The values ofΔ m^2 imply that all of the neutrino masses must be within about 0.1 eV of one another. However, these results do not give the absolute scale of neutrino masses. They also do not give the ordering of the levels. There are two possibilities, called the *normal* and *inverted* hierarchy.

$$m \Bigg\uparrow \quad \overline{} 3 \qquad \overline{\overline{}} 1,2$$
$$\qquad\qquad \underset{or}{}$$
$$\qquad \overline{\overline{}} 1,2 \qquad \overline{} 3 \tag{20.37}$$

Normal and *inverted* neutrino mass hierarchy.

In each case, the isolated mass eigenstate is an almost pure combination of ν_μ and ν_τ, while the two closely spaced states mix ν_e with the orthogonal linear combination of ν_μ and ν_τ.

It is possible in principle to distinguish these possibilities by observing the effect on neutrino mixing of neutrino interactions with matter as the neutrinos pass through the earth over hundreds of km. So far, the issue has not been resolved.

The next question we might address is that of whether ν_3 contains some admixture of ν_e. This mixing is controlled by the third mixing angle in the PMNS matrix. It can be detected by looking for an oscillation of reactor antineutrinos ($\bar{\nu}_e$) at the oscillation Δm^2 of the atmospheric neutrino oscillation, which corresponds to a km wavelength for neutrinos of MeV energy. This was finally observed in 2012 by the reactor experiments Daya Bay, in China, and RENO, in Korea (An *et al.* 2012, Ahn *et al.* 2012). These experiments constructed closely matched pairs of detectors and contrasted the rate observed in a "far" detector with that predicted from the rate observed in a "near" detector. Figure 20.6 shows the comparison of near and far detector fluxes at Daya Bay. The value of the third neutrino mixing angle is

Direct measurement of the third PMNS mixing angle.

$$\sin^2 \theta_{13} = 0.0241 \pm 0.0025 . \qquad (20.38)$$

The question remains of whether the possible phase in the PMNS matrix is nonzero. There is room for this CP-violating term in the neutrino mass matrix. Still, it is a fundamental question whether the couplings of the neutrinos violate CP and T. In principle, CP violation in the neutrino system can be measured by observing asymmetries such as

$$\text{Prob}(\nu_\mu \to \nu_e) \neq \text{Prob}(\bar{\nu}_\mu \to \bar{\nu}_e) \qquad (20.39)$$

However, we do not have the answer yet.

Exercises

(20.1) Estimate quantitatively the neutrino flight path required for neutrino oscillations.

    (a) Consider first the oscillation, mainly between ν_μ and ν_τ, mediated by θ_{23}. Assume a pure ν_μ source. Using the parameters of this oscillation given in the text, compute the position of the first maximum for ν_τ appearance and the position of the succeeding zero, for ν_μ energies of 1 GeV and 20 GeV (for neutrinos from an accelerator source).

    (b) Now consider the oscillation between ν_e and other species that gives rise to the oscillation of solar neutrinos. Compute the position of the first maximum for ν_μ appearance (or maximal ν_e disappearance) and the position of the succeeding zero, for ν_e energies of 1 MeV (for reactor or solar neutrinos) and 1 GeV and 20 GeV.

(20.2) Compute the cross section for neutrino-electron elastic scattering. Assume that $s \gg m_e^2$, so that, in the center of mass frame, both the electron and the neutrino can be treated as massless. For this

very short-time interaction, you can neglect neutrino flavor mixing. Also, assume that $s \ll m_W^2$, so the interaction can be treated as pointlike.

    (a) Consider first ν_μ-e scattering. This process occurs through the first Feynman diagram in (20.35). Write the spinors for the initial- and final-state particles. Compute the scattering amplitudes associated with this diagram for $\nu_L e_L^-$ and $\nu_L e_R^-$ scattering.

    (b) Square the amplitudes, integrate over phase space, and compute the cross section. The initial electron should be averaged over polarizations; the initial neutrino is, of couse, purely left-handed. You should find

$$\sigma(\nu_\mu e \to \nu_\mu e) = \frac{G_F^2 s}{\pi} \left(\frac{1}{4} - s_w^2 + \frac{4}{3} s_w^4 \right) . \qquad (20.40)$$

    (c) For ν_e-e scattering, both diagrams shown in (20.35) contribute. Notice that the charged-current diagram is present only for e_L^-. For this case, compute the scattering amplitude associated with the second diagram. Use the same spinors as you used in part (a) and keep careful track of the relative sign between this amplitude and the amplitude for the first diagram in (20.35). You should find that the relative sign is *positive*.

(d) However, there is one more contribution to the relative sign of the two amplitudes. Between the first and second diagrams in (20.35), there is an exchange of positions of two fermions. This gives an extra factor of (-1). With this factor included, show that the two diagrams interfere *destructively*.

(e) Compute the full cross section for ν_e-e scattering, averaged over the electron spin. By what factor is this cross section larger than that for ν_μ or ν_τ scattering?

(20.3) This problem concerns the effect of propagation through matter on neutrino flavor mixing. This problem gives an application of the formula (15.76) at the end of Exercise 15.1. So, it might be worth reviewing (or working through) that problem before attempting this one.

(a) Write down the terms in the Lagrangian of the Standard Model that include the W and Z fields. Cross out the $(F_{\mu\nu})^2$ terms that involve W and Z. This is equivalent to the approximation $q^2 \ll m_W^2$. Now the Lagrangian contains only the W and Z mass terms and the interactions with fermions, with no derivatives. This structure is very simple. Write the simplified field equations for W and Z, and solve them. Plug the results back into the Lagrangian. Show that this gives a term in the Lagrangian

$$\Delta\mathcal{L} = -\frac{4G_F}{\sqrt{2}}(j_L^{+\mu}j_{\mu L}^- + (j^{\mu 3} - s_w^2 j_Q^\mu)^2) .$$
(20.41)

This is actually a derivation of the weak interaction matrix element the we have seen before in (16.64). However, we have now obtained the overall sign in front, in a form that we can compare to other terms in the Standard Model that involve the quarks and leptons.

(b) For a massive neutrino of momentum p, with $p \gg m$, simplify the kinetic energy term by applying the time derivative to the neutrino

wavefunction. Show that this gives

$$\nu_L^\dagger i\bar\sigma^0\partial_0\nu = (p + \frac{m^2}{2p} + \cdots)\nu_L^\dagger\nu_L .$$
(20.42)

The factor in parentheses is the phase accumulated by a neutrino per unit time of propagation.

(c) If a neutrino flies through matter, it can interact with the background matter. This can be represented by taking $\langle j^{03}\rangle$ and $\langle j_Q^0\rangle$ to be nonzero, so that the term

$$-4\sqrt{2}G_F(\nu^\dagger\nu)\langle j^{\mu 3} - s_w^2 j_Q^\mu\rangle .$$
(20.43)

must be added to the Lagrangian. In principle, this contributes to the neutrino phase. Show that this contribution is identical for the three species of neutrino, so it does not affect the flavor mixing.

(d) Let the background matter density be n (baryons/cm^3). Assume that the matter is composed of atoms of light elements with equal numbers of proton and neutrons. In this approximation, evaluate the contribution from protons, neutrons, and electrons to $\langle j^{03}\rangle$ and $\langle j_Q^0\rangle$.

(e) Specifically for ν_e, there is another contribution. Apply the Fierz identity to the charged-current term and show that this yields a term proportional to

$$\nu_e^\dagger\nu_e \, e_L^\dagger e_L .$$
(20.44)

Show that $\langle e_L^\dagger e_L\rangle = n/4$, where n is the background baryon density. Then we can interpret this term as a shift of the (diagonal) ν_e mass. Evaluate this term and show that this shift is

$$\Delta m_e^2 = -\sqrt{2}G_F \, n \cdot p .$$
(20.45)

(f) The central density of the sun is approximately

$$\rho = 150 \text{ g/cm}^3 .$$
(20.46)

For neutrinos of 1 MeV, compute the m^2 shift numerically and compare to the δm^2 of the solar oscillation. When the mass shift due to the matter effect is much greater than the Δm^2 in vacuum, flavor mixing is turned off and the ν_e propagates as an independent species.

The Higgs Boson

21

There is one more particle of the Standard Model of particle physics that we still have not discussed—the Higgs boson. In Chapters 16 and 18, I have emphasized that the masses of all quarks, leptons, and vector bosons arise from the spontaneous symmetry breaking of $SU(2) \times U(1)$ gauge symmetry. In this chapter, I will describe the predictions of the Standard Model for the properties of the Higgs boson and the extent to which those predictions have been verified experimentally.

In Chapter 14, we saw that a phenomenon analogous to electroweak symmetry breaking, the spontaneous breaking of the chiral symmetry of QCD, has a dynamical explanation in terms of the attraction and pair condensation of light quarks. It would be wonderful if there were a physical mechanism that allowed us to understand qualitatively why the $SU(2) \times U(1)$ symmetry of the weak interaction is spontaneously broken. Today, we have no such understanding.

The Standard Model gives a simpler explanation for this symmetry breaking. It postulates a scalar field, the Higgs field, with the potential (16.29) and gauge and Yukawa couplings allowed by symmetry. The potential has the correct shape for $SU(2) \times U(1)$ symmetry breaking because it has a minus sign in front of the μ^2 term. This explanation is too *ad hoc* to be a final physics explanation. However, the Standard Model at least gives us a definite theory that makes precise reference predictions for the properties of the Higgs boson. Perhaps by measuring the couplings of the Higgs boson and testing these predictions are precisely as possible, we can obtain hints toward a deeper explanation. That program has now begun.

21.1 Constraints on the Higgs field from the weak interaction

Before entering into the specifics of the Standard Model theory of the Higgs boson, I would like to point out two aspects of the Higgs field theory that are fixed by aspects of the weak interaction that we have already studied. Most of the tests of the $SU(2) \times U(1)$ gauge theory that we have discussed so far are independent of the nature of the Higgs field. They involve experiments using light quarks and leptons, whose

Concepts of Elementary Particle Physics. Michael E. Peskin.
© Michael E. Peskin 2019. Published in 2019 by Oxford University Press.
DOI: 10.1093/oso/9780198812180.001.0001

couplings are fixed by $SU(2) \times U(1)$ invariance alone. However, the properties of the W and Z bosons do depend on the Higgs field.

In particular, there is one prediction involving the W and Z bosons that depends on the mechanism of electroweak symmetry breaking and lets us glimpse into its properties. This is the relation (16.47),

$$m_W = m_Z c_w \ . \tag{21.1}$$

In Chapter 16, we saw that this was a specific outcome of the Standard Model, but it is interesting to inquire further. Thinking more generally, the relation (21.1) comes from the fact that the gauge boson mass matrix, in the original $SU(2) \times U(1)$ basis, has the form

$$m^2 = \frac{1}{4} \begin{pmatrix} g^2 v^2 & & & \\ & g^2 v^2 & & \\ & & g^2 v^2 & -gg'v^2 \\ & & -gg'v^2 & g'^2 v^2 \end{pmatrix} \tag{21.2}$$

Custodial $SO(3)$ symmetry, a symmetry of the interaction that breaks the weak interaction gauge symmetry.

acting on the vector $(A_\mu^1, A_\mu^2, A_\mu^3, B_\mu)$. This structure does not require every detail of the Standard Model, but it requires assumptions beyond those of $SU(2) \times U(1)$ gauge symmetry. In particular, it follows from the two assumptions: (1) The symmetry breaking leaves invariant an $SO(3)$ symmetry acting on $A_\mu^1, A_\mu^2, A_\mu^3$, which requires that the first three diagonal elements are equal; (2) The symmetry breaking leaves invariant a $U(1)$ gauge symmetry, which requires that the matrix m^2 have a zero eigenvalue. We saw in Section 16.3 that these assumptions are satisfied in the model of $SU(2) \times U(1)$ symmetry breaking by one $I = \frac{1}{2}$ scalar field. However, these assumptions are also true in some more complex models of electroweak symmetry breaking (Sikivie, Susskind, Voloshin, and Zakharov 1980). The $SO(3)$ symmetry, called *custodial symmetry*, should be a property of any more advanced model that we might propose.

Another aspect of the physics of W and Z bosons that bears directly on the Higgs field is the behavior of their interactions in the limit of very high energy. When high-energy W and Z bosons are emitted in an elementary particle reaction, it is possible to see the presence of the Higgs boson in the quantum states of the massive W and Z. An illustrative example is found in the theory of the decay of the top quark. Working out this theory using only the V−A structure of the weak interaction, we find a Higgs field-like behavior. I will quote the main results here; you can derive them in Exercise 21.3 .

The top quark is sufficiently heavy that it can decay to a b quark and an on-shell W boson. Starting from the standard weak interaction coupling

$$\frac{g}{\sqrt{2}} W_\mu^- b_L^\dagger \bar{\sigma}^\mu t_L \tag{21.3}$$

we find for the top quark decay rate.

$$\Gamma_t = \frac{g^2}{64\pi} \frac{m_t^3}{m_W^2} \left(1 + 2\frac{m_W^2}{m_t^2}\right)\left(1 - \frac{m_W^2}{m_t^2}\right)^2 \ . \tag{21.4}$$

This formula behaves oddly. We do not find the expected behavior

$$\Gamma_t \sim \alpha_w m_t \qquad (21.5)$$

but, instead,

$$\Gamma_t \sim \alpha_w m_t \cdot \frac{m_t^2}{m_W^2} . \qquad (21.6)$$

The enhancement is associated with the decay to a longitudinally polarized (helicity 0) W boson. It is important to remember that this state exists in the first place only if the $SU(2) \times U(1)$ symmetry is broken and the massless W fields eat the resulting Goldstone bosons. It is instructive to compare the result (21.4) to the result in a theory with no W bosons but with the Higgs Yukawa coupling to the top quark. In that theory, the top quark would decay to a b quark and a Goldstone boson. The predicted decay rate would be

A helicity 0 W or Z boson couples like the Higgs boson state that the vector boson ate to become massive.

$$\Gamma_t = \frac{y_t^2}{32\pi} m_t . \qquad (21.7)$$

Using the relation between y_t and m_t, we can convert this into a form similar to (21.4). Since

$$\frac{y_t^2}{2} = \frac{m_t^2}{v^2} = \frac{g^2}{4} \frac{m_t^2}{m_W^2} , \qquad (21.8)$$

the prediction (21.7) is equal to

$$\frac{g^2}{64} m_t \frac{m_t^2}{m_W^2} , \qquad (21.9)$$

which precisely reproduces the leading term in (21.4). This understanding of the enhancement also implies that most of the W bosons produced in top quark decays should be longitudinally polarized. This prediction is verified by experiment (Khachatryan *et al.* 2016).

Apparently, the massive W boson automatically knew that it needed to contain a Goldstone boson from symmetry breaking as a part of its structure. There are many other examples in the physics of W and Z bosons at high energy that illustrate this point (Chanowitz and Galliard 1985).

21.2 Expected properties of the Higgs boson

Now we look into the more specific properties of the Higgs field as predicted by the Standard Model. The Higgs field of the Standard Model contains only 4 degrees of freedom. We saw below (16.34) that 3 of these are Goldstone bosons that are eaten as the W and Z bosons obtain mass. What remains is only 1 dynamical field, the Higgs field $h(x)$.

In the Standard Model Lagrangian $h(x)$ always appears together with the Higgs field vacuum expectation value v. Then the couplings of the Higgs boson are generated by the replacement

$$v \to v + h(x) \ . \tag{21.10}$$

The couplings of h are then associated with the Standard Model mass terms. The Higgs interaction terms in the Standard Model Lagrangian are

$$\Delta \mathcal{L} = -\sum_f m_f \bar{f} f \frac{h(x)}{v} + 2 m_W^2 W_\mu^+ W^{-\mu} \frac{h(x)}{v} + m_Z^2 Z_\mu Z^\mu \frac{h(x)}{v}$$

Couplings of the Higgs boson to Standard Model particles.

$$-3 m_h^2 \frac{h(x)}{v} + \mathcal{O}(h^2) \tag{21.11}$$

These terms are all P and C conserving, so the Higgs boson is a spin 0 particle with $P = +1$. In (17.14), we found that v has the value

$$v = 246 \text{ GeV} \ . \tag{21.12}$$

So all of the couplings of the Higgs boson are highly suppressed, except for the couplings to W, Z, and t. More general models of $SU(2) \times U(1)$ symmetry breaking also have this problem. Either W, Z, or t must be involved in the relevant processes, or the expected rates of Higgs boson processes are extremely small.

I will now discuss the processes that we can use to observe the Higgs boson. We must discuss both the production and decay processes. I will start with the decays. If m_h were greater than $2 m_W$ and $2 m_Z$, the dominant decays would be the decays to these particles

$$h \to W^+ W^- \ , \qquad h \to ZZ \ . \tag{21.13}$$

These decays have been searched for at the LHC, but the only result has been to put strong limits on the production cross sections (Khachatryan *et al.* 2015). Thus, the mass of the Higgs boson must be below the threshold for decay into WW.

Decay of the Higgs boson to quarks and leptons.

In this case, the Higgs boson would decay dominantly into the next lightest Standard Model particle, the b quark. Using methods discussed in this book, it is not so difficult to work out the decay rate for Higgs boson decay to $b\bar{b}$. The calculation is described in Exercise 21.1. The result is

$$\Gamma(h \to b\bar{b}) = 3 \frac{\alpha_w}{8} m_h \frac{m_b^2}{m_W^2} \left(1 - \frac{4 m_b^2}{m_h^2}\right)^{3/2} \ . \tag{21.14}$$

The quark mass should be evaluated at $Q \approx m_h$, giving a value of about 3 GeV for m_b. Then, for a Higgs boson of mass 125 GeV, we find

$$\Gamma(h \to b\bar{b}) \approx 2 \text{ MeV} \tag{21.15}$$

Recall that the width of the Z boson is about 2.5 GeV, a thousand times larger. So the Higgs boson is very narrow, so narrow that it will be difficult to measure the width directly. Other relevant decays to quarks and leptons

$$h \to \tau^+\tau^- , \qquad h \to c\bar{c} \qquad (21.16)$$

give decay rates about 10 times smaller than the decay rate to $b\bar{b}$.

Because the decay to $b\bar{b}$ is so highly suppressed, higher-order decay processes can compete with it. First, although $h \to WW, ZZ$ are forbidden, it is possible that the Higgs boson can decay through a diagram in which an off-mass-shell W or Z appears as a resonance,

Decay of the Higgs boson to off-mass-shell W and Z pairs.

$$(21.17)$$

These decays are called $h \to WW^*$, $h \to ZZ^*$. The suppression from multi-body phase space and the tail of the Breit-Wigner distribution is comparable to the suppression seen above from the small size of $(m_b/v)^2$.

It is also possible for a Higgs boson to decay through higher-order processes involving virtual top quarks or W bosons. This gives decays to two gluons,

Decay of the Higgs boson to gg and $\gamma\gamma$.

$$(21.18)$$

to two photons

$$(21.19)$$

and to γZ. For a 125 GeV Higgs boson, the rate for $h \to 2g$ is comparable to the rate for $h \to \tau^+\tau^-$ and $h \to WW^*$. The rate for $h \to \gamma\gamma$ is about a factor of 50 smaller.

A full set of predictions for the branching ratios of the Higgs boson within the Standard Model, as a function of the Higgs boson mass, is shown in Fig 21.1 (Heinemeyer *et al.*2013). These predictions of the Standard Model do not involve any parameters other than those that I have already discussed in this book. Thus, the predictions can be highly precise. Does nature agree with these results?

Finally, we find that, for a Higgs boson of mass 125 GeV, the Standard Model predicts 9 distinct decay modes with branching ratios larger than 10^{-4} that are potentially observable.

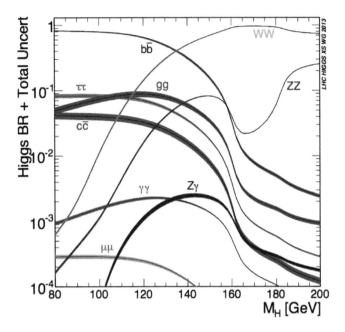

Fig. 21.1 Branching ratios for decays of the Higgs boson as a function of the Higgs boson mass, predicted in the Standard Model, from (Heinemeyer *et al.* 2013).

21.3 Measurements of Higgs boson properties at the LHC

Reversing the decay processes, we find processes for producing the Higgs bosons in high energy collisions. An obvious production process is

$$b\bar{b} \to h \ . \tag{21.20}$$

However, at the LHC, the cross section for this process is multiplied by the very small b quark pdf in the proton. The most promising production mode at the LHC turns out to be $gg \to h$, or *gluon-gluon fusion*,

Reactions for the production of Higgs bosons at the LHC.

$$\tag{21.21}$$

using the Higgs coupling to two gluons shown above. The intrinsic strength of the interaction is smaller, but the initial gluons can be taken from the very large gluon pdf in the proton. At the 13 TeV LHC, a gluon momentum fraction of $x \sim 0.01$ is all that is required.

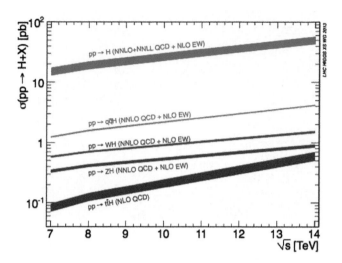

Fig. 21.2 Cross sections for production of the Higgs boson at the LHC as a function of center of mass energy, predicted in the Standard Model, from (Carena, Grojean, Kado, and Sharma 2014).

Another important production process is *vector boson fusion*

$$(21.22)$$

in which high-x quarks in the proton create virtual W or Z bosons that then combine to produce a Higgs boson. Notice that this process results in a Higgs boson and two high-energy jets emitted in the forward direction. The presence of the forward jets can then be used to enhance the Higgs boson signal.

A third important reaction is production of a Higgs boson in association with a W or Z boson. This process can be imagined as $q\bar{q}$ annihilation to the weak boson followed by radiation of a Higgs boson using the relatively large Higgs coupling to these particles.

$$(21.23)$$

Predictions for these and other Higgs boson production processes at the LHC are shown in Fig. 21.2 (Carena, Grojean, Kado, and Sharma

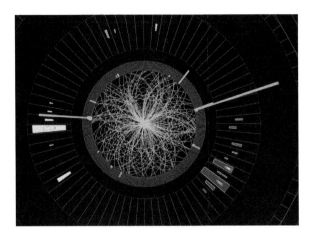

Fig. 21.3 Candidate $pp \to h \to \gamma\gamma$ event observed by the CMS experiment at the LHC (figure courtesy of CERN and the CMS Collaboration).

2014). For $m_h = 125$ GeV and an LHC center of mass energy of 13 TeV, the cross sections are

$$\sigma(pp \to gg \to h) = 50 \text{ pb}$$
$$\sigma(pp \to WW \to h) = 4 \text{ pb}$$
$$\sigma(pp \to Wh, Zh) = 2 \text{ pb} \tag{21.24}$$

These results should be compared with the proton-proton total cross section of about 100 mb, which is higher by a factor of 2×10^9!

At the LHC, we do not observe the total rate for Higgs production; rather, we reconstruct the Higgs boson in a particular decay mode. The quantity that we measure has the form of a cross section times branching ratio, $\sigma \cdot BR$, for example,

$$\sigma(gg \to h) \cdot BR(h \to b\bar{b}) \ . \tag{21.25}$$

In general, a separate selection must be used for each separate decay mode.

Unfortunately, many of the most important Higgs boson decay modes are difficult to observe at the LHC. For example, the process

Difficulty of observing hadronic Higgs decays at the LHC.

$$gg \to h \to b\bar{b} \tag{21.26}$$

results in two b quark jets. However, the QCD process

$$gg \to b\bar{b} \tag{21.27}$$

also produces pairs of b quark jets, with jet pair masses at and above the Higgs boson mass, at a rate about a million times greater. In the decays $h \to WW^*$ and $h \to ZZ^*$, events with hadronic decays of the W and Z are difficult to recognize for the same reason.

To discover the Higgs boson, the ATLAS and CMS experiments at the LHC concentrated their efforts on decay modes of the Higgs boson

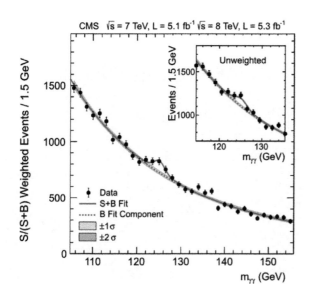

Fig. 21.4 Mass distribution of γ pairs measured by the CMS experiment at the LHC, from (Chatrchyan *et al.* 2012). In main plot, events more likely to be well-reconstructed $pp \to \gamma\gamma$ events are given higher weight.

to photons and leptons in which all final state particle would be visible. These modes are

$$h \to \gamma\gamma \, BR \qquad = 0.23\%$$
$$h \to ZZ* \to 4\ell \qquad BR = 0.016\% \, . \qquad (21.28)$$

In all, Higgs boson production and decay into these modes occurs in about 1 in 2×10^{12} pp collisions. It was quite a feat to collect such events in the presence of enormous numbers of more ordinary LHC collisions.

By collecting a very large data set, the LHC experiments were able to identify the Higgs boson in these channels. Figure 21.3 shows a candidate $h \to \gamma\gamma$ event from CMS. Figure 21.4 shows the distribution of $pp \to \gamma\gamma$ events found by CMS as of June 2012 as a function of the invariant mass of the $\gamma\gamma$ pair. There is a clear resonance on the expected smooth background at a mass of about 125 GeV. Figure 21.5 shows a candidate $h \to e^+e^-\mu^+\mu^-$ event collected by the ATLAS experiment. Figure 21.6 shows the 4-lepton events seen by the ATLAS experiment as of June 2012, plotted as a function of the 4-lepton invariant mass. A signficant resonance signal is seen at the same mass of 125 GeV. On July 4, 2012, both experiments showed significant signals in both of these channels, presenting strong evidence for the appearance of this particle.

Discovery of the Higgs boson at the LHC using the decays $h \to \gamma\gamma$ and $h \to ZZ^*$.

With the new particle identified, we can ask whether it indeed has the properties expected for the Higgs boson. First, is it a particle with $J^P = 0^+$, as the Standard Model predicts? We showed in Exercise 3.4 that, if the new resonance decays to $\gamma\gamma$, it cannot be a particle of spin 1. However, the possibilities of spin greater than 1, and of $P = -1$, would still be open. These hypotheses can be addressed using $h \to$

Tests of the 0^+ spin-parity of the Higgs boson.

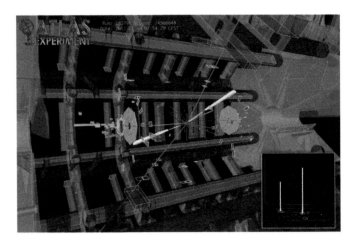

Fig. 21.5 Candidate $pp \to h \to e^+e^-\mu^+\mu^-$ event observed by the ATLAS experiment at the LHC (figure courtesy of CERN and the ATLAS Collaboration).

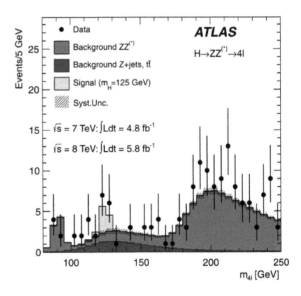

Fig. 21.6 Mass distribution in four-lepton events measured by the ATLAS experiment at the LHC, from (Aad *et al.* 2012).

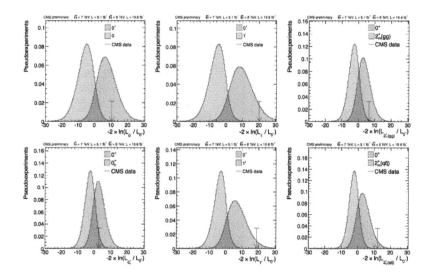

Fig. 21.7 Comparison of hypotheses for the spin and parity of the 125 GeV resonance from event distributions in $h \to 4\ell$, from (CMS Collaboration 2013). What is shown in each plot is the distribution of a test statistic expected for the Standard Model 0^+ hypothesis (yellow) and for an alternative hypothesis (blue). The arrow shows the value of the test statistic given by the data.

$ZZ^* \to 4$ lepton events. The relative orientations of the leptons in these events give information on the polarizations of the Z bosons in $h \to ZZ^*$. Also, they allow tests of whether the particle production and decay is independent of orientation, as would be expected for a spin 0 particle. Figure 21.7 shows tests of the various spin and parity hypotheses relative to the hypothesis of $J^P = 0^+$. In all cases, the 0^+ hypothesis is favored. In most cases, this hypothesis is strongly favored already with this sample of about 25 events.

The decay $h \to W^+W^-$ is more difficult to observe. The specific process visible at the LHC is

Observation of the decay $h \to WW$.

$$h \to W^+W^- \to \ell^+\nu\ell^-\overline{\nu} \qquad (21.29)$$

where ℓ is e or μ. That is, one looks for events with minimal jet activity, two leptons of opposite sign, and unbalanced momentum carried off by the neutrinos. The process $pp \to W^+W^-$, where both W bosons are on mass shell, is an obvious background that cannot be cleanly distinguished from the Higgs events. These processes differ in some details; the leptons from Higgs decay tend to have lower invariant mass and a smaller spread in angle. Figure 21.8 shows the distributions of $\ell^+\ell^-$ invariant mass for events with e or μ, unbalanced momentum, and 0 or 1 jet (Aad *et al.* 2015). The small but significant excess over the expectation from other Standard Model processes is due to the Higgs boson.

If the new particle is the Higgs boson that gives rise to the masses of quarks and leptons, we should be able to discover events in which

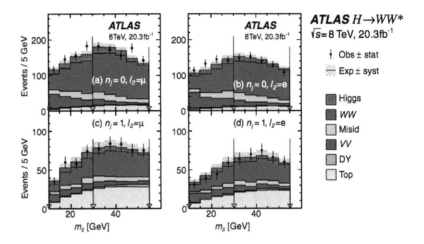

Fig. 21.8 Distribution of $\ell^+\ell^-$ invariant mass in LHC events at 8 TeV collected by the ATLAS experiment, with $e\mu$, unbalanced momentum, and 0 or 1 jet, from (Aad *et al.* 2015). The excess of events in red is attributed to the Higgs boson decaying to WW^*.

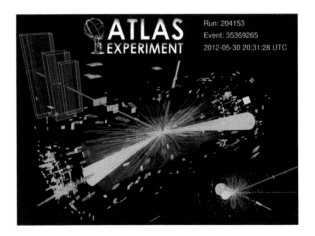

Fig. 21.9 Candidate $pp \to h \to \tau^+\tau^-$ event observed by the ATLAS experiment at the LHC, from (ATLAS Collaboration 2013).

the Higgs boson decays to quarks and leptons. The highest branching ratios correspond to the heaviest available fermions, the b quark and the τ lepton. I will first discuss the evidence for $h \to \tau^+\tau^-$.

Higgs decays to $\tau^+\tau^-$ are generally not sufficiently characteristic that they can be identified in the main LHC reaction of gluon fusion. The WW fusion process, in which events contain additional forward jets for tagging and in which the main competition to Higgs events comes from electroweak reactions, provides a much better setting for this search. A candidate event is shown in Fig. 21.9 (ATLAS Collaboration 2013). This event contains $\tau \to e$ and $\tau \to \mu$ decays, two forward jets as expected from the WW fusion process, and unbalanced momentum consistent with the neutrinos emitted along the τ directions. Events of this type can easily be faked by Standard Model reactions that do not involve the Higgs boson. The most important examples are

Observation of the decay $h \to \tau^+\tau^-$.

$$pp \to Z \to \tau^+\tau^- \ ,$$
$$pp \to W + \text{jet} \to \tau\nu + \pi \text{ faking } \tau \ ,$$
$$pp \to Z \to 2 \text{ jets} \ , \text{ jets faking } \tau \ . \tag{21.30}$$

Very recently, the CMS Collaboration demonstrated the presence of the $h \to \tau^+\tau^-$ decay with a high degree of statistical significance, based mainly on the analysis of WW fusion events (Sirunyan *et al.* 2017a). The background rate, about 10 times the signal rate in the final sample, was estimated by the study of related processes such as $pp \to Z \to \mu^+\mu^-$ and extrapolation from kinematic regions outside the Higgs boson signal region.

The observation of $h \to b\bar{b}$ presents an even more challenging problem. In this case, the most promising reaction is Higgs production in association with a W or Z boson. In this setting, though, the reaction $pp \to W + h$, $h \to b\bar{b}$ is difficult to distinguish from Standard Model reactions without a Higgs boson

Observation of the decay $h \to b\bar{b}$.

$$pp \to W + Z \ , \quad Z \to b\bar{b} \ ,$$
$$pp \to W + g \ , \quad g \to b\bar{b} \ . \tag{21.31}$$

In the second reaction, the gluon is radiated off-shell from the initial quark or antiquark and splits to a quark-antiquark pair through the parton shower physics that we discussed in Chapter 12. Very recently, the ATLAS and CMS Collaborations presented strong evidence for the $h \to b\bar{b}$ decay based on very complex analyses that relied on the $b\bar{b}$ mass distribution and more subtle features of the events. The final separation of signal and background was done using machine learning technniques (Aaboud *et al.* 2017, Sirunyan *et al.* 2017b).

Figure 21.10 shows the status of $\sigma \cdot BR$ rate measurements as summarized by the CMS collaboration at the end of 2017 (CMS Collaboration 2017). The quantity plotted on the horizontal axis is

$$\mu = \frac{\sigma \cdot BR(\text{observed})}{\sigma \cdot BR(\text{SM})} \ , \tag{21.32}$$

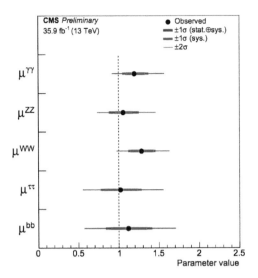

Fig. 21.10 Ratios μ of the rates for Higgs boson production measured by the CMS experiment in a variety of decay channels to the predictions of the Standard Model, from (CMS Collaboration 2017). In the figure $\mu = 0$ would indicate no signal from the Higgs boson, and $\mu = 1$ indicates the Standard Model prediction.

All of the major decay modes predicted for the Higgs boson have now been observed for the 125 GeV resonance.

the ratio of the observed rate to the Standard Model prediction. The figure shows clear evidence for all of the major decay modes of the Higgs boson predicted by the Standard Model. With the new 2017 results, we can now say that the whole pattern of Higgs boson decays predicted by the Standard Model is well supported by data from the LHC. The measurements agree with the predictions to the 20–30% level.

Eventually, we will see measurements of the Higgs boson couplings to 1% accuracy. It is possible that nature will still follow the predictions of the Standard Model. But it is also possible, consistent with all of our knowledge, that these measurements will reveal additional contributions from new interactions outside that Standard Model that have not yet been discovered.

Exercises

(21.1) This problem works through the computation of the partial width of the Higgs boson to $b\bar{b}$.

    (a) Draw the Feynman diagram for the $h \to b\bar{b}$ decay. Write the term in the Standard Model Lagrangian that gives the vertex in this diagram.

    (b) The b quark mass appears in the Yukawa cou-

pling y_b, but the b and $\bar{b}$ produced in the decay are very relativistic. Thus, it is a good approximation to neglect the b quark mass everywhere else in this calculation, and, in particular, to use massless spinors for the b and $\bar{b}$. Write the appropriate massless spinors and use them to compute the decay amplitude.

(c) Square this amplitude, integrate over phase space, and verify (21.14) to the leading order in the m_b.

(d) Repeat this calculation using the spinors of the massive Dirac equation. Verify (21.14) in full.

(21.2) Figure 21.1 shows that, if the Higgs mass were greater than the W and Z boson masses, the decays $h \to WW$ and $h \to ZZ$ would dominate. Compute these decay rates assuming $m_h > 2m_W, 2m_Z$.

(a) Draw the Feynman diagrams that lead to the $h \to WW$ and $h \to ZZ$ decays, and identify the terms in the Standard Model Lagrangian that generate these vertices.

(b) Show that, since the Higgs boson has spin 0, the final-state vector mesons must have equal helicities. We must sum over three helicity states, $h = +1, 0, -1$.

(c) Consider first $h \to WW$. Choose coordinates such that the W^+ boson moves in the $+\hat{3}$ direction. For a W boson at rest, the three W polarization vectors are the three spacelike unit vectors. Choosing vectors of definite angular momentum, we can write these as

helicity $+1$: $\quad \epsilon_+^\mu = (0, 1, i, 0)^\mu/\sqrt{2}$
helicity 0 : $\quad \epsilon_0^\mu = (0, 0, 0, 1)^\mu$
helicity -1 : $\quad \epsilon_-^\mu = (0, 1, -i, 0)^\mu/\sqrt{2}$.
(21.33)

Boost these to the W momentum $k^\mu = (E, 0, 0, k)^\mu$, where $E = m_h/2$. Show that the helicity $+1$ and -1 vectors are unchanged, while the helicity 0 vector boosts to

$$\epsilon_0^\mu = (k, 0, 0, E)^\mu/m_W .$$ (21.34)

(d) Construct the polarization vectors for the W^- by rotating the vectors found in part (c) by $180°$.

(e) Compute the three nonzero decay amplitudes to W boson pairs of definite helicity.

(f) Square these amplitudes and integrate over phase space to obtain a total decay rate. Show that, for $m_h > 2m_W$,

$$\Gamma(h \to WW) = \frac{\alpha_w}{16}\frac{m_h^3}{m_W^2}\left[1 - 4\frac{m_W^2}{m_h^2} + 12\frac{m_W^4}{m_h^4}\right]$$
$$\cdot(1 - 4\frac{m_W^2}{m_h^2})^{1/2} .$$ (21.35)

(g) Find the corresponding expression for $\Gamma(h \to ZZ)$ for $m_h > 2m_Z$.

(h) The growth of these decay rates proportional to m_h^3 is a surprise. Which helicity amplitudes are responsible for this growth? Exercise 21.3 might shed some additional light on this phenomenon.

(21.3) The top quark is so heavy that it can decay to an on-shell W boson and a bottom quark. The decay matrix element is

$$\mathcal{M}(t \to bW^+) = \frac{g}{\sqrt{2}}u_L(b)^\dagger \bar\sigma^\mu u_L(t)\epsilon^{\mu*}(W)$$
(21.36)

Ignore the b quark mass. In this limit, the b is always left-handed. The spinor $u_L(t)$ is the top 2 components of the top quark Dirac spinor. In the following, work in the top quark rest frame, and assume for definiteness that the top quark is polarized in the $+\hat{3}$ direction.

(a) Write formulae for the energies and momenta of the final state W and b in terms of m_t, m_W. Assuming that the W^+ momentum is at an angle θ with respect to the $\hat{3}$ axis,

$$p_W = (E_W, p_W \sin\theta, 0, p_W \cos\theta)$$ (21.37)

write the b quark momentum vector and the b spinor $u_L(p_b)$.

(b) Compute the partial width $\Gamma(t \to b_L W_-^+)$, to a W boson of helicity (-1). The appropriate W polarization vector is

$$\epsilon(W) = \frac{1}{\sqrt{2}}(0, \cos\theta, -i, -\sin\theta)$$ (21.38)

This is a rotation of the polarization vector found in Exercise 21.2, part (c).

(c) Compute the partial width $\Gamma(t \to b_L W_+^+)$, to a W with positive helicity. You should get zero for the result. Why is this process forbidden?

(d) As was seen in Exercise 21.2, part (c), the polarization vector for a helicity-0 (longitudinally-polarized) boson boosted along the $\hat{3}$ axis is

$$\epsilon(W) = \left(\frac{p_W}{m_W}, 0, 0, \frac{E_W}{m_W}\right) .$$ (21.39)

Rotate this polarization vector appropriately, and compute the partial width $\Gamma(t \to b_L W_0^+)$, to a W with helicity zero.

(e) Compute the total width of the top quark and derive the formula (21.4).

(f) Compute the ratio of rates for top quark decays to transverse and longitudinal W bosons. Which mode accounts for the enhancement of top decays discussed in Section 21.1?

Part IV

Epilogue

Epilogue

In this book, I have described the structure of the Standard Model of particle physics and its correspondence with experiment. We have seen that the amazing variety of elementary particle phenomena described in this book can be accounted for quantitatively by the Standard Model Lagrangian (18.37). It is remarkable that we have achieved this state of knowledge of the fundamental interactions at very short distances.

The Standard Model is so powerful that many people assert that this is the end of the story of elementary particles. But there are good reasons to think that it is not. Though the Standard Model is a synthesis of what came before it, it still lacks the simple and self-contained character of, for example, Maxwell's equations or Einstein's theory of general relativity. It is not what Steven Weinberg imagined as a "final theory" of fundamental forces (Weinberg 1993). In this book, I have concentrated on the questions in particle physics whose answers are known. But now it is time to discuss what is not known, and what is yet to be discovered. I will organize this discussion as a series of questions. For a few of these questions, the Standard Model gives answers that are known to be incorrect. For most of them, it is incapable of giving any answer at all.

The questions are of three types. The first set of questions concerns the structure of the Standard Model itself. The second set concerns the relation of our understanding of particle physics to the picture of the universe in the large that has recently emerged from astrophysics. The third set concerns the relation of the Standard Model to grand questions about the nature of space and time.

First, I will consider the questions that the Standard Model raises about its own structure and leaves unanswered:

How do the pieces of the Standard Model fit together? The Standard Model Lagrangian contains 3 different gauge symmetry groups and 15 different fermion representations. In a final theory, shouldn't there just be one gauge symmetry and one type of matter?

For the unification of the gauge symmetries, there is an attractive hypothesis. The group $SU(3) \times SU(2) \times U(1)$ can be considered as a subgroup of the Lie group $SU(5)$ or $SO(10)$. Each generation of fermions fills out an $SU(5)$ or $SO(10)$ representation. The case of $SO(10)$ is particularly elegant, since in this case, the quantum numbers of a com-

Are the $SU(3)$, $SU(2)$, and $U(1)$ symmetries of the Standard Model pieces of a "grand unification" symmetry group?

Concepts of Elementary Particle Physics. Michael E. Peskin.
© Michael E. Peskin 2019. Published in 2019 by Oxford University Press.
DOI: 10.1093/oso/9780198812180.001.0001

plete generation of quark and leptons (including a right-handed neutrino) are contained in a 16-dimensional irreducible representation of $SO(10)$. The idea that there is a single fundamental gauge group, and that this symmetry group is spontaneously broken at short distances to $SU(3) \times SU(2) \times U(1)$, is called *grand unification*.

The grand unified theory is a Yang-Mills theory with a single coupling constant. This seems to contradict our knowledge that the $SU(3)$, $SU(2)$, and $U(1)$ coupling constants are very different, as expressed in (11.73) and (17.12). However, since the QCD coupling becomes smaller at short distances while a $U(1)$ coupling becomes larger (as we discussed in Sections 11.6 and 11.7), it is not hard to imagine that the three Standard Model couplings meet at some small distance scale where the spontaneous breaking takes place. Expressed as an energy scale, the location of the symmetry breaking turns out to be close to 10^{16} GeV. The theory predicts the value of g'/g. The theory also predicts that the baryon number is violated and baryons decay, however, with a long lifetime exceeding 10^{33} years. However, proton decay has not yet been observed. Also, while the qualitative relation of the Standard Model couplings is explained by grand unification, the precise values of these couplings differ somewhat from the grand unification predictions. There is not yet an attractive grand unified theory that explains the presence of three generations of fermions. More information about the theory of grand unification and its experimental status can be found in (Ross 1984) and (Raby 2006).

Perhaps the unification of the elements of the Standard Model requires more ambitious ideas. I present some possibilities below.

Why do the quark and lepton masses vary over such a large range? One of the most striking features of the Standard Model is that it accomodates a top quark of mass about 170 GeV and an u quark whose mass is 100,000 times smaller. In each case, the mass of the quark is given by the size of the corresponding Higgs boson Yukawa coupling. These coupling constants are inputs to the Standard Model. In principle, any input would be acceptable, so the Standard Model gives no explanation for the large mass ratios between the various quarks and leptons.

Actually, the Standard Model does not even give insight into what might seem to be an easier question: Is the top quark a "heavy" quark, while the other quarks and leptons have more ordinary values, or does the top quark have a "normal" value for its mass, while the masses of all other fermions are for some reason suppressed? Since the mass of the top quark is of the same order of magnitude as the masses of the W, Z, and Higgs bosons, it is tempting to say that the top quark mass is of the expected magnitude for fermion masses. However, there are theories created to explain aspects of the Standard Model in which the large size of the top quark Yukawa coupling plays an essential role. So even this simpler question has, at the moment, no definite answer.

Why is the weak interaction gauge symmetry $SU(2) \times U(1)$ **spontaneously broken?** In the Standard Model, we postulate one

Grand unification can qualitatively explain the sizes of the $SU(3)$, $SU(2)$, and $U(1)$ coupling constants.

Is the top quark a "heavy" quark or a "normal" quark?

multiplet of Higgs scalar fields and assume that these fields have the potential (16.29) with the parameter μ^2 assumed to have a negative value. This does explain the breaking of $SU(2) \times U(1)$ symmetry, but this explanation seems to be rather *ad hoc*.

In Section 14.2, we discussed the spontaneous breaking of the chiral $SU(2) \times SU(2)$ symmetry of QCD. We gave an intuitive explanation for the symmetry-breaking based on the properties of light quarks with strong attractive interactions. That explanation was modelled on the well-understood explanation for superconductivity in metals put forward by Bardeen, Cooper, and Schrieffer. In my presentation of the Higgs mechanism in Section 16.2, I emphasized that the original papers of Higgs, Englert and Brout, and Guralnik, Hagen, and Kibble all referred to the theory of superconductivity as the inspiration for their proposals. There are many other condensed matter system with spontaneously broken symmetries, and, in each case, there is a definite physical explanation for the fact that an asymmetric ground state of the Hamiltonian has the lowest energy. These explanations differ for superconductors, superfluids, magnets, liquid crystals, and other systems presented, for example, in (Sethna 2006). But, in each case, there is fascinating physics there. Why should this not be true also for the large physical system that we call the universe?

Is there a physics explanation for the spontaneous breaking of the weak interaction $SU(2) \times U(1)$ symmetry?

Many theories have been put forward to provide an underlying explanation for the shape of the potential energy of the Higgs field and its preference for an asymmetric vacuum state. I have reviewed this subject recently in (Peskin 2016). The models proposed are of many different types, some relying on weak interactions and special symmetries of the underlying theory, some on new strong interactions at short distances and, possibly, composite structure of the Higgs bosons and the fermions. The one feature that these models have in common is that the Standard Model is not enough. New particles and new interactions must be added to it.

Any such explanation requires new fundamental forces that have not yet been discovered.

I personally find it fascinating that, if one dismisses the Standard Model Higgs potential as too simplistic and looks for a physics-based explanation for the symmetry-breaking of $SU(2) \times U(1)$, it is unavoidable that there exist new fundamental interactions still to be discovered at higher energies or shorter distances.

Next, I will discuss the relation of the Standard Model to the picture of the universe revealed by astrophysical observations.

This book has focussed almost completely on particle physics proper, without even glances at the implications of particle physics for astrophysics and cosmology. In truth, though, the subjects of particle physics and cosmology are closely related. We have strong evidence from astrophysics that the universe originated in a state of very high temperature, the "Big Bang". At this temperature, all of the particles of the Standard Model would have been created by thermal pair production and would have been present in large numbers. As the universe cooled, particles and antiparticles would have annihilated, leaving us with the universe with empty space and clumps of stable matter that we see today. In principle, we should be able to understand the composition of the universe and the growth of cosmic structure such as galaxies and galaxy clusters by taking the high-temperature state as the initial condition and evolving from that point using the equations of the Standard Model.

An excellent introduction to cosmology that emphasizes its connection to fundamental particle physics is (Dodelson 2003). As this reference emphasizes, the same period since the 1970's in which the Standard Model was validated in particle experiments showed tremendous improvements in our observational knowledge about the universe. The interpretation of these observations might have been another triumph for the Standard Model. But, in fact, the comparison of the current universe to that expected from the Standard Model alone has revealed essential gaps and highlighted additional ingredients that need to be included. Thus, when we are interested in the defects of the Standard Model rather than its successes, we need to look to astrophysics for important evidence of what is missing. Thus, we turn to the questions:

Why is the universe not uniform, but, rather, full of structure? If the universe were born in a uniform thermal state, it would stay uniform, and there would be no galaxies, galaxy clusters, or other cosmic structures. To grow these structures, we need, first, seeds given by small density inhomogeneities, and, second, a mechanism for these inhomogeneities to grow as the universe evolves. The growth of structure can be accomplished by gravity, with small excesses of matter attracting additional matter and growing into large density excesses. But how did the small excess arise in the first place? The snapshot of the early universe provided by the cosmic microwave background radiation tells us that the original density inhomogeneities had some specific special properties: Their statistical distribution was close to Gaussian, approximately scale-invariant, and essentially identical in widely separated parts of the universe that were not causally connected. In particular, the temperature of the cosmic microwave background is observed to be the same at opposite poles of the sky, even though this radiation was created at 100,000 years after the Big Bang, when these regions were separated by 27 billion light-years.

In 1981, Alan Guth proposed an explanation of these features from a model called the *inflationary universe* (Guth 1981). In this picture, the universe began its evolution containing a scalar field with a very large positive value of its potential energy. The coupling of this scalar field to gravity leads to an exponential expansion of every small patch of the universe. In the inflationary model, a patch of a few cm in size expands to the size of the current universe. Inflation is terminated by the transition of the scalar field to its ground state with much lower potential energy. The conversion of the original potential energy into heat provides the thermal energy of the Big Bang. A more detailed review of inflation and its solutions to the problems listed in the previous paragraph can be found in (Olive 1990) and in (Dodelson 2003).

So, already, in order to create the correct initial conditions for the universe, we need to postulate at least one additional scalar field that is

The application of the Standard Model to the physics of the universe has revealed gaps that the model does not explain.

Why did large scale structure—galaxies and clusters of galaxies—form in the early universe?

The "inflationary universe" can explain at the qualitative properties of the large-scale structure of the universe.

not contained in the Standard Model.

Why does the universe contain more matter than antimatter? An obvious property of the observed universe is that it is full of matter (protons, neutrons, and electrons), with very little antimatter. In principle, the universe could have begun with an initial small excess of matter over antimatter. When quarks and antiquarks annihilated as the temperature of the universe fell below 1 GeV, the excess quarks would have been left over. An initial excess of only 1 part in 10^{10} is needed.

However, if we accept the idea that the initial conditions of the universe came from a period of inflation, this explanation cannot be valid. The dramatic expansion required by inflation emptied the universe of particles and set the initial matter-antimatter asymmetry to zero. Then the needed asymmetry must have developed in the evolution of the universe after the Big Bang.

In principle, we can compute the evolution of the components of the universe from the equations of the Standard Model. In particular, in order to create a nonzero asymmetry between the numbers densities of matter and antimatter, these equations must be asymmetric between matter and antimatter, violating CP symmetry. The Standard Model contains a CP-violating parameter, the CKM phase, and this does produce a matter-antimatter asymmetry in the early universe. However, it turns out that this asymmetry is too small by a factor of 10^8 to produce today's known matter density of the universe. The influence of the CKM phase in the early universe is proportional to the product of the light quark Yukawa couplings, and so is very small. Then, another source of CP-violation is needed (Riotto and Trodden 1999).

The excess of matter over antimatter in the universe requires new CP-violating interactions not yet observed in particle physics.

Looking at models more general than the Standard Model, there are many possibilities for new CP-violating interactions. A model with two Higgs field multiplets can contain an additional CP-violating phase. There are many more complex possibilities. Most of the models discussed under the previous question as providing explanations of $SU(2) \times U(1)$ symmetry breaking offer the possibility of new CP-violating parameters. Unfortunately, these models also allow other new flavor-dependent terms, threatening some of the beautiful conclusions presented in Section 18.4.

One interesting suggestion is that the new source of CP violation is the Majorana mass (20.24) for right-handed neutrinos (Fukugita and Yanagida 1986). This neutrino mechanism for the production of a baryon asymmetry is called *leptogenesis*. I have emphasized at the end of Chapter 20 that the presence of CP violation in the neutrino sector is highly suggested from the structure of the Standard Model but is not yet experimentally established. Unfortunately, the CP-violating phase that would lead to leptogenesis is not directly observable in neutrino mixing experiments.

What is the "dark matter" of the universe? Another aspect of modern cosmology that challenges the Standard Model is the accumulating evidence that atoms made of Standard Model particles are not the only type of matter in the universe. In fact, the internal dynamics

Most of the matter in the universe is "dark matter", a type of matter not accounted for by the Standard Model.

of galaxies and of clusters of galaxies require that these objects contain large amounts of invisible and weakly interacting matter that interacts gravitationally with the atoms. These observations are corroborated by observations of the cosmic microwave background. In addition, we now know that the growth of structure in the universe since the Big Bang would be too slow if the gravitational clumping of matter were driven by the gravity of atomic matter only. From all of these sources, we deduce that 85% of the non-relativistically moving matter in the universe is of this new type, called *dark matter* (Bertone 2010).

Dark matter must be made of particles, but those particles cannot be any of the particles in the Standard Model. For particle physicists, this is a supreme embarassment: With all of our knowledge, we are ignorant of the origin of most of the matter in the universe.

Models proposed to extend the Standard Model provide many possible candidates for the particle of dark matter. Each proposal offers new experiments that might discover the dark matter particle, and, at the same time, give evidence for new particles and forces beyond those of the Standard Model (Feng 2010). It is possible to search for dark matter particles not only by production at accelerators but also by searching for collisions of cosmic dark matter particles with detectors on Earth. Unfortunately, so far, none of these experiments has given a positive signal. Still, the dark matter is there in our astrophysical observations; only its identity is missing.

What is the "dark energy" of the universe? In 1998, another mysterious ingredient was added to this picture. From measurements of the red shifts of distant supernovae, two groups of observers demonstrated that the universe is in a phase of exponential expansion even today (Riess *et al.* 1998, Perlmutter *et al.* 1999). This expansion would be accounted for by a small potential energy in each unit volume of empty space, due either to another new scalar field or to quantum effects from known and unknown elementary particles. This ingredient is called *dark energy*. A contemporary summary of the resulting three-part picture of the universe, with two mysterious ingredients, can be found in (Bahcall *et al.* 1999).

The universe contains a small but nonzero vacuum energy. Its origin is unknown, and its size defies all attempts to predict it.

In principle, quantum effects of the Standard Model can lead to a vacuum energy that accounts for the dark energy. However, it is not known how to compute the Standard Model contribution to the energy of the vacuum state of space. The application of obvious methods leads to a result 120 orders of magnitude too large. It is also mysterious why the energy density of the vacuum today can be so much smaller—by a similar number of orders of magnitude—than the energy density present during the period of inflation.

It is important the students of particle physics should be aware of this compelling evidence from astrophysics that our current fundamental understanding of nature is incomplete. At the same time, it is important that both particle physicists and astrophysicists realize that the answers to these questions cannot come purely from astronomical observations. At some point, the initial conditions from astrophysics must be put together with a full dynamical model—at the level of the underlying particles and fields—to explain the evolution of the universe to its current state.

Finally, we come to the questions about the relation of the Standard Model of particle physics to deep questions about the nature of space and time:

Are there higher symmetries of nature that lead to new particles and interactions? Many of the questions that we have already considered—in particular, the missing explanations for $SU(2) \times U(1)$ symmetry breaking, the generation of the observed matter-antimatter asymmetry, and the dark matter of the universe—call for new particles and interactions that must be added to the Standard Model. It is a very attractive idea that these new interactions might have a fundamental basis. Perhaps, by extending the space-time symmetries of the Standard Model, these new ingredients might naturally appear.

There is in fact a unique extension of the group of space-time symmetries, the translations, rotations, and Lorentz boosts. This extended group adds *supersymmetries*, operations that change the total spin by $\frac{1}{2}$ unit, transforming bosons into fermions and fermions into bosons. The square of a supersymmetry operation is an infinitesimal translation (Haag, Lopszanski, and Sohnius 1975).

"Supersymmetry" is the most natural extension of the Lorentz symmetry of space-time. But, is supersymmetry actually present in nature?

Supersymmetry doubles the number of elementary particles. For each known boson, it predicts a new fermion. For each quark or lepton, it predicts a new boson. This gives ample material to propose solutions to all of the questions that I have outlined above. In particular, the supersymmetric partner of the photon is a fermion with zero charge and very weak interactions that is an excellent candidate for the particle of dark matter. The extension of the Standard Model to a supersymmetric theory has been worked out in great detail. Descriptions of this theory can be found in the review articles (Martin 1997) and (Peskin 2008) and in the books (Drees, Godbole, and Roy 2004) and (Baer and Tata 2006).

The supersymmetric extension of the Standard Model gives the most robust explanations for $SU(2) \times U(1)$ symmetry breaking and dark matter if the supersymmetry partners of the gluon and the top quark have masses of several hundred GeV. The ATLAS and CMS experiments at the LHC have searched diligently for these particles and have now essentially ruled out the possibility that these particles have masses below 1000 GeV (Adam 2017). That was a signficant blow to proponents of supersymmetric models. The search for supersymmetric particles continues, but models of this type no longer have the pride of place that they held before the start of the LHC experiments.

It is possible that nature is supersymmetric at extremely short distances, far above the TeV energies accessible to the LHC. This possibility plays into the answers to the questions to follow.

How does gravity fit together with the Standard Model? A truly final theory should incorporate all of the known forces of nature, including gravity. The description of gravity at the classical level is given by Einstein's theory of general relativity. This is a very well-tested the-

352 *Epilogue*

The quantum theory of gravity is incomplete. How is gravity related to the interactions of the Standard Model?

"String theory" is a possible framework for unifying gravity with the Standard Model. Does this idea have unique observable consequences?

Properties of black holes challenge the basic notion of continuous space-time.

ory. General relativity has a quantum version. In that theory, the gravitational force is mediated by a massless spin-2 particle, the graviton. This quantum theory of gravity has a formalism of Feynman diagrams, similar to those of the Standard Model, from which scattering amplitudes can be computed. From this point of view, the quantum theory of gravity can simply be appended as another component of the Standard Model, though its unification with the other forces is not explained.

There, is, however, a more serious problem with quantum gravity. The quantum theory of gravity differs from the gauge theories of the Standard Model in having a dimensionful coupling constant, Newton's constant G_N. Expressed as an energy, Newton's constant gives a mass parameter, the Planck scale m_{Pl}, equal to 10^{19} GeV. At energies below m_{Pl}, the Feynman diagram expansion makes sense, but at energies approaching m_{Pl} the theory becomes strongly coupled and our methods of calculation fail catastrophically. Speaking roughly, Einstein's theory predicts that, at distances of $1/m_{Pl}$, space-time itself becomes singular due to quantum fluctuations.

Solutions to this problem have been proposed, but none are yet completely successful. The most interesting of these is *string theory*, which discards the notion of point particles moving in continuous space-time in favor of a picture in which all elementary particles, including the graviton, are extended 1-dimensional objects (Zwiebach 2004, Polchinski 2005). For reasons too subtle to explain here, string theory removes all distances shorter than $1/m_{Pl}$ while at the same time retaining continous translation and Lorentz symmetries. In addition to the massless spin-2 graviton, string theory contains massless spin-1 particles with the properties of Yang-Mills gauge bosons. Thus, string theory can be the setting for a complete unified theory of all of the forces of nature. However, it has not yet been possible to identify the Standard Model gauge symmetry group as a unique consequence of string theory, or to use string theory to give definitive solutions to the other questions that I have posed in this chapter. Some approaches to these issues involve extending string theory to a supersymmetric model, bringing in all of the virtues and challenges described for supersymmetry in the discussion above.

Is space-time a fundamental concept that will survive in the final theory? The difficulties of formulating a quantum theory of gravity valid at all energies suggest the idea that continuous space-time itself is an approximate notion that will be replaced in a more fundamental theory. At currently explored energies, up to the energies probed by the LHC, Lorentz invariance, the continuity of space, and the locality of quantum field theory interactions are all extremely well tested (Kostelecky and Russell 2011). But it is a long way from TeV energies to the Planck scale. Many surprises and new concepts might make themselves apparent between here and there.

Even now, theoretical investigations of black holes have challenged the idea that space-time is ultimately continuous. The quantum theory of black holes is partially understood, and what we know leads to paradoxes. These issues are reviewed lucidly in (Polchinski 2016). Among these is the fact that the number of quantum degrees of freedom of a black hole is proportional to its area, not its volume, so a large black hole has many fewer degrees of freedom than one might expect in a description based on quantum field theory. These ideas have led to the *holographic principle*, the idea that the quantum degrees of freedom in any 3-dimensional volume are encoded on a 2-dimensional boundary. This idea is fundamentally incompatible with quantum field theory except as an approximation at energies much lower than the Planck scale (Bousso 2002).

An idea pushing in the other direction is that of the existence of additional dimensions of space beyond those seen in our common experience. These extra dimensions would need to be curled up to a small size R. Then particle physics experiments at energies of order $1/R$ could possibly access them (Hewett and Spiropulu 2002). String theory actually requires the existence of extra space dimensions and, in a sense, blurs the distinction between the presence of dimensions of space and the presence of extra quantum fields added to the Lagrangian.

> Are there extra dimensions of space visible at very short distances?

It is fun to think about such dramatic modifications of our ideas of space-time. It may be that new concepts drawn from these ideas are needed to address the open questions of particle physics. On the other hand, it is equally likely that the most pressing of problems that we have highlighted—$SU(2) \times U(1)$ symmetry breaking, CP violation, and the nature of dark matter—will be solved with new interactions compatible with quantum field theory that are present at energies just beyond our current reach. There are many aspects of quantum field theory that we do not yet understand, especially for theories that are strongly coupled. New concepts as profound and unusual as asymptotic freedom and quark confinement could drive the behavior of these new interactions. Just as was true for the insights that led to the structure of QCD and the electroweak theory, it may take new accelerator experiments at higher energies to bring them to the surface.

> New ideas are needed to answer all of these questions. Can you discover them?

The completion of the Standard Model with the discovery of the Higgs boson closes this book on the concepts of elementary particle physics. However, many question remain, enough to fill another book, perhaps many more. Today, we have few clues to address the next level of questions about the fundamental interactions of physics. But, always, we build our knowledge level by level. I hope that the concepts I have presented in this book will aid you in confronting this new set of open questions and, by discovering new principles, in pushing forward the quest for an ultimate understanding of nature.

Notation

Units

Throughout this book, I use *natural units* in which

$$\hbar = c = 1 . \tag{A.1}$$

All masses and momenta are measured in energy units, typically in GeV. Distances and times are computed in inverse energy units (GeV^{-1}). For example, I use the symbol m_e (the mass of the electron) to represent all of the following quantities:

$$m_e = 9.10938 \times 10^{-28} \text{ g} = 0.510999 \text{ MeV}$$
$$= (3.86156 \times 10^{-11} \text{ cm})^{-1} = (1.28809 \times 10^{-21} \text{ sec})^{-1} . \tag{A.2}$$

The conversion of quantities in natural units to quantities in more familiar units is discussed in Section 2.2.

Vectors and tensors

Vectors in 3 dimensions are notated with arrows or Latin indices. Vectors in Minkowski space are denoted with Greek indices. Thus,

$$x^\mu = (x^0, x^i)^\mu \text{ or } (x^0, \vec{x})^\mu . \tag{A.3}$$

Sometimes, I write $x^0 = t$ or, for momenta, $p^0 = E$. Distances and momenta naturally carry raised indices. Greek indices are raised and lowered with the metric tensor of special relativity

$$\eta^{\mu\nu} = \eta_{\mu\nu} = \begin{pmatrix} 1 & 0 & 0 & 0 \\ 0 & -1 & 0 & 0 \\ 0 & 0 & -1 & 0 \\ 0 & 0 & 0 & -1 \end{pmatrix} . \tag{A.4}$$

Then the Lorentz-invariant product of vectors is written

$$x \cdot y = x^\mu \eta_{\mu\nu} y^\nu = x^\mu y_\mu = x^0 y^0 - x^i y^i = x^0 y^0 - \vec{x} \cdot \vec{y} . \tag{A.5}$$

The Lorentz-invariant interval is

$$(x - y)^2 = (x^0 - y^0)^2 - (\vec{x} - \vec{y})^2 . \tag{A.6}$$

The derivative operator naturally carries a lowered index,

$$\partial_\mu = (\partial/\partial x^0, \partial/\partial x^i) , \tag{A.7}$$

This is set up so that

$$x^\mu \partial_\mu \tag{A.8}$$

is Lorentz-invariant. Note that

$$\partial^\mu = \eta^{\mu\nu} \partial_\nu = (\partial/\partial x^0, -\partial/\partial x^i) . \tag{A.9}$$

The D'Alembertian operator

$$\partial^\mu \partial_\mu = \partial^2/\partial(x^0)^2 - \partial^2/\partial(x^i)^2 = \partial^2/\partial t^2 - (\vec{\nabla})^2 \tag{A.10}$$

is Lorentz-invariant.

The totally antisymmetric symbols ϵ_{ab}, ϵ^{ijk}, $\epsilon^{\mu\nu\lambda\sigma}$ satisfy

$$\epsilon_{12} = +1 , \quad \epsilon^{123} = +1 , \quad \epsilon^{0123} = +1 \tag{A.11}$$

Note that this implies $\epsilon_{0123} = -1$ after index lowering.

Momentum vectors

A particle of mass m has a momentum vector satisfying

$$p^2 = (p^0)^2 - (\vec{p})^2 = m^2 . \tag{A.12}$$

The quantity E_p is defined in the text to be a function of $\vec{p}$ equal to

$$E_p = [(\vec{p})^2 + m^2]^{1/2} ; \tag{A.13}$$

it is the energy of a particle (on mass shell) with momentum $\vec{p}$.

Basic quantum-mechanical operators

I write the energy and momentum operators acting on Schrödinger wavefunctions as

$$E = i\frac{\partial}{\partial t} , \qquad \vec{p} = -i\frac{\partial}{\partial \vec{x}} = -i\vec{\nabla} . \tag{A.14}$$

Note that, with (A.9), these combine into a 4-vector operator

$$p^\mu = i\partial^\mu . \tag{A.15}$$

The plane wave with 4-momentum k has wavefunction $e^{-ik\cdot x}$, since

$$p^\mu \, e^{-ik\cdot x} = i\partial^\mu \, e^{-ik\cdot x} = k^\mu \, e^{-ik\cdot x} . \tag{A.16}$$

The Pauli sigma matrices are

$$\sigma^1 = \begin{pmatrix} 0 & 1 \\ 1 & 0 \end{pmatrix} , \quad \sigma^2 = \begin{pmatrix} 0 & -i \\ i & 0 \end{pmatrix} , \quad \sigma^3 = \begin{pmatrix} 1 & 0 \\ 0 & -1 \end{pmatrix} . \tag{A.17}$$

These satisfy

$$(\sigma^i)^2 = 1 \qquad \sigma^i \sigma^j = \epsilon^{ijk} \sigma^k . \tag{A.18}$$

Fourier transforms and distributions

The Dirac delta function in d dimensions is denoted $\delta^{(d)}(x)$. This is a distribution that equal zero for $x \neq 0$ and satisfies

$$\int d^d x \, \delta^{(d)}(x) = 1 . \tag{A.19}$$

In writing Fourier transforms, I always associate a factor 2π with the momentum integral. In Minkowski space,

$$f(x) = \int \frac{d^4 k}{(2\pi)^4} e^{-ik \cdot x} \tilde{f}(k) \qquad \tilde{f}(k) = \int d^4 x \, e^{+ik \cdot x} f(x) \tag{A.20}$$

Factors of 2π will also appear from

$$\int d^4 x \, e^{ik \cdot x} = (2\pi)^4 \delta^{(4)}(x) . \tag{A.21}$$

Electrodynamics

I use standard SI notation for electrodynamics but with $\epsilon_0 = \mu_0 = 1$ in natural units. The constant e is a positive number, equal to the electric charge of the proton. In general, I write the electric charge of a particle as Qe, with $Q = -1$ for an electron and $Q = +1$ for a proton.

The Coulomb potential of a point charge is written

$$V(r) = \frac{Qe}{4\pi r} . \tag{A.22}$$

The electrodynamic potential and the vector potential form a 4-vector

$$A^\mu(x) = (-\Phi(x), -\vec{A}(x))^\mu . \tag{A.23}$$

The electromagnetic fields are contained in the tensor

$$F_{\mu\nu} = \partial_\mu A_\nu - \partial_\nu A_\mu . \tag{A.24}$$

It is a nice exercise to show that the components of this tensor form the correct expressions for the $\vec{E}$ and $\vec{B}$ fields,

$$F^{0i} = E^i , \qquad F^{ij} = \epsilon^{ijk} B^k . \tag{A.25}$$

Conversion factors and physical constants

Conversion factors

$$c \quad = 2.99729 \times 10^8 \text{ m/sec}$$
$$c^{-2} \quad = 1.78266 \times 10^{-30} \text{ kg/MeV}$$
$$\hbar \quad = 6.582119 \times 10^{-22} \text{ MeV-sec}$$
$$\hbar c \quad = 197.327 \text{ MeV-fm}$$

(1 fm $= 10^{-15}$ m; 1 barn $= 10^{-24}$ cm^2.)

$$\alpha = e^2/4\pi\epsilon_0\hbar c \qquad = 1/137.03560$$
$$r_e = e^2/4\pi\epsilon_0 m_e c^2 \qquad = 2.817940 \times 10^{-15} \text{ m}$$
$$Ry = e^4 m_e/2(4\pi\epsilon_0)^2\hbar^2 \ = 13.6057 \text{ eV}$$

Standard magnetic moments and cyclotron frequencies

Use these relations to convert from MeV and sec to magnetic field strengths.

$$\mu_B = e\hbar/2m_e = 5.78838 \times 10^{-11} \text{ MeV/T}$$
$$\omega_e/B = e/m_e = 1.75882 \times 10^{11} \text{ /sec/T}$$
$$\mu_N = e\hbar/2m_p = 3.15245 \times 10^{-14} \text{ MeV/T}$$
$$\omega_p/B = e/m_p = 9.5788 \times 10^7 \text{ /sec/T}$$

(1 Tesla $= 10^4$ gauss)

Masses of leptons

$$m(e) \quad = 0.510999 \text{ MeV}$$
$$m(\mu) \quad = 105.658 \text{ MeV}$$
$$m(\tau) \quad = 1776.9 \text{ MeV}$$

Masses of baryons

$$
\begin{array}{lll}
m(p) & = 938.272 \text{ MeV} & m(n) \quad = 939.565 \text{MeV} & m(\Lambda) \quad = 1115.68 \text{ MeV} \\
m(\Sigma^+) = 1189.37 \text{ MeV} & m(\Sigma^0) \quad = 1192.64 \text{ MeV} & m(\Sigma^-) \; = 1197.45 \text{ MeV} \\
m(\Xi^0) \; = 1314.86 \text{ MeV} & m(\Xi^-) \quad = 1321.71 \text{ MeV} \\
m(\Delta) \; = 1232. \text{ MeV} & m(\Omega^-) \; = 1672.5 \text{ MeV} \\
m(\Lambda_c) \; = 2286.5 \text{ MeV} & m(\Lambda_b) \quad = 5619.5 \text{ MeV}
\end{array}
$$

Masses of mesons

$$
\begin{array}{lll}
m(\pi^+) \; = 139.570 \text{ MeV} & m(\pi^0) \quad = 134.977 \text{ MeV} \\
m(K^+) = 493.68 \text{ MeV} & m(K^0) \quad = 497.61 \text{ MeV} \\
m(\eta) \quad = 547.86 \text{ MeV} & m(\eta') \quad = 957.78 \text{ MeV} \\
m(\rho^+) \; = 775.11 \text{ MeV} & m(\rho^0) \quad = 775.26 \text{ MeV} \\
m(K^{*+}) = 891.6 \text{ MeV} & m(K^{*0}) = 895.8 \text{ MeV} \\
m(\omega) \; = 782.66 \text{ MeV} & m(\phi) \quad = 1019.46 \text{ MeV} \\
m(D^+) = 1869.6 \text{ MeV} & m(D^0) \quad = 1864.8 \text{ MeV} & m(D_s^+) \; = 1968.2 \text{ MeV} \\
m(\eta_c) \; = 2983.4 \text{ MeV} & m(J/\psi) = 3096.9 \text{ MeV} & m(\psi') \quad = 3686.1 \text{ MeV} \\
m(B^+) = 5279.3 \text{ MeV} & m(B^0) \quad = 5279.6 \text{ MeV} & m(B_s^0) \; = 5366.8 \text{ MeV} \\
m(\eta_b) \; = 9399.0 \text{ MeV} & m(\Upsilon) \quad = 9460.3 \text{ MeV} & m(\Upsilon') \; = 10023.3 \text{ Me}
\end{array}
$$

Masses of weak-interaction bosons

$$
m(W) = 80.385 \text{ GeV} \quad m(Z) = 91.1876 \text{ GeV}
$$

Strengths of the fundamental interactions at $Q = 91.$ GeV

$$
\begin{array}{ll}
\alpha = e^2/4\pi \; = 1/129 & \alpha_s = g_s^2/4\pi = 1/8.5 \\
\alpha_w = g^2/4\pi = 1/29.8 & \alpha' = g'^2/4\pi = 1/99.1
\end{array}
$$

All quantities in this appendix except for the final values of fundamental interaction strengths are taken from the summary tables in (Patrignani *et al.* 2016).

Formulae for the creation and destruction of elementary particles

Spin 0

$$\langle 0|\,\phi(x)\,|\varphi(p)\rangle = e^{-ip\cdot x}\ , \qquad \langle\varphi(p)|\,\phi(x)\,|0\rangle = e^{+ip\cdot x}\ . \qquad (\text{C.1})$$

Spin 1/2

massive fermions:

$$\langle 0|\,\Psi(x)\,|f^s(p)\rangle = U^s(p)e^{-ip\cdot x}\ , \qquad \langle f^s(p)|\,\overline{\Psi}(x)\,|0\rangle = \overline{U}^s(p)e^{+ip\cdot x}\ ,$$
$$\langle 0|\,\overline{\Psi}(x)\,\big|\overline{f}^s(p)\big\rangle = \overline{V}^s(p)e^{-ip\cdot x}\ , \qquad \big\langle\overline{f}^s(p)\big|\,\Psi(x)\,|0\rangle = V^s(p)e^{+ip\cdot x}\ ,$$
$$(\text{C.2})$$

where $\Psi(x)$ is a 4-component spinor field and $U^s(p)$, $V^s(p)$ are 4-component spinors, with s indicating the spin direction.

massless, chiral fermions:

$$\langle 0|\,\psi_L(x)\,|f_L(p)\rangle = u_L(p)e^{-ip\cdot x}\ , \qquad \langle f_L(p)|\,\psi_L^\dagger(x)\,|0\rangle = u_L^\dagger(p)e^{+ip\cdot x}\ ,$$
$$\langle 0|\,\psi_L^\dagger(x)\,\big|\overline{f}_R(p)\big\rangle = v_R(p)e^{-ip\cdot x}\ , \qquad \big\langle\overline{f}_R(p)\big|\,\psi_L^\dagger(x)\,|0\rangle = v_R(p)e^{+ip\cdot x}\ ,$$
$$\langle 0|\,\psi_R(x)\,|f_R(p)\rangle = u_R(p)e^{-ip\cdot x}\ , \qquad \langle f_R(p)|\,\psi_R^\dagger(x)\,|0\rangle = u_R^\dagger p)e^{+ip\cdot x}\ ,$$
$$\langle 0|\,\psi_L(x)\,\big|\overline{f}_R p)\big\rangle = v_L^\dagger(p)e^{-ip\cdot x}\ , \qquad \big\langle\overline{f}_L(p)\big|\,\psi_R^\dagger(x)\,|0\rangle = v_L(p)e^{+ip\cdot x}\ ,$$
$$(\text{C.3})$$

where $\psi_L(x)$, $\psi_R(x)$ are a 2-component spinor fields and $u_L(p)$, $u_R(p)$, $v_L(p)$, and $v_R(p)$ are 2-component spinors. For $\vec{p}\parallel\hat{3}$,

$$u_L(p) = \sqrt{2E}\begin{pmatrix}0\\1\end{pmatrix} = v_R(p)\ ,$$
$$u_R(p) = \sqrt{2E}\begin{pmatrix}1\\0\end{pmatrix} = v_L(p)\ . \qquad (\text{C.4})$$

For fermions moving in the direction $\hat{p} = \cos\theta\hat{3} + \sin\theta\hat{1}$,

$$u_L(p) = \sqrt{2E}\begin{pmatrix} -\sin\theta/2 \\ \cos\theta/2 \end{pmatrix} = v_R(p)$$

$$u_R(p) = \sqrt{2E}\begin{pmatrix} \cos\theta/2 \\ \sin\theta/2 \end{pmatrix} = v_L(p) \ . \tag{C.5}$$

In particular, for fermions moving in the $-\hat{3}$ direction,

$$u_L(p) = \sqrt{2E}\begin{pmatrix} -1 \\ 0 \end{pmatrix} = v_R(p)$$

$$u_R(p) = \sqrt{2E}\begin{pmatrix} 0 \\ 1 \end{pmatrix} = v_L(p) \ . \tag{C.6}$$

Spin 1

$$\langle 0|\, A^\mu(x)\, |V_s(p)\rangle = \epsilon_s^\mu(p)e^{-ip\cdot x} \ , \qquad \langle V_s(p)|\, A^\mu(x)\, |0\rangle = \epsilon_s^{*\mu}(p)e^{+ip\cdot x} \ , \tag{C.7}$$

where s indicates the spin direction. The vectors $\epsilon_s^\mu(p)$ satisfy

$$p\cdot\epsilon_s = 0 \ . \tag{C.8}$$

For vector bosons moving in the $\hat{3}$ direction, these vectors are

$$\epsilon_+^\mu = \frac{1}{\sqrt{2}}(0,1,i,0)^\mu \ , \qquad \epsilon_-^\mu = \frac{1}{\sqrt{2}}(0,1,-i,0)^\mu \ ,$$

$$\epsilon_0^\mu = (\frac{p}{m},0,0,\frac{E}{m})^\mu \ . \tag{C.9}$$

For vector bosons moving in the direction $\hat{p} = \cos\theta\hat{3} + \sin\theta\hat{1}$, these vectors are

$$\epsilon_+^\mu = \frac{1}{\sqrt{2}}(0,\cos\theta,i,-\sin\theta)^\mu \ , \qquad \epsilon_-^\mu = \frac{1}{\sqrt{2}}(0,\cos\theta,-i,-\sin\theta)^\mu \ ,$$

$$\epsilon_0^\mu = (\frac{p}{m},\frac{E}{m}\sin\theta,0,\frac{E}{m}\cos\theta)^\mu \ . \tag{C.10}$$

For massless vector bosons (*e.g.*, photons), the polarization state ϵ_0^μ is absent, and only the polarizations $\epsilon_\pm^\mu$ correspond to physical states.

Master formulae for the computation of cross sections and partial widths

D

Partial widths

For decays of a particle X of mass m_X,

$$\Gamma(X \to A_1 + \cdots + A_n) = \frac{1}{2m_X} \int d\Pi_n \, |\mathcal{M}(X \to A_1 + \cdots + A_n)|^2 \, , \quad \text{(D.1)}$$

summed over final spins and, when appropriate, averaged over the spin of X. $d\Pi_n$ is the integral over phase space, below.

Cross sections

For the cross section of a reactions of particles A and B with initial energies E_A, E_B and velocities v_A, v_B,

$$\sigma(A + B \to C_1 + \cdots + C_n)$$
$$= \frac{1}{E_A E_B |v_A - v_B|} \int d\Pi_n \, |\mathcal{M}(A + B \to C_1 + \cdots + C_n)|^2 \, ,$$
$$\text{(D.2)}$$

summed over final spins and, when appropriate, averaged over the spins of A and B.

Phase space

For n-body phase space,

$$\int d\Pi_n = \prod_{i=1}^{n} \int \frac{d^3 p_i}{(2\pi)^3 2E_i} \, (2\pi)^4 \delta^{(4)} \left(P_{CM} - \sum_i p_i \right) \quad \text{(D.3)}$$

2-body phase space

For a system with center of mass energy E_{CM}, in the center of mass frame, 2-body phase space takes the form

$$\int d\Pi_2 = \frac{1}{8\pi} \left(\frac{2k}{E_{CM}} \right) \int \frac{d\Omega}{4\pi} \, , \quad \text{(D.4)}$$

where k is the momentum of each of the two products and $d\Omega$ is the integral over their angular distribution.

3-body phase space

For a system with center of mass energy $E_{CM}^2 = s$ and energies of the three products in the CM frame given by E_1, E_2, E_3, let

$$x_1 = \frac{2E_1}{\sqrt{s}} \,, \quad x_2 = \frac{2E_2}{\sqrt{s}} \,, \quad x_3 = \frac{2E_3}{\sqrt{s}} \,. \tag{D.5}$$

In this frame $\vec{p}_1 + \vec{p}_2 + \vec{p}_3 = 0$. The three final momentum vectors lie in a plane. Then 3-body phase space, integrated over the orientation of this plane, takes the form

$$\int d\Pi_3 = \frac{s}{128\pi^3} \int dx_1 \, dx_2 \,. \tag{D.6}$$

Since $x_1 + x_2 + x_3 = 2$, any pair of x_i can be used as the integration variables.

The masses of the 2-particle combinations, for example, $m_{12}^2 = (p_1 + p_2)^2$, are given by

$$m_{12}^2 = s(1 - x_3) + m_3^2 \,, \quad m_{23}^2 = s(1 - x_1) + m_1^2 \,, \quad m_{31}^2 = s(1 - x_2) + m_2^2 \,, \tag{D.7}$$

so (D.6) can also be written as

$$\int d\Pi_3 = \frac{1}{128\pi^3 s} \int dm_{12}^2 \, dm_{23}^2 \,. \tag{D.8}$$

As in (D.6), any two of m_{12}^2, m_{23}^2, m_{31}^2 can be used as integration variables.

QCD formulae for hadron collisions

Parton model formula for cross sections

$$\sigma(pp \to X) = \int_0^1 dx_1 \, dx_2 \sum_{f_1, f_2} f_{f_1}(x_1, Q) \, f_{f_2}(x_2, Q) \, \sigma(f_1 f_2 \to X)$$

$$(\text{E.1})$$

Altarelli-Parisi splitting functions

$$P_{g \leftarrow q}(z) = \frac{4}{3} \frac{1 + (1 - z)^2}{z}$$

$$P_{q \leftarrow q}(z) = \frac{4}{3} \Big[\frac{1 + z^2}{(1 - z)} + A\delta(z - 1) \Big]$$

$$P_{q \leftarrow g}(z) = \frac{1}{2}\big(z^2 + (1 - z)^2\big)$$

$$P_{g \leftarrow g}(z) = 3 \Big[\frac{1 + z^4 + (1 - z)^4}{z(1 - z)} + B\delta(z - 1) \Big] \qquad (\text{E.2})$$

Differential cross sections for parton-parton scattering

$$\frac{d\sigma}{d\cos\theta_*}(ud \to ud) = \frac{2\pi\alpha_s^2}{9s} \Big[\frac{s^2 + u^2}{t^2} \Big]$$

$$\frac{d\sigma}{d\cos\theta_*}(uu \to uu) = \frac{2\pi\alpha_s^2}{9s} \Big[\frac{s^2 + u^2}{t^2} + \frac{s^2 + t^2}{u^2} - \frac{2}{3}\frac{s^2}{tu} \Big]$$

$$\frac{d\sigma}{d\cos\theta_*}(u\bar{u} \to d\bar{d}) = \frac{2\pi\alpha_s^2}{9s} \Big[\frac{t^2 + u^2}{s^2} \Big]$$

$$\frac{d\sigma}{d\cos\theta_*}(u\bar{u} \to u\bar{u}) = \frac{2\pi\alpha_s^2}{9s} \Big[\frac{s^2 + u^2}{t^2} + \frac{t^2 + u^2}{s^2} - \frac{2}{3}\frac{u^2}{st} \Big]$$

$$\frac{d\sigma}{d\cos\theta_*}(u\bar{u} \to gg) = \frac{16\pi\alpha_s^2}{27s} \Big[\frac{u}{t} + \frac{t}{u} - \frac{9}{4}\frac{t^2 + u^2}{s^2} \Big]$$

$$\frac{d\sigma}{d\cos\theta_*}(ug \to ug) = \frac{2\pi\alpha_s^2}{9s} \Big[-\frac{u}{s} - \frac{s}{u} + \frac{9}{4}\frac{s^2 + u^2}{t^2} \Big]$$

$$\frac{d\sigma}{d\cos\theta_*}(gg \to u\bar{u}) = \frac{\pi\alpha_s^2}{12s}\left[\frac{u}{t} + \frac{t}{u} - \frac{9}{4}\frac{t^2 + u^2}{s^2}\right]$$

$$\frac{d\sigma}{d\cos\theta_*}(gg \to gg) = \frac{9\pi\alpha_s^2}{4s}\left[3 - \frac{tu}{s^2} - \frac{su}{t^2} - \frac{st}{u^2}\right]$$

(E.3)

An exceptionally swift and easy derivation of these cross section formulae can be found in (Peskin 2011). That derivation makes use of some abstract, but simple and quite fascinating, concepts from quantum field theory.

References

Here are some especially useful references on elementary particle physics. The material in these books will supplement that in this textbook. They also provide an introduction to the deeper aspects of the subject that are beyond the level of this text.

Among books at the general level of this text that provide alternative viewpoints on particle physics, I recommend

Thomson, M., *Modern Particle Physics* (Cambridge University Press, 2013).
Griffiths, D., *Introduction to Elementary Particles* (Wiley-VCH, 2008.)
Seiden, A., *Particle Physics: a Comprehensive Introduction* (Addison Wesley, 2005).

Three texts that are now out of date, but are classics with much insight into the problems they discuss, are

Källén, G., *Elementary Particle Physics* (Addison-Wesley, 1964).
Gasiorowicz, S., *Elementary Particle Physics* (Wiley, 1966).
Feynman, R. P., *Photon-Hadron Interactions* (Benjamin, 1972).

Among particle physics texts at a higher level, that assume knowledge of quantum field theory, I recommend

Halzen, F. and Martin, A. D., *Quarks and Leptons: an Introductory Course in Modern Particle Physics* (Wiley, 1984).

Valuable monographs on the Standard Model and the structure of the strong and weak interaction are

Donoghue, J. F., Golowich, E., and Holstein, B. R. *Dynamics of the Standard Model* (Cambridge University Press, 1992).
Commins, E. D. and Bucksbaum, P. H. *Weak Interactions of Leptons and Quarks* (Cambridge University Press, 1983).
Ellis, R. K., Stirling, W. J., and Webber, B. R. *QCD and Collider Physics* (Cambridge University Press, 2003).

A collection of the most important papers in the history of particle physics, with insightful commentary, is found in

Cahn, R. N. and Goldhaber, G., *The Experimental Foundations of Particle Physics* (Cambridge University Press, 2009).

To study a little quantum field theory, I recommend

Peskin, M. E. and Schroeder, D. V., *An Introduction to Quantum Field Theory* (Westview Press, 1995).
Schwartz, M. D., *Quantum Field Theory and the Standard Model* (Cambridge University Press, 2014).

A beautiful introduction to this subject can also be found in

Feynman, R. P., *QED: the Strange Theory of Light and Matter* (Princeton University Press, 2006).

A very readable introduction to Lie groups and their application to particle physics is

Georgi, H., *Lie Algebras in Particle Physics* (Perseus Books, 1999).

For more detailed discussion of particle detectors, two very useful books are

Green, D., *The Physics of Particle Detectors* (Cambridge University Press, 2005).
Grupen, C. and Shwartz, B., *Particle Detectors*, 2nd ed. (Cambridge University Press, 2008).

A topic not covered in this book is the treatment of uncertainty in particle physics experiments. Two excellent books on this topic are

Cowan, G., *Statistical Data Analysis* (Oxford University Press, 1998).
Lyons, L., *Statistics for Nuclear and Particle Physicists*, (Cambridge University Press, 1989).

The web site of the Particle Data Group—`pdg.lbl.gov/`—is an invaluable resource. There you will find up to date summaries of the properties of all elementary particles and somewhat technical but comprehensive reviews of many areas of theory and experimental techniques. The current static printed version of this data compilation, the *Review of Particle Physics*, is given in (Patrignani *et al.* 2016).

Finally, like all textbooks produced by mortal men and women, this book will have errors that need correction. Please send any needed corrections to me at: mpeskin@slac.stanford.edu. I will post a list of errata on the web at

`www.slac.stanford.edu/~mpeskin/ConceptsBook.html` .

References to specific information in the text are listed below in alphabetical order by first author. References with arXiv numbers can be found at `arxiv.org/`. The bibliographic service `inspirehep.net` provides full-text links to almost all of the papers listed here. References in the list of recommended references above are marked with a *.

Aaboud, M. *et al.* (2017). arXiv:1708.03299 [hep-ex].

Aad, G. *et al.*. (2008). *JINST* **3**, 508003.

Aad, G. *et al.*(2012). *Phys. Lett.* **B716**, 1.

Aad, G. *et al.*. (2012). *Phys. Rev.* **D86**, 014022.

Aad, G. *et al.*, (2015). *Phys. Rev.* **D92**, 012006.

Abbiendi, G. *et al.* (2001). *Eur. Phys. J.* **C19**, 587.

Abe, K. *et al.* (1997). *Phys. Rev.* **D55**, 2533.

Abe, K. *et al.* (1998). *Phys. Rev. Lett.* **81**, 942.

Abe, K. *et al.* (2001a). *Phys. Rev. Lett.* **86**, 1162.

Abe, K. *et al.* (2001b). *Phys. Rev. Lett.* **87**, 091802.

Abe, S. *et al.* (2008). *Phys. Rev. Lett.* **100**, 221803.

Abele, A. *et al.* (1999). *Phys. Lett.* **B469**, 270.

Abrams, G. S. *et al.* (1974). *Phys. Rev. Lett.* **33**, 1453.

Abreu, P. *et al.* (1999). *Eur. Phys. J.* **C11**, 383.

Ackerman, N. *et al.* (2011). *Phys. Rev. Lett.* **107**, 212501.

Acosta, D. *et al.* (2005). *Phys. Rev.* **D71**, 112002.

Adam, W. (2017). *Proceedings of Science – ICHEP2016*, 017.

Ade, P. A. R. *et al.* (2016). *Astron. Astrophys.* **594**, A14.

Ahmad, Q. R. *et al.* (2002). *Phys. Rev. Lett.* **89**, 011301.

Ahn, M. H. *et al.* (2006). *Phys. Rev.* **D74** 072003.

Ahn, J. K. *et al.* (2012). *Phys. Rev. Lett.* **108**, 191802.

Aid, S. *et al.* (1996). *Nucl. Phys.* **B470**, 3;

Aktas, A. *et al.* (2006). *Eur. Phys. J.* **C45**, 23.

Altarelli, G., and Parisi, G. (1977) *Nucl. Phys.* **B126**, 298.

Amaldi, U. *et al.* (1987). *Phys. Rev.* **D36**, 1385.

Ambrosino, F. *et al.* (2008). *JHEP* **0804**, 059.

An, F. P. *et al.* (2012). *Phys. Rev. Lett.* **108**, 171803.

Anderson, P. W. (1963). *Phys. Rev.* **130**, 439.

Andreotti, M. *et al.* (2005). *Phys. Rev.* **D72**, 032001.

Appelquist, T., Dine, M. and Muzinich, I. J. (1977). *Phys. Lett* **69B**, 231.

Arnison, G. *et al.* (1983a). *Phys. Lett.* **122B**, 103.

Arnison, G. *et al.* (1983b). *Phys. Lett.* **126B**, 398.

Assev, V. N. *et al.* (2011). *Phys. Rev.* **D84**, 112003.

Ashie, Y. (2005). *Phys. Rev.* **D71**, 112005.

ATLAS Collaboration (2013). ATLAS-CONF-2013-108.

Aubert, J. J. *et al.* (1974). *Phys. Rev. Lett.* **33**, 1404.

Aubert, B. *et al.* (2001). *Phys. Rev. Lett.* **87**, 091801.

Aubert, B. *et al.* (2002). *Phys. Rev. Lett.* **89**, 201802.

Augustin, J.-E. *et al.* (1974). *Phys. Rev. Lett.* **33**, 1406.

Augustin, J.-E. *et al.*(1975). *Phys. Rev. Lett.* **34**, 233.

Bahcall, J. N. (1964) *Phys. Rev. Lett.* **12**, 300.

Bahcall, N. A., Ostriker, J. P., Perlmutter, S., and Steinhardt, P. J. (1999). *Science* **284**, 1481.

Bahr, M. *et al.* (2008). *Eur. Phys. J.* **C58**, 639.

Ball, R. D. *et al.* (2015). *JHEP*, **1504**, 40.

Banner, M. *et al.* (1983a). *Phys. Lett.* **122B**, 476.

Banner, M. *et al.* (1983b). *Phys. Lett.* **129B**, 130.

Bardeen, J., Cooper, L. N., and Schrieffer, J. R. (1957) *Phys. Rev.* **106**, 162, **108**, 1175.

Bardon, M. *et al.* (1965). *Phys. Rev. Lett.* **14**, 449.

Baer, H. and Tata, X. (2006). *Weak Scale Supersymmetry*. Cambridge University Press.

Barnes, V. E., *et al.* (1964). *Phys. Rev. Lett.* **12**, 204.

Baron, J. *et al.* (2014). *Science* **343**, 269.

Barr, G. D. *et al.* (1993). *Phys. Lett.* **B317**, 233.

Benvenuti, A. *et al.* (1973). *Phys. Rev. Lett.* **30**, 1084.

Berko, S. and Pendleton, H. N. (1980). *Ann. Rev. Nucl. Part. Sci.* **30**, 543.

Bertone, G., ed. (2010). *Particle Dark Matter*. Cambridge University Press.

Bethe, H. A. (1930). *Ann. Phys.* **397**, 325.

Bethke, S., Dissertori, G., and Salam, G. P. (2016), in in (Patrignani *et al.* 2016).

Bevan, A. J. *et al.* (2014). *Eur. Phys. J.* **C74**, 3026.

Bichsel, H., Groom, D. E., and Klein, S. R. (2016), in (Patrignani *et al.* 2016).

Bigi, I. I. and Sanda, A. I. (1981). *Nucl. Phys.* **B193**, 85.

Bjorken, J. D. (1966). *Phys. Rev.* **148**, 1467.

Bjorken, J. D. and Dunietz, I. (1987). *Phys. Rev.* **D36**, 2109.

Bloom, E. *et al.* (1969). *Phys. Rev. Lett.* **23**, 930.

Bousso, R. (2002). *Rev. Mod. Phys.* **74**, 825.

Brandelik, R. *et al.* (1979). *Phys. Lett.* **86B**, 243.

Braunschweig, W. *et al.* (1975) *Phys. Lett.* **57B**, 407.

Breidenbach, M. *et al.* (1969). *Phys. Rev. Lett.* **23**, 935.

Buckley, A. *et al.* (2015) *Eur. Phys. J.* **C75**, 132.

Cabibbo, N. *Phys. Rev. Lett.* **10**, 531.

Cacciari, M., Salam, G. P., and Soyez, G. (2008). *JHEP* **0804**, 063.

* Cahn, R. N. and Goldhaber, G. (1989). *The Experimental Foundations of Particle Physics*. Cambridge University Press.

Campbell, J. M. *et al.* (2013). arXiv:1310.5189 [hep-ph].

Carena, M., Grojean, C., Kado, M., and Sharma, V. (2014), in K.A. Olive *et al.* (Particle Data Group), *Chin. Phys.* **C38**, 090001.

Carena, M., Grojean, C., Kado, M., and Sharma, V. (2016), in (Patrignani *et al.* 2016).

Carithers, W. C. *et al.* (1975). *Phys. Rev. Lett.* **34**, 1244.

Carter, A. B. and Sanda, A. I. *Phys. Rev. Lett.* **45**, 952.

Chanowitz, M. and Gaillard, M. K. (1985). *Nucl. Phys.* **B261**, 379.

Charles, J. *et al.* (2005). *Eur. Phys. J.* **C41**, 1. Updated results and plots from this collaboration can be found at `http://ckmfitter.in2p3.fr`.

Chatrchyan, S. *et al.* (2012). *Phys. Lett.* **B716**, 30.

Christenson, J. H., Cronin, J. W., Fitch, V. L., and Turlay, R. (1964). *Phys. Rev. Lett.* **13**, 138.

CMS Collaboration (2013). CMS PAS HIG-13-002.

CMS Collaboration (2017). CMS PAS HIG-17-031.

* Commins, E. D. and Bucksbaum, P. H. (1983). *Weak Interactions of Leptons and Quarks*. Cambridge University Press.

* Cowan, G. (1998). *Statistical Data Analysis*. Claredon Press, Oxford.

Davis, R., Harmer, D. S., and Hoffman, K. C. (1968). *Phys. Rev. Lett.* **20**, 1205.

de Groot, J. G. H. *et al.* (1979). *Z. Phys.* **C1**, 143.

Decamp, D. *et al.* (1990). *Z. Phys.* **C 48**, 365.

Derrick, M. *et al.* (1986). *Phys. Rev.* **D34**, 3286.

Dine, M. (2000). "TASI Lectures on the Strong CP Problem", arXiv:hep-ph/0011376.

Dodelson, S. (2003). *Modern Cosmology*. Academic Press.

Dokshitzer, Y. L. (1977). *Sov. Phys. JETP* **46**, 641.

* Donoghue, J. F., Golowich, E., and Holstein, B. R. (1992). *Dynamics of the Standard Model*. Cambridge University Press.

Drees, M., Godbole, R., and Roy, P. (2004). *Theory and Phenomenology of Sparticles*. World Scientific Press.

Drell, S. D. and Yan, T.-M. (1970). *Phys. Rev. Lett.* **25**, 902.

Eichten, E., Godfrey, S., Mahlke, H., and Rosner, J. L. (2008). *Rev. Mod. Phys.* **80**, 1161.

* Ellis, R. K., Stirling, W. J., and Webber, B. R. (2003). *QCD and Collider Physics*. Cambridge University Press.

Englert, F. and Brout, R. (1964). *Phys. Rev. Lett.* **13**, 321.

Feldman, G. J. *et al.* (1975). *Phys. Rev. Lett.* **35**, 821; (E) **35** 1184.

Feng, J. (2010). *Ann. Rev. Astron. Astrophys.* **48**, 495.

Fermi, E. (1934). *Nuov. Cim.* **11**, 1; *Z. Phys.* **88**, 161.

* Feynman, R. P. (1972). *Photon-Hadron Interactions*. W. A. Benjamin.

* Feynman, R. P. (2006). *QED: the Strange Theory of Light and Matter*. Princeton University Press.

Feynman, R. P. and Gell-Mann, M. (1958). *Phys. Rev.* **109**, 193.

Foster, B., Martin, A. D., Thorne, R. S., and Vincter, M. G. (2016), in (Patrignani *et al.* 2016).

Friedman, J. I. and Telegdi, V. (1957). *Phys. Rev.* **105**, 1681.

Fritzsch, H. (1977). *Phys. Lett.* **B 70**, 436 (1977).

Froidevaux, D. and Sphicas, P. (2006). *Ann. Rev. Nucl. Part. Sci.* **56**, 375.

Fukuda, Y. *et al.* (1998). *Phys. Rev. Lett.* **81**, 1562.

Fukugita, M. and Yanagida, T. (1986). *Phys. Lett* **B 174**, 45.

Garwin, R. L., Lederman, L. M., and Weinrich, M. (1957). *Phys. Rev.* **105**, 1415.

* Gasiorowicz, S. (1966). *Elementary Particle Physics*. Wiley.

Gell-Mann, M. (1964). *Phys. Lett* **8**, 214.

Gell-Mann, M., Ramond, P., and Slansky, R. (1979). In, *Supergravity*, P. van Nieuwenhuizen and D. Z. Freedman, eds. North-Holland. This paper is now available as: arXiv:1306.4669.

* Georgi, H. (1999). *Lie Algebras in Particle Physics*. Perseus Books.

Georgi, H. and Glashow, S. L. (1972). *Phys. Rev. Lett.* **28**, 1494.

Geweniger, C. *et al.* (1974). *Phys. Lett.* **48B**, 487.

Ginzburg, V. L. and Landau, L. D. (1950). *Zh. Eksp. Teor. Fiz.* **20**, 1064.

Gjesdal, S. *et al.* (1974). *Phys. Lett.* **52B**, 113.

Glashow, S. L. (1961). *Nucl. Phys.* **22**, 579.

Glashow, S. L., Iliopoulos, J., and Maiani, L. (1970). *Phys. Rev.* **D2**, 1285.

Gleisberg, T. *et al.* (2009). *JHEP* **0902**, 007.

Goldberger, M. L. and Treiman, S. B. (1958). *Phys. Rev.* *110*, 1178.

Goldstone, J. (1961). *Nuov. Cim.* **19**, 154.

Goncharov, M. *et al.* (2001). *Phys. Rev.* **D64**, 112006.

* Green, D. (2006). *The Physics of Particle Detectors*. Cambridge University Press.

* Griffiths, D. (2008). *Introduction to Elementary Particles*. Wiley-VCH.

Gross, D. J. and Wilczek, F. (1973). *Phys. Rev. Lett.* **30**, 1343.

* Grupen, C. and Shwartz, B. (2008). *Particle Detectors*, 2nd ed. Cambridge University Press.

Guralnik, G. S., Hagen, C. R., and Kibble, T. W. B. (1964). *Phys. Rev. Lett.* **13**, 585.

Guth, A. H. (1981). *Phys. Rev.* **D23**, 347.

Haag, R., Lopuszanski, J. T., and Sohnius, M. (1975) *Nucl. Phys.* **B88**, 257.

* Halzen, F. and Martin, A. D. (1984). *Quarks and Leptons: an Introductory Course in Modern Particle Physics*. Wiley.

Han, M. Y. and Nambu, Y. (1965). *Phys. Rev.* **139** B1006.

Hanson, G. *et al.* (1975). *Phys. Rev. Lett.* **35**, 1609.

Hardy, J. C. and Towner, I. S. (2009). *Phys. Rev.* **C70**, 055502.

Heinemeyer, S. *et al.* (2013). *Handbook of Higgs Cross Sections: 3*. CERN Yellow Report 2013-004. CERN.

Heisenberg, W. (1932). *Z. für Phys.*, **77**, 1.

Heister, A. *et al.* (2001). *Eur. Phys. J.* **C20**, 401.

Heister, A. *et al.* (2004). *Eur. Phys. J.* **C35**, 457.

Herb, S. W. *et al.* (1977) *Phys.Rev.Lett.* **39**, 252.

Hewett, J. and Spiropulu, M. (2002). *Ann. Rev. Nucl. Part. Sci.* **52**, 397.

Higgs, P. W. (1964). *Phys. Lett.* **12**, 132; *Phys. Rev. Lett.* **13**, 508.

Hill, C. T. (1995). *Phys. Lett.* **B 345**, 483.

Hofstadter, R. (1957). *Ann. Rev. Nucl. Part. Sci.* **7** 231.

Jackson, J. D. (1998). *Classical Electrodynamics*, 3rd ed. Wiley. (first edition: 1962).

* Källén, G. (1964). *Elementary Particle Physics*. Addison-Wesley.

Khachatryan, V. *et al.* (2015). *JHEP* **1510**, 144.

Khachatryan, V. *et al.* (2016). *Phys. Lett.* **B762**, 512.

Kinoshita, T. (1990). *Quantum Electrodynamics*. World Scientific Press.

Kobayashi, M. and Maskawa, T. (1973). *Prog. Theor. Phys.* **49**, 652.

Koks, F. W. and van Klinken, J. (1976). *Nucl. Phys.* **A272**, 61.

Kopp, S. *et al.* (2001) *Phys. Rev.* **D63**, 092001.

Kronfeld, A. S. (2012). *Ann. Rev. Nucl. Part. Sci.* **62**, 265.

Kraus, Ch. *et al.* (2005). *Eur. Phys. J* **C40**, 447.

Kostelecky, V. A. and Russell, N. (2011). *Rev. Mod. Phys.* **83**, 11.

Kurie, F. N. D., Richardson, J. R., and Paxton, H. C. (1936). *Phys. Rev.* **49**, 368.

Landau, L. D. (1944). *J. Exp. Phys. (USSR)*, **8**, 201.

Lattes, C. M. G., Occhialini, G. P. S., and Powell, C. F. (1947). *Nature* **160**, 453.

Lee, T. D. and Yang, C. N. (1956). *Phys. Rev.* **104**, 254, (E) **106**, 1371.

Lees, J. P. *et al.* (2012). *Phys. Rev. Lett.* **109**, 211801.

Llewellyn Smith, C. H. (1983). *Nucl. Phys.* **B228**, 205.

* Lyons, L. (1989). *Statistics for Nuclear and Particle Physicists*. Cambridge University Press.

Majorana, E. (1937) *Nuov. Cim.* **14**, 171.

Maki, Z., Nakagawa, M., and Sakata, S. (1962). *Prog. Theor. Phys.* **28**, 870.

Manohar, A. V., Sachrajda, C. T., and Barnett, R. M. (2016), in (Patrignani *et al.* 2016).

Marshak, R. E. and Sudarshan, E. C. G. (1958). *Phys. Rev.* **109**, 1860.

Martin, S. (1997). "A Supersymmetry Primer", arXiv:hep-ph/9709356.

Michael, D. G. *et al.* (2006). *Phys. Rev. Lett.* **97**, 191801.

Minkowski, P. (1977). *Phys. Lett.* **67B**, 421.

Nambu, Y. (1960). *Phys. Rev.* **117**, 648.

Nambu, Y. and Jona-Lasinio, G. (1961). *Phys. Rev.* **122**, 345.

Nanopoulos, D. (1973). *Lett. Nuov. Cim.* **8**, 873.

Oddone, P. (1989). *Ann. N. Y. Acad. Sci.* **578**, 237.

Olive, K. (1990). *Phys. Repts.* **190**, 307.

Pais, A. (1986) *Inward Bound*. Clarendon Press, Oxford.

Pais, A. and Piccioni, O. (1955). *Phys. Rev.* **100**, 1487.

* Patrignani, C. *et al.* (Particle Data Group) (2016). *Chinese Physics* **C40**, 100001. See also the web site of the Particle Data Group, `pdg.lbl.gov/`.

Pauli, W. (1930). Unpublished letter to a conference on radioactivity, reproduced in (Pais 1986) and other sources. You must read this letter.

Pendlebury, J. M. *et al.* (2015). *Phys. Rev.* **D92**, 092003.

Perl, M. L. *et al.* (2001). *Int. J. Mod. Phys.* **A16** 2137.

Perlmutter, S. *et al.* (1999). *Ap. J.* **517**, 565.

Peskin, M. E. (2008). "Supersymmetry in Elementary Particle Physics", arXiv:0801.1928.

Peskin, M. E. (2011). "Simplifying Multi-Jet QCD Computation", arXiv:1101.2414.

Peskin, M. E. (2016). *Annalen Phys.* **528**, 20.

* Peskin, M. E. and Schroeder, D. V. (1995). *An Introduction to Quantum Field Theory*. Westview Press.

Polchinski, J. (2005) *String Theory*, vols. I and II. Cambridge Univesity Press.

Polchinski, J. (2016). The Black Hole Information Problem, arXiv:1609.04036.

Politzer, H. D. (1973). *Phys. Rev. Lett.* **30**, 1346.

Pomeranchuk, I. Ya. (1958). *Sov. Phys. JETP* **7**, 499.

Pontecorvo, B. (1958), *Sov. Phys. JETP* **7**, 172.

Raby, S. (2006). "Grand Unified Theories", arXiv:hep-ph/0608183.

Riess, A. G. *et al.* (1998). *Astron. J.* **116**, 1009.

Riotto, A. and Trodden, M. (1999) *Ann. Rev. Nucl. Part. Sci.*, **49**, 35.

Rojo, J. *et al.* (2015). *J. Phys.* **G42**, 103103.

Rosner, J. L. *et al.* (2005). *Phys. Rev. Lett.* **95**, 102003.

Ross, G. G. (1984). *Grand Unified Theories.* Westview Press.

Salam, A. (1968), in *Proceedings of the 8th Nobel Symposium*, N. Svarholm, ed. Almqvist and Wiksell.

Salam, G. (2010). *Eur. Phys. J.* **C67**, 637.

Schael, S. *et al.* (2006). *Phys. Repts.* **427**, 257.

Scharre, D. L. (1981), in *Proceedings of the 1981 International Symposium on Lepton and Photon Interactions*, W. Pfeil, ed. Bonn University.

* Schwartz, M. D. (2014). *Quantum Field Theory and the Standard Model.* Cambridge University Press.

* Seiden, A. (2005). *Particle Physics: a Comprehensive Introduction.* Addison Wesley.

Serebrov, A. P. *et al.* (2015). *Phys. Rev.* **C92**, 055501.

Serenelli, A. M., Haxton, W. C., and Pena-Garay, C. (2011). *Astrophys. J.* **743**, 24.

Sethna, J. (2006) *Statistical Mechanics: Entropy, Order Parameters, and Complexity.* Clarendon Press, Oxford.

Sikivie, P., Susskind, L., Voloshin, M. B., and Zakharov, V. I. (1980). *Nucl. Phys.* **B173**, 189.

Silverman A. (1981), in *Proceedings of the 1981 International Symposium on Lepton and Photon Interactions*, W. Pfeil, ed. Bonn University.

Simmons-Duffin, D. (2017), "The Conformal Bootstrap". arXiv:1602.07982 [hep-th].

Sirunyan, A. M. *et al.* (2017a). arXiv:1708.00373 [hep-ex].

Sirunyan, A. M. *et al.* (2017b). arXiv:1709.07497 [hep-ex].

Sjöstrand, T. *et al.* (2015). *Comput. Phys. Commun.*, **191**, 159.

Stoker, D. P. *et al.* (1985). *Phys. Rev. Lett.* **54**, 1887.

Streater, R. F. and Wightman, A. S. (2000). *PCT, Spin and Statistics, and All That*, 5th ed. Princeton Unversity Press. (first edition: Benjamin-Cummings, 1964.)

Tannenbaum W. *et al.* (1978). *Phys. Rev.* **D17**, 1731.

Testa, M. (2007). *Proceedings of Science – KAON'07*, 042.

't Hooft, G. (1972), in *Colloqium on Renormalization of Yang-Mills Fields*, C. P. Korthals-Altes, ed. Univesity of Marseille.

't Hooft, G. (1976). *Phys. Rev. Lett.* **37**, 8.

Tinkham, M. (1966) *Introduction to Superconductivity.* McGraw-Hill.

* Thomson, M. (2013). *Modern Particle Physics.* Cambridge University Press.

Vasiliev, M. A. (2000). *Phys. Lett.* **B243**, 378.

Weinberg, S. (1967). *Phys. Rev. Lett.* **19**, 1264.

Weinberg, S. (1972). *Phys. Rev.* **D5**, 1962.

Weinberg, S. (1973). *Phys. Rev. Lett.* **31**, 494.

Weinberg S. (1993). *Dreams of a Final Theory.* Pantheon Press.

Wilson, K. (1974). *Phys. Rev.* **D10** 2445.

Wolfenstein, L. (1983) *Phys. Rev. Lett.* **51**, 1945.

Wu, C. S., *et al.* (1957). *Phys. Rev.* **105**, 1413.

Yanagida, T. (1979). *Prog. Theor. Phys.* **64**, 1103.

Yang, C. N. (1962). *Rev. Mod. Phys.* **34**, 694.

Yang, C. N. and Mills, R. (1954). *Phys. Rev.* **96**, 191.

Yukawa, H. (1935) *Proc. Phys. Math. Soc. Japan* **17**, 48.

Zweig, G. (1964). CERN report CERN-TH-401.

Zwiebach, B. (2004). *A First Course in String Theory.* Cambrige University Press.

Index